MASTERING NATURE IN THE MEDIEVAL ARABIC AND LATIN WORLDS

CONTACT AND TRANSMISSION
INTERCULTURAL ENCOUNTERS FROM
LATE ANTIQUITY TO THE EARLY MODERN PERIOD

VOLUME 4

General Editors
Görge K. Hasselhoff, Technische Universität Dortmund
Ann Giletti, University of Oxford

Editorial Board
Charles Burnett, Warburg Institute, University of London
Ulisse Cecini, Universitat Autònoma de Barcelona
Harvey Hames, Ben-Gurion University of the Negev
Beate Ulrike La Sala, Humboldt Universität, Berlin
Frans van Liere, Calvin University, Grand Rapids

Mastering Nature in the Medieval Arabic and Latin Worlds

Studies in Heritage and Transfer of Arabic Science in Honour of Charles Burnett

Edited by
ANN GILETTI *and* DAG NIKOLAUS HASSE

BREPOLS

British Library Cataloguing in Publication Data
A catalogue record for this book is available from the British Library.

ISBN: 978-2-503-60448-0
e-ISBN: 978-2-503-60449-7
ISSN: 2736-6952
e-ISSN: 2736-6960
DOI: 10.1484/M.CAT-EB.5.132219

Printed in the EU on acid-free paper.

D/2023/0095/122

Portrait of Charles Burnett. Photo by I. Bavington Jones.

Table of Contents

List of Illustrations 9

Introduction 11

Abū Maʿshar and the Tradition of
Planetary Lots in Astrology
Dorian Gieseler Greenbaum 19

The Ikhwān al-Ṣafāʾ on the *Ṣūrat al-Arḍ*:
A Geography in Motion
Godefroid de Callataÿ 57

Adelard of Bath on Climates and the Elements:
An Adaptive View on Nature
Pedro Mantas-España 83

Avicenna's *On Floods* (*De diluviis*) in Latin Translation:
Analysis and Critical Edition with an English Translation
of the Arabic
Dag Nikolaus Hasse 107

Latin Scholastics on the Eternity of the World
and Eternal Creation on the Part of the Creature:
Did They Amount to the Same Thing?
Ann M. Giletti 143

Whitewash for 'Black Magic': Justifications and
Arguments in Favour of Magic in the Latin *Picatrix*
David Porreca 177

Censorship, *maleficia*, and the Medieval Readers of the *Liber vaccae*
Sophie Page 207

The Transmission of Materialized Knowledge: A Medieval Saphea with Islamic Projections, Re-engraved in the Renaissance
Koenraad Van Cleempoel 231

Bibliography of Works by Charles S. F. Burnett 253

Index of Names 287

List of Illustrations

Portrait of Charles Burnett. Photo by I. Bavington Jones. 5

Portrait of Charles Burnett. Drawing by Ken Burnett. 18

Dorian Gieseler Greenbaum

Figure 1.1. Calculating the Lot of Fortune for a Day Birth. 21

Figure 1.2. The Lots of Fortune, Daimon, Eros,
and Necessity forming a double mirror image. 26

Figure 1.3. Birth chart of 10 August 787, 10:12 p. m. 37

Table 1.1. 'Planetary' Lots in Hellenistic Astrology.
Lots according to Vettius Valens (b. 120 AD);
Lots according to Paulus of Alexandria (fl. 378 AD). 22

Table 1.2. Planetary Lots in the *Great Introduction*:
Their Names and Day Formulae 32

Table 1.3. Orders of Lots and Planets, Abū Maʿshar
and Paulus of Alexandria/Olympiodorus 34

Table 1.4. Lots in the *Abbreviation*: Name, Day Formula
and Associated Planet 35

Godefroid de Callataÿ

Table 2.1. Networks of Associations among Climes,
Planets, and Skin Colour 62

Table 2.2. Correspondences between Climes and Regions 63

Pedro Mantas-España

Figure 3.1. Petrus Alfonsi, *Dialogus contra iudaeos,* Oxford, Bodl. Lib., MS Laud Misc. 356 (14th c.), fol. 120^{r}. 87

Figure 3.2. Petrus Alfonsi, *Dialogus contra iudaeos,* Oxford, Bodl. Lib., MS Laud Misc. 356 (14th c.), fol. 120^{r}. 87

Figure 3.3. 'The Illustration of the Encompassing Sphere and the Manner in Which It Embraces All Existence, and Its Extent', from *Kitāb Gharā'ib al-funūn wa-mulaḥ al-ʿuyūn,* Oxford, Bodl. Lib., MS Arab. c. 90 (dated 1190–1210), fols 2^{b}–3^{a}. 90

Figure 3.4. 'The Waq-Waq tree' from *Kitāb al-Bulhān,* Oxford, Bodl. Lib. MS Or. 133 (15th c.), fol. 41^{b}. 94

Koenraad Van Cleempoel

Figure 8.1. Oxford *Saphea, c.* 1450 (probably France) and *c.* 1580 (Louvain or Liège), Side A. Oxford, History of Science Museum (inv. 14645). 233

Figure 8.2. Oxford *Saphea, c.* 1450 and *c.* 1580, Side A (detail). Oxford, History of Science Museum (inv. 14645). 234

Figure 8.3. Oxford *Saphea, c.* 1450 and *c.* 1580, Side B. Oxford, History of Science Museum (inv. 14645). 235

Figure 8.4. Oxford *Saphea, c.* 1450 and *c.* 1580, Side B (detail). Oxford, History of Science Museum (inv. 14645). 237

Figure 8.5. 'Sicilian' Astrolabe, *c.* 1300. Oxford, History of Science Museum (inv. 40829). 242

Figure 8.6. *De Astrolabio universalis et Saphea sine cursore,* Munich, Bayerische Staatsbibliothek, MS lat. 25026 (16th c.), fol. 14^{v}. 244

Figure 8.7. Adrian Zeelst and Gerard Stempel, *Utriusque astrolabii tam particularis quam universalis fabrica et usus* (Liège: Ouwerx, 1602), unnumbered fol. Oxford, History of Science Museum (inv. 55945). 247

Figure 8.8. Astrolabe by Adrian Zeelst, *c.* 1590 (Cologne or Louvain). Cologne, Kölnisches Stadtmuseum (inv. 1.184). 248

Introduction

If we can speak of a collective endeavour of medieval Arabic scientific experts, it was to discover, compile and possess information about the great phenomena and minute details of the natural world. The effort, for many of them, was to master nature. Gaining mastery did not mean that they could control nature, but equipped with this knowledge they could situate human beings in relation to the surrounding world, navigate through it, and predict, prepare for and mitigate the harsher impacts of nature and of fates held in the stars. Even the science and practice of magic, which sometimes intervened in nature through manipulations, did not generally seek to control it. Yet while experts dominated nature only inasmuch as they possessed knowledge of it, this in itself was a great power. Latin scholars inheriting this perspective repeated a saying derived from Arabic sources: 'Sapiens dominabitur astris' (The wise man will dominate the stars). The benefit to humankind was explained by the unknown author of the much-read *Centiloquium*, the Arabic *Kitāb al-thamara*: 'The astrologer is able to prevent many effects of the stars, if he knows the nature of what affects him, and he may remedy the effect before it occurs in a way that he can bear it'.[1]

Several of the articles in this volume show how the stars were thought to influence a variety of sublunar phenomena, not only the fates of human beings but also the surface of the earth, climate, catastrophic events such as floods and fire storms, and the rise and fall of civilizations. Stars, however, were only one of many factors studied by medieval scientists and scholars, who also sought to master nature by understanding the causal role of the four elements, bodily humours and spirits, and by analysing the world through the concepts of matter and form, action and passion, time and place, and finitude and infinity. These undertakings unfolded in various disciplines of the Arabic-Latin tradition including, as discussed in this volume, astrology

1 Anonymous (Ps.-Ptolemy), *Kitāb al-thamara*, in Aḥmad ibn Yūsuf ibn al-Dāya, *Commento al Centiloquio Tolemaico*, ed. by Franco Martorello and Giuseppe Bezza (Milan: Mimesis, 2013), Ch. 5, p. 62:

قد يقدر المنجّم على دفع كثير من أفعال النجوم إذا كان عالمًا بطبيعة
ما تؤثّر فيه وربّما أصلح للفعل قبل وقوعه قابلاً يحتمله.

For the Latin translation of Plato of Tivoli, see: Anonymous (Ps.-Ptolemy), *Centiloquium*, in ibid., *Quadripartitum. Centiloquium cum commento Hali* (Venice: Erhard Ratdolt, 1484), Ch. 5, fol. 7vb: 'Optimus astrologus multum malum prohibere poterit, quod secundum stellas venturum est, cum earum naturam praesciuerit. Sic enim praemuniet eum, cui malum futurum est, ut possit illud pati'.

Mastering Nature in the Medieval Arabic and Latin Worlds: Studies in Heritage and Transfer of Arabic Science in Honour of Charles Burnett, ed. by Ann Giletti and Dag Nikolaus Hasse, CAT 4 (Turnhout: Brepols, 2023), pp. 11–18 BREPOLS PUBLISHERS 10.1484/.CAT-EB.5.134024

and astronomy, geography, climatology and meteorology, physics and metaphysics. In contrast with these sciences, magic was a more interventative method to master nature. While some Arabic magical texts translated into Latin, such as the *Picatrix*, aimed to harness the powers of stars and celestial spirits, other forms of magic were meant to interfere directly in the structure of the natural world, such as through recipes to produce living beings from inanimate matter, as set out in the *Book of Laws*.

Within this tradition, the experts were masters. In the classical Arabic world, the expert in a particular science was often called *ṣāḥib*, master of a science, such as *ṣāḥib ʿilm al-nujūm* (master of the science of the stars), *ṣāḥib ʿilm al-manāẓir* (master of the science of optics), and *ṣāḥib ʿilm al-handasa* (master of the science of geometry). In some Latin texts, the term *ṣāḥib* was translated as *magister* or *auctor*, such as *magister scientiae iudiciorum astrorum* (master of the science of judgements from the stars). Some of these experts engaged not only in study but also in design and manufacture of devices and instruments for measuring nature, that is, with 'the science of the instruments of the stars' (*ṣināʿat ālāt nujūmiyya*),[2] a prime example being the astrolabe.

Arabic-Latin natural sciences did not constitute a body of knowledge unified by a consensus of experts within their sciences or across natural science as a whole. None of the natural sciences in Arabic and Latin cultures was immune to controversy: the contributions in this volume point to alternative explanations and to competing theories and positions. These sciences were rich and complex, and issues of disagreement among the masters proved to be a constant engine of intellectual productivity. However, negative reception of some ideas and practices went beyond scientific disagreement. In both Arabic and Latin cultures, magic was met with suspicion and criticism, as reflected by the extensive apologetics of magic in the *Picatrix* and signs of severe censorship in Latin manuscripts of the *Book of Laws*. Certain practices of magic and scientific theories such as the eternity of the world were met with suspicion because religious authorities saw them as conflicting with their faith. Negative reactions took the form of scribal censorship or the prohibition of ideas. Scientific specialists grappling with these perceptions sometimes offered justifications, such as the author of the *Picatrix*, who argued that God's power was at the root of magic; and some Latin scholastics proposed ways of accepting challenging theories that did not contradict their faith.

The articles in this volume discussing these diverse scientific topics are offered in honour of Charles Burnett. He himself is a master of the complex field of research treated here. The bibliography of his books and articles included at the end of this volume gives testimony to this, and shows his contributions concerning additional Arabic and Latin traditions in sciences not covered here, notably music, arithmetic, geometry, medicine and psychology. In this

2 For example, al-Fārābī called it this: al-Fārābī, *La statistique des sciences*, ed. by Osman Amine (Cairo: Anglo-Egyptian Library, 1968), p. 109, l. 15.

bibliography the reader will also find many suggestions for future research. Burnett's studies exemplify cross-cultural and multi-disciplinary research that views the Arabic and Latin worlds as part of a vast cultural area spreading across West Asia, North Africa and Europe, and bridged by inquisitive scholars and learned translators. This volume is a small tribute to the chapter of the past that Charles Burnett contributes so much to uncovering.

The collection begins with a contribution on astrology by **Dorian Gieseler Greenbaum** in Chapter 1. She examines the work of the great Persian astrologer Abū Maʿshar (787–886) on the subject of planetary lots, and shows how he drew on and adapted lot traditions in Hellenistic astrology. Lots are an astrological technique that address human concerns, such as health and fortune (Lot of Fortune), and love and desire (Lot of Eros). Each lot has a formula for making a distance measurement and projection in relation to celestial bodies (sun, moon, planets), which points to a zodiacal sign that is then interpreted in regard to the topic of interest. Greenbaum explains how lots are calculated, as well as what 'planetary lots' are. Each is associated with one of the seven classical planets — for instance, the Lot of Fortune with the moon — and that planet is part of its formula. She traces the history of planetary lots in the works of experts from the second century BC to the tenth century AD, and situates Abū Maʿshar in relation to the early, Hellenistic tradition by making a detailed analysis of the lot formulae in three works by him, *Kitāb al-mudkhal al-kabīr ilā ʿilm aḥkām al-nujūm* ('The Great Introduction to Astrology'), *Kitāb al-mudkhal al-ṣaghīr* ('The Abbreviation of the Introduction to Astrology'), and *Kitāb fī taḥāwil sinī al-mawālīd* ('On the Revolutions of the Years of Nativities'). She discovers that Abū Maʿshar combined two lot traditions in Hellenistic astrology. One, that she designates as 'Egyptian', had four lots and did not associate them with planets; the other, called 'Hermetic', had seven lots and did make this association. So, while Abū Maʿshar referred to his lots as 'planetary lots', his formulae sometimes do not fit this description. Greenbaum carefully identifies variations in his formulae, and traces possible sources in earlier thinkers' works. She rounds off the study by examining the afterlife of Abū Maʿshar's doctrine, presenting examples of later medieval astrologers who were deeply dependent on his system.

Godefroid de Callataÿ introduces us in Chapter 2 to geography, the Ikhwān al-Ṣafāʾ ('Brethren of Purity'), and their doctrine that features of the earth's surface move as a result of cyclical motions of the heavens. This is a startling divergence from other medieval Muslim geographers, who perceived the earth's surface as fixed, anchored by mountains. The Ikhwān, a group of anonymous authors in Iraq, set out the doctrine in their encyclopaedic *Rasāʾil Ikhwān al-Ṣafāʾ* ('Epistles of the Brethren of Purity') in the ninth/tenth centuries. De Callataÿ analyses Epistle 4, *On Geography*, initially examining the general geographical concepts it treats, including climes, and then focusing on its intriguing final section. The Ikhwān explain here that celestial cycles are responsible for recurring growth or fall in the proliferation of plants, animals, human welfare, scholars, justice, buildings, cities, and even rises in civilisation and good people, or descent into corrupt morals and wicked people. This expanded view of geography — adding time and moral qualities — departs from the technical data of the rest of the Epistle. De Callataÿ shows that the cycle discussion's meaning and purpose can be understood through other passages in the *Rasāʾil*. Passages elaborating on the role of time and celestial cycles in geography explain how features characterising the earth's four regions — desert, sea, mountains, or human civilisation — shift among regions over time, influenced by astronomical cycles and conjunctions. Other passages provide insight indicating a religious purpose, associating the Ikhwān with the Qurʾānic Seven Sleepers story. De Callataÿ shows how the Ikhwān link it with astronomical-millenary interpretations of the sequence of prophets, and the duty of a group of distinguished people living during a cycle of wicked people to preserve the purity of the message they bear. He concludes that the Ikhwān's ideological convictions and identification with this role drive the cycle discussion in Epistle 4.

Pedro Mantas-España moves in Chapter 3 to the subjects of climate and the four elements in the thinking of Adelard of Bath (d. 1152), knowledge that Adelard learned from Arab experts while travelling in the Mediterranean. Mantas-España outlines the Hellenistic and Arab/Persian roots of the geographical division of the earth into climes, which are zones identified by their climates. The 'first clime' is the zone of the Mediterranean and Mesopotamia which are characterised by comfortable climatic conditions considered ideal for philosophers, and the further climes are gradually colder. He examines what Adelard taught in his *Quaestiones naturales* ('Questions on Natural Science') and *De opere astrolapsus* ('Treatise on the Astrolabe') concerning climes, their habitability, and the influence of their respective climates on the conditions and activity of people living in them. He compares Adelard's view with that of the Spanish *converso* Petrus Alfonsi (d. 1140) in his *Dialogus contra iudaeos* ('Dialogue against the Jews'), and finds that they agree on a connection between mild, healthy climates and human physical prosperity and thriving intelligence. Mantas-España shows that Adelard's view builds on his understanding of the four elements and how everything in nature draws on what is appropriate to it, such that plants draw on 'earthiness' to grow, while human intelligence thrives on a moist brain. A balanced climate makes for stable elements and bodily humours, resulting in a balanced environment and human physical and intellectual prosperity. However, Adelard does see the possibility of equilibrium in other climes and, unlike Petrus Alfonsi, accepts that inhabitants in challenging climates can nevertheless enjoy wellbeing and productive intellectual lives.

Dag Nikolaus Hasse turns in Chapter 4 to the subject of meteorology and natural catastrophes in the work of Avicenna (d. 1037). In *On Floods*, Avicenna discusses water floods along with fire storms, earthy deluges and destructive winds, explaining the four kinds of flood as excesses of one of the four elements, caused by certain conjunctions of the planets in combination with elemental dispositions. Avicenna also speaks of the possibility of extinction of life through floods and spontaneous generation of life afterwards. Hasse presents a study of the work, an English translation of the original Arabic, and a critical edition of the medieval Latin translation. The original text is the last chapter of Avicenna's *Meteorology* in his summa *al-Shifāʾ* ('The Cure'), bearing the title *Fī l-ḥawādith al-kibār allatī taḥduthu fī l-ʿālam* ('On the Great Events which Happen in the World'). It was translated from Arabic into Latin before the mid-thirteenth century. The English and Latin translations are set out here on facing pages, revealing how the shorter, Latin version is an abbreviation of the original work. In his introductory study, Hasse analyses the scientific content of the work, explaining the underlying theories and identifying possible sources informing the thinking, and he sets the Latin translation in the context of scholastic reception of other translations of Arabic meteorological texts. He is also able to attribute the Latin version to Michael Scot, famous for his Latin translations of Averroes and as the court astrologer of Emperor Frederick II Hohenstaufen. This attribution is based on strong evidence of both the use of certain expressions characteristic of Michael's lexicon, and the abbreviation technique typical of Michael's translations, which omits phrases and sentences rather than paraphrase, and otherwise preserves the text as an almost word-for-word translation.

Ann Giletti looks into questions about the origin of the world in Chapter 5, and the issue of transgression of science against religion. She examines two theories claiming to prove that the world is eternal, which had been controversial in the Arabic philosophical tradition for conflicting with the belief in Creation. Giletti analyses the theories' late medieval Latin reception in the context of this faith-reason problem. One theory was set out in Aristotle's *Physics*, and was based on natural philosophical principles of time, matter and motion. The other theory, Eternal Creation, came from Neoplatonic and Islamic metaphysics traditions, and was based on the principle of God's eternal, unchanging power. It was known to Latin scholastics through Augustine, who criticised it, and Avicenna, who accepted it in his *Metaphysics*. Through a range of examples in scholastic philosophical and theological texts, Giletti shows that, while the theories were produced by separate fields of knowledge, principles and arguments, the authors used a single case — one set of philosophical arguments — to try to prove that both were impossible. While this approach may seem coherent with the theories' shared conclusion that the world is eternal, there were different implications regarding their transgressive nature. Aristotle's theory contradicted the Bible and was potentially heretical, while Eternal Creation arguably was not. The danger lay in holding that this set of arguments could not disprove that the world was eternal. This position was held by Latin scholars influenced by Avicenna and his rebuttals favouring Eternal Creation. However, in accepting the position for Eternal Creation, they also implicitly accepted it for Aristotle's theory, which more clearly denied Christian doctrine.

David Porreca takes up the subject of magic in Chapter 6, and looks with a different perspective at the matters of scientific validity and transgression. He explores how, in responding to common criticisms, texts on magic justified to their readers that their rituals were effective and acceptable in moral and religious terms. He looks at the Latin version of *Ghāyat al-Ḥakīm* ('The Goal of the Sage'), known as *Picatrix*, a representative example of ceremonial and astral magic in the Latin Middle Ages and Renaissance. The original work, composed in Arabic in the 950s by the Andalusian *ḥadīth* scholar Maslama ibn Qāsim al-Qurṭubī, drew on over 200 sources in the Arabic, Babylonian, Egyptian, Greek, Indian, Persian and Syrian traditions. It was translated into Spanish in the 1250s, and soon after from Spanish into Latin. Much of the Spanish version has not survived. The Latin version shows divergences from the Arabic — missing or simplified passages, and significant interpolations — introduced somewhere in the trilingual translation process. The pro-magic rhetoric was present in the original Arabic, reacting to mistrust of magic in the Islamic world, and was retained in the translations for a similar purpose. Porreca identifies twelve themes serving to defend magic, such as: claims of its effectiveness and references to *auctoritates* who were successful in conducting magic; appropriation of positive reputations of other scientific disciplines and emphasis on magic's coherence and rationality; and assertions that God's power is at the root of magic and that practitioners must be pure and devout. His analysis of each theme shows that they repeat through the work and give it the overall tone of a pro-magic apologetic, to ward off scepticism and make magic a legitimate part of knowledge. Nevertheless, the significant number of Renaissance manuscripts but absence of an edition at this time indicate that, while the *Picatrix* attracted substantial interest, its topic remained controversial.

Sophie Page continues with the themes of magic and transgression in Chapter 7, conveying us through the late Middle Ages to the beginnings of the Renaissance. She analyses the content and reception of the *Liber vaccae*, a twelfth-century Latin translation of a ninth-century Arabic magic work, *Kitāb al-nawāmīs* ('Book of Laws'), and she identifies strategies of scribal censorship in surviving manuscripts. The work is composed of two books containing 85 experiments (recipes for magic). By examining manuscripts, Page shows that, while modern scholarship has focused on the transgressive nature of Book I and has characterised Book II as containing benign domestic magic or parlour tricks, Book II in fact was not considered innocuous but was subjected to censorship. It presents experiments such as: how to plant seeds to make them grow instantly to maturity, using human blood; how to make a wick that causes women to dance in a frenzy, using animal blood; and how to make a lamp that creates the illusion that a house is full of snakes, using a funeral cloth. Page analyses the methods of censorship, ranging from omission by the scribe of part or all of the text, to omission or later erasure of one or two elements or ingredients, rendering the experiments inoperable. While magic texts from the Arabic, Jewish and Greco-Roman traditions circulating among Latin readership generally involved the powers of stars, celestial spirits, invocations and talismans, the *Liber vaccae* offered magic employing interventions into nature. This manipulation of natural substances — to produce such things as new living beings and transformation of people into apes — caused critics to regard the *Liber vaccae* as *maleficia* (harmful magic). Page shows how this kind of 'organic' magic and its characterisation as *maleficia* were associated with female practitioners in medieval literature and scholarship, and how this would become a central element of witchcraft mythologies starting in the fifteenth century.

In the final contribution, **Koenrad van Cleempoel** takes us in Chapter 8 to material culture and scientific instruments. He presents findings on a unique astrolabe, the Oxford *Saphea* (inv. 14645, History of Science Museum, Oxford), whose story spans 450 years, linking medieval Toledo and Renaissance Louvain. Astrolabes originated with the ancient Greeks, developed further in the Islamic world, and passed into the medieval Latin West. These brass instruments are for calculating astronomical positions and other applications such as surveying heights. They represent in two dimensions the earth's surface and the heavens. Their construction includes a plate with engravings corresponding to a particular latitude, over which a rete (skeletal frame), indicating stars and the zodiac, can rotate. Astrolabes can have multiple plates for perspectives at various latitudes, but what van Cleempoel presents is a *saphea*, one consisting of a single plate engraved with a universal projection. Universal projection for astrolabes was developed in Toledo by astronomers at the court of al-Maʾmūn (1037–1075). From there, the universal projection of Abū Isḥāq Ibrāhīm al-Zarqālī, known as Azarquiel or Arzachel (fl. 1048–1087), entered the Latin West thanks to translations of his writings. Van Cleempoel explains the parts and engravings of astrolabes, and details the remarkable features of the Oxford *Saphea*. He shows that it was designed and engraved *c.* 1450 in France with the eleventh-century Islamic projection, and that it received additional engraving *c.* 1600. He is able to attribute the Renaissance engraving to the Louvain mathematician Adrian Zeelst. The attribution is founded on striking evidence that Zeelst used this medieval *saphea* as the basis for an illustrated treatise on astrolabes commissioned by Prince-Bishop Ernst of Bavaria, and that Zeelst also used it as the model to make another, spectacular astrolabe now in Cologne.

The studies in this volume are based on papers presented at a conference at the Institute of Philosophy of Würzburg University, which former PhD students of Charles Burnett organized in his honour. He was the only participant who knew nothing of the existence of the conference and the gathering of his former doctoral students before he entered the lecture room. Some of us who attended the conference were unable to contribute to this volume, but join in expressing their gratitude: Anne McLaughlin, Bink Hallum, and David Brancaleone. We could not have found a better supervisor than Charles Burnett. He set an example for us of humble wisdom, rigorous scholarship, joyful curiosity, and kindness and collegiality. His breadth and depth of knowledge are an inspiration, and following his approach is an impulse we hope we will never lose. We are privileged and delighted to offer these studies in honour of our teacher, mentor and friend, Charles Burnett.

In preparing this volume, we were fortunate to have excellent critical help from anonymous peer reviewers. We would like to thank them, as well as Johathan Maier for preparing the index. We also thank, with great affection, Ken Burnett and Ian Jones for their kind contributions of portraits of Charles Burnett for this volume.

Ann Giletti and Dag Nikolaus Hasse

Portrait of Charles Burnett.
Drawing by Ken Burnett.

DORIAN GIESELER GREENBAUM

Abū Maʿshar and the Tradition of Planetary Lots in Astrology

Abū Maʿshar, one of the great medieval Arabic astrologers, is known for his encyclopaedic knowledge of the theory, practice, and philosophy of astrology. His knowledge extended to all of astrology's branches: nativities, for interpreting the astrological chart for the time of birth; general, for interpreting events pertaining to countries, cities or states, and their populations; and katarchic, for interpreting for individuals specific events and timing of events occurring outside of natal astrology.[1] He was also skilled in predictive techniques (in scholarship sometimes called 'continuous' astrology), which examine effects of movements of the stars through real or symbolic time.[2] A subject which greatly interested Abū Maʿshar was lots, one of the oldest techniques in astrology for assessing topics of human concern such as fortune, family and health. Each lot represents one of these topics, whose significance for a particular astrological chart is determined through a technique involving measurements of ecliptic degree-arcs between celestial bodies (such as planets) as they were at the

* I am deeply grateful for help and guidance from Charles Burnett, especially with Arabic questions, transliterations and translations, and access to pre-publication versions of *The Great Introduction to Astrology*. I also warmly thank Liana Saif for stimulating discussions on Arabic words in the text, Stephan Heilen for help with material from Rhetorius, Rosa Comes for obtaining an Escorial folio for me, and Juan Acevedo for Arabic transcription. Finally, I appreciate this volume's editors and the anonymous peer reviewer for their careful attention to this contribution. Needless to say, all errors remaining are my own.

1 Katarchic astrology includes these related sub-branches: elections (choosing the auspicious astrological moment for beginning something), events (interpreting the chart for an event after the fact), interrogations (predicting an outcome based on the time a question is asked) and decumbitures (prediction based on the time someone becomes ill). For more information on these and other branches of astrology, see: Greenbaum, 'The Hellenistic Horoscope', pp. 448–50; Greenbaum, 'Divination and Decumbiture', pp. 115–16.

2 For example, David Pingree uses the term 'continuous astrology' in *Astral Omens to Astrology*, p. 21. The time employed can be either the real passage of time and planetary movements at that time (for example Jupiter at 23° of Aquarius on 25 September 2021), or symbolic, e.g. using a day in real time to represent a year symbolically (for example, the first day after a birth could represent the first year of life).

Dorian Gieseler Greenbaum (Duxbury, MA, USA and London, UK) is a historian of astrology in Greco-Roman and later Antiquity and the Middle Ages and its transmission from Hellenistic to Arab thinkers.

Mastering Nature in the Medieval Arabic and Latin Worlds: Studies in Heritage and Transfer of Arabic Science in Honour of Charles Burnett, ed. by Ann Giletti and Dag Nikolaus Hasse, CAT 4 (Turnhout: Brepols, 2023), pp. 19–56 BREPOLS PUBLISHERS 10.1484/.CAT-EB.5.134025

moment the chart records. Abū Maʿshar devoted the last book of his *Great Introduction to Astrology* and Ch. 6 of his *Abbreviation of the Introduction to Astrology* to lots. In the former (viii, 2.3), he introduces the 'authenticated' lots used by 'the Ancients': the 'first kind' of these authenticated lots are the 'lots of the seven planets'.[3] That these lots form the first group demonstrates their importance in Abū Maʿshar's mind. This article will first explore the relevant lot traditions that preceded Abū Maʿshar's use of the phrase 'lots of the planets'; then demonstrate how Abū Maʿshar valued and synthesised these traditions; and finally examine their use, meanings, and transmission into medieval astrology.

1. What is a Lot?

Lots in astrology are a very old technique, mentioned in some of the pseudepigraphical fragments of Nechepso and Petosiris (second–first centuries BC), among the earliest extant texts on what is generally known as Hellenistic astrology.[4] Lots are used, along with other techniques, to interpret many human concerns, such as physical health and fortune (Lot of Fortune), spiritual and mental topics (Lot of Daimon), parents (Lots of Father and Mother), siblings (Lot of Siblings), marriage (various lots covering the topic of marriage from different perspectives such as gender), and love and desire (Lot of Eros). Because one lot can have many meanings depending on the context in which it is being assessed, its interpretive range is great.

Lots are found by: (1) taking the ecliptic degree-arc[5] (measuring the distance in degrees) between two celestial bodies or points (the 'natural and fixed' indicators); (2) projecting that arc (the same measurement) from a third point, usually the Ascendant (the 'movable indicator'),[6] which is the degree of the zodiacal sign ascending on the eastern horizon at the moment the chart records; and (3) noting the zodiacal sign at the projection's end point. This position and its relationship to other components of the chart allow it to be interpreted within that milieu. The components used to create the lot are based on, and reflect their connection with, the specific topic desired.

3 Abū Maʿshar, *The Great Introduction to Astrology*, ed. and trans. by Yamamoto and Burnett, with an edition of the Greek version by David Pingree (hereafter *Great Introduction*), i, pp. 824–25 (Arabic and English).

4 See the original edition of Nechepso and Petosiris, ed. by E. Riess, *Fragmenta magica*, e.g. fragments 13, 19, 19a; additional fragments in Heilen, 'Metrical Fragments', fragments +3, +5a-b, +12a-b, +23, +27.

5 Abū Maʿshar calls this degree-arc 'the distance between the two indicators' in *Great Introduction*, viii, 1.8, vol. i, pp. 820–21 (Arabic and English). All references to and quotations from this text (in English translation, Arabic transliteration, or Greek) will be from this edition unless otherwise noted.

6 *Great Introduction*, viii, 1.8, vol. i, pp. 822–23 (Arabic and English).

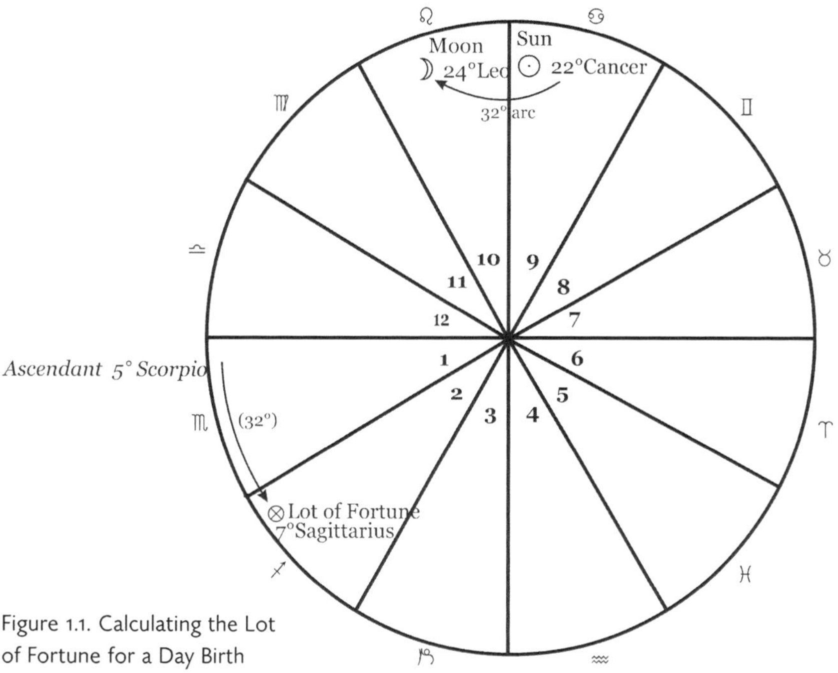

Figure 1.1. Calculating the Lot of Fortune for a Day Birth (diagram by the author).

What is arguably the most important, and most used in practice, is the Lot of Fortune.[7] As an example of how a lot is calculated, the Lot of Fortune is shown in Fig. 1.1.

The calculation of the lot differs for a person born during the day and one born at night. By day the degree-arc is taken from the Sun to the Moon, and by night from the Moon to the Sun. The arc should always be calculated in the order of the zodiac, so in the illustration, which uses a day chart (Sun above the horizon line, i.e. Ascendant), the arc from Sun to Moon is projected in zodiacal order from the Ascendant to arrive at the position of the Lot of Fortune. For a night chart (Sun below the horizon), the arc is taken from the Moon to the Sun and projected from the Ascendant. The same method is used to calculate any lot.

7 It, and probably its mirror-image companion, the Lot of Daimon, are mentioned in Nechepso and Petosiris (see above, Riess and Heilen in n. 4). See also: commentary in Heilen, *Hadriani genitura*, II, pp. 1158–82; Greenbaum, *Daimon in Hellenistic Astrology*, pp. 332–35. The Lots of Fortune and Daimon are the two most frequently used lots in documentary and literary evidence: see Greenbaum, 'Lots of Fortune and Daemon in Extant Charts'.

Table 1.1. 'Planetary' Lots in Hellenistic Astrology. Lots according to Vettius Valens (b. 120 AD) = italic font; Lots according to Paulus of Alexandria (fl. 378 AD) = Roman font.

Name of Lot	**Planet**	**Formula** (arc always projected from Ascendant; day formula given here)
Lot of Fortune	Moon	*From Sun to Moon* / From Sun to Moon (reverse at night)
Lot of Daimon	Sun	*From Moon to Sun* / From Moon to Sun (reverse at night)
Lot of Eros		1. *From Fortune to Daimon* (reverse at night)
	Venus	2. From Daimon to Venus (reverse at night)
Lot of Necessity		1. *From Daimon to Fortune* (reverse at night)
	Mercury	2. From Mercury to Fortune (reverse at night)
Lot of Courage	Mars	From Mars to Fortune (reverse at night)
Lot of Victory	Jupiter	From Daimon to Jupiter (reverse at night)
Lot of Nemesis	Saturn	From Saturn to Fortune (reverse at night)

2. What is a Planetary Lot?

This phrase is generally employed to describe a group of lots that are based on the association of each of the seven classical planets (including the Sun and Moon) to a specific lot in that group. This kind of grouping is not seen in the extant literature until the fourth century AD, nor does it appear to have been known to astrologers before then.[8] Thus in Table 1.1 we may note that for the second-century AD author Vettius Valens, the

8 Paulus of Alexandria (fl. 378 AD), whose chapter title in the transmitted manuscripts cites a book called the *Panaretos*, and his commentator Olympiodorus (fl. 564 AD), mention and describe a group of seven lots associated with the seven planets. In a *Compendium* of earlier writers' work said to be written by Rhetorius (sixth century AD), these lots are described in a similar fashion to Paulus, in a section attributed to the earlier author Antiochus of Athens (*c.* second century AD). László, 'Rhetorius, Zeno's Astrologer', has recently (2020) analysed the author attributions, dates and content of this *Compendium*, supporting a later date of introduction (fourth century) for these planetary lots). The one extant original papyrus

Lots of Eros and Necessity, associated respectively with Venus and Mercury, are not linked in that way by him, and their formulae do not use those planets.[9] However, they are so linked, and formulated, in the fourth-century writer Paulus of Alexandria.[10]

This discrepancy arises from the use of two different lot traditions in Hellenistic astrology, which I designate as 'Egyptian' and 'Hermetic'.[11] In this list we see that the Lots of Eros and Necessity can be calculated with two different sets of astrological positions;[12] furthermore, the Lots of Courage, Victory, and Nemesis do not appear in the historical record at all until the fourth century AD. Thus the first, Egyptian, grouping of lots (seen in the second century AD, in both texts and original documents of nativities) consists of the Lots of Fortune, Daimon, Eros, and Necessity. The second, Hermetic, grouping adds the Lots of Courage, Victory, and Nemesis to the original four, and associates each of these seven lots to a planet. Vettius Valens and Paulus of Alexandria exemplify, respectively, these different Egyptian and Hermetic traditions. These historical circumstances come to play a large part in the history of the transmission of lot doctrine, especially in the work of Abū Maʿshar. Thus some attention to these different traditions and their history is in order.

3. An Abbreviated History of 'Planetary' Lots

This history will concentrate on practices and methods that have an impact on transmission of this tradition into the medieval period. Each subsection below presents the major and minor players in this history, listed in rough chronological order:

birth chart that contains all of these seven lots in formulae according to Paulus is cast for a date in 319 AD; see *editio princeps* in Greenbaum and Jones, 'P. Berl. 9825'.

9 The formulae for these lots are found in Vettius Valens, *Anthologies*, IV, 25.13 and 16, ed. by Pingree, p. 192.22–24 and 30–31. The editor has bracketed them as not genuine, but neglects the historical evidence for these formulae. See Greenbaum, *Daimon in Hellenistic Astrology*, pp. 360–67, esp. 361 and n. 105, 364 and n. 109 for discussion of this issue.

10 Paulus of Alexandria, *Introduction to Astrology*, Ch. 23, ed. by Boer, pp. 47.15–49.10; trans. by Greenbaum, pp. 41–42.

11 This designation is based on a Scholion to Hephaestio's *Apotelesmatica* which mentions two lot traditions. See below, Section 3-B and n. 22. For more, see Greenbaum, *Daimon in Hellenistic Astrology*, pp. 366–78. The 'Hermetic' tradition does not appear in extant texts until the fourth century AD.

12 A variant on Valens's formulae for the Lots of Eros and Necessity is offered by Firmicus Maternus (fl. 338 AD), reversing the positions of the Lots of Daimon and Fortune in Valens's formulae, so that the Lot of Eros in Valens = the Lot of Necessity in Firmicus and the Lot of Necessity in Valens = the Lot of Eros in Firmicus.

Nechepso and Petosiris
Dorotheus of Sidon
Vettius Valens
Antiochus of Athens
Claudius Ptolemy
Julius Firmicus Maternus
Paulus of Alexandria and Olympiodorus
Rhetorius
Documentary evidence of relevant lots in horoscopes

Though some of these have more of an impact than others, it is important to situate all of them contextually within the historical lot tradition, as in the following subsections. As noted above, lots associated as a group with the seven planets are not mentioned in extant texts or original documents until the fourth century.[13]

A. Nechepso and Petosiris (second–first centuries BC)

Only fragments quoted by later astrological writers remain of these pseudepigraphical authors. The relevant fragments are 19, 19a, +12a, and +12b.[14] Until recently only the Lot of Fortune was identified in fragments, but Stephan Heilen's recent, persuasive proposal gives evidence that the Lot of Daimon, by implication, must also be inferred in their work.[15]

B. Dorotheus of Sidon (first century AD)

Dorotheus of Sidon's text, written in verse and known variously as the *Carmen Astrologicum* or the *Pentateuch*, is mostly complete only in an extant Arabic version.[16] Greek fragments of his work exist in several places, especially in the *Apotelesmatica* of Hephaestio of Thebes (early fifth century AD). In direct quotation, Dorotheus mentions only the Lots of Fortune, 'Demon',[17]

---

13 For more commentary on the origins of the planetary lots in Late Antiquity, see Greenbaum and Jones, 'P.Berl. 9825', paragraphs 205–10.

14 Riess, *Fragmenta magica*, pp. 363–65; Heilen, 'Metrical Fragments', pp. 58–61; Heilen, *Hadriani genitura*, II, pp. 1161–63 and 1170–79.

15 Heilen, *Hadriani genitura*, II, pp. 1170–79; see also Greenbaum, *Daimon in Hellenistic Astrology*, pp. 332–34.

16 Dorotheus of Sidon, *Carmen astrologicum*, ed. and trans. by Pingree.

17 Pingree's translation (n. 16 above). However, the two (only extant) manuscripts (Berlin, MS Berol. or. oct. 2663 and Istanbul, MS Yeni Cami 784) show '*dīn*', 'religion', here (see p. 10 of the Arabic). Pingree has emended the Arabic text by inserting an 'm' to make the word an Arabic transliteration of 'demon'; but other Arabic authors (e.g., Sahl ibn Bishr and al-Bīrūnī) also use 'dīn' for the lot of absence (as well as *ghayb*, absent or hidden), so

and Eros, but does not call them planetary lots. The one Greek fragment of Dorotheus that mentions the formula for the Lot of Eros matches the one used by Vettius Valens.[18] Hephaestio quotes Dorotheus on the Lot of Eros, but provides no formula for it;[19] a related passage said to be Dorothean in *De triginta sex decanis* (also known as the *Liber Hermetis*) also mentions the 'Part of Cupid'.[20] In addition, Hephaestio also mentions the four lots, 'Fortune, Daimon, Necessity, Eros', in katarchic astrology.[21] A scholion to this passage provides a later gloss, mentioning two lot traditions, one associated with Hermes Trismegistus and one with 'the Egyptians'.[22]

C. Vettius Valens of Antioch (second century AD)

Vettius Valens of Antioch uses the Lots of Fortune, Daimon, Eros, and Necessity in his textbook, *Anthologies*. He does not call them planetary lots, and their formulae are all related: Fortune and Daimon, as we have seen above (in Section 1; see esp. n. 7), are mirror images of one another; and Eros and Necessity are formed from the arc between the Lots of Fortune and Daimon, making them also mirror images of one another. Thus the set of lots forms double mirror images, as Figure 1.2 shows.

Vettius Valens is arguably the astrologer most interested in the interpretation of these lots; he uses a number of examples from client charts to illustrate these interpretations. He also provides detailed descriptions of the lots.[23] When Valens mentions these lots as a group, he uses the order Fortune, Daimon, Eros, Necessity.

Pingree's emendation may be contentious. Benjamin Dykes's recent translation of the Arabic Dorotheus calls it 'the Lot of Religion' (Dorotheus of Sidon, *Carmen astrologicum*, trans. by Dykes, p. 72).

18 In Hugo of Santalla, *Liber Aristotilis*, ed. by Burnett and Pingree, p. 206, Appendix II, Fr. XVI, 6; the same Fr. XVI in English translation in Dorotheus of Sidon, *Carmen astrologicum*, trans. by Gramaglia, p. 338. Other Dorothean fragments naming the Lot of Eros do not give the formula (see Dorotheus of Sidon, *Carmen astrologicum*, ed. and trans. by Pingree, pp. 432–34).

19 Hephaestio of Thebes, *Apotelesmatica*, II, 23.10–11 and 16, ed. by Pingree, I, pp. 183.17–20 and 184.3–5.

20 Dorotheus of Sidon, *Carmen astrologicum*, ed. and trans. by Pingree, p. 433; Ch. 21, 13, Hermes Trismegistus, *De triginta sex decanis*, ed. by Feraboli, p. 73.41–44.

21 Hephaestio of Thebes, *Apotesmatica*, III, 6.11, ed. by Pingree, I, pp. 253.20–254.4.

22 Dorotheus of Sidon, *Carmen astrologicum*, ed. and trans. by Pingree, Scholium ad Heph., III, 6.11, pp. 433.14–434.1. This is why I called the two traditions 'Egyptian' and 'Hermetic'. Christian Bull's translation of this passage (*The Tradition of Hermes Trismegistus*, p. 37) should be viewed with caution, as it is flawed and shows no awareness of lot doctrine.

23 Vettius Valens, *Anthologies*, e.g. at IV, 25, ed. by Pingree, pp. 191–92.

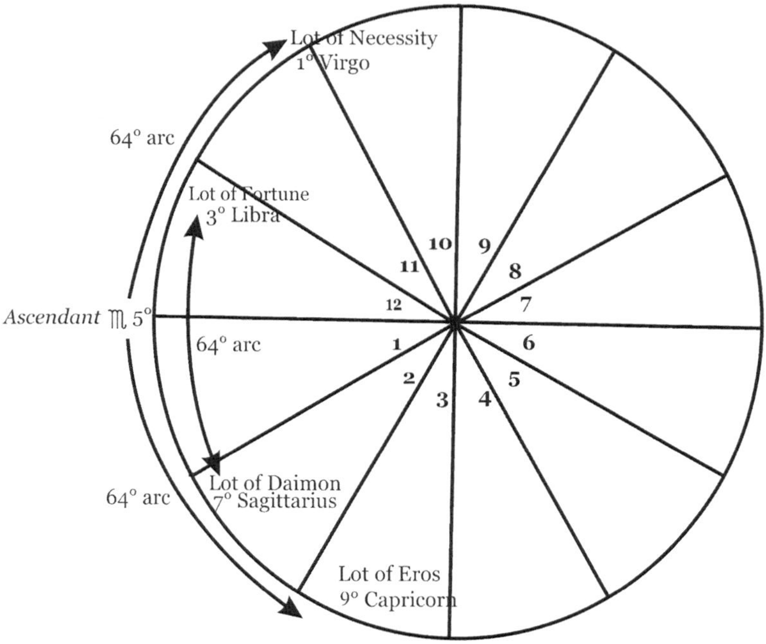

Figure 1.2. The Lots of Fortune, Daimon, Eros, and Necessity forming a double mirror image (diagram by the author).

D. Antiochus of Athens (second century AD?)

Though his dates likely lie in the second century AD,[24] we know the work of Antiochus of Athens mainly from sixth-century AD quotations of his *Thesauri* (*Treasuries*) in an epitome by Rhetorius (see below, Section 3-H), and fragments of an *Isagoge* (*Introduction*) preserved in a summary. Some quotations on lots (Ch. 47) involve planetary lots — though not identified as a group by that name[25] — since they mention, in order, the 'Moon's Lot' (Fortune) and the 'Sun's Lot' (Daimon), shortly followed by 'Kronos's Lot' (Nemesis), 'Zeus's Lot' (Victory), 'Ares' Lot' (Courage), 'Aphrodite's Lot' (Eros), and 'Hermes' Lot' (Necessity). He interprets the Lots of Fortune and Daimon in the *Thesauri*, Ch. 48.[26] The prose of Ch. 47 is often quite similar

24 See Pingree, 'Antiochus and Rhetorius', p. 204; Beck, *Religion of the Mithras Cult*, pp. 253–55; Greenbaum, *The Daimon in Hellenistic Astrology*, pp. 183–84.

25 Antiochus of Athens, *Thesauri*, Ch. 47, ed. by Franz Boll, *Catalogus Codicum Astrologorum Graecorum*, I, p. 160.11–28. This is part of Rhetorius's *Compendium astrologicum*, Book V, Ch. 47, ed. by Pingree and Heilen, in preparation.

26 Antiochus of Athens, *Thesauri*, Ch. 47, ed. by Franz Boll, *Catalogus Codicum Astrologorum Graecorum*, I, p. 161.1–21 and 23–25.

to the words used by Paulus in his 378 AD *Introduction to Astrology*, Ch. 23.[27] Could Antiochus have been a common source for Paulus and Rhetorius; or were Antiochus and Paulus drawing on a common source, later quoted by Rhetorius as being from Antiochus; or could the section on lots actually have been excerpted from Paulus and was not original to Antiochus?[28] Unfortunately, these passages provide no formulae for any of the lots mentioned.

E. Claudius Ptolemy (second century AD)

Although Ptolemy's work (both genuine and not, cf. the *Centiloquium*) was a major influence on many medieval astrologers writing in Arabic, his lot doctrine and philosophy appear to have had little impact on practice at this time. Ptolemy used only the Lot of Fortune in its day formula, which he describes as a 'σεληνιακὸς ὡροσκόπος' (lunar horoscope)[29] and has no use for 'κλήρων καὶ ἀριθμῶν ἀναιτιολογήτων' (lots and numbers for which no cause can be reckoned).[30]

F. Julius Firmicus Maternus (fourth century AD)

In *Mathesis*, written *c.* 338 AD, Julius Firmicus Maternus mentions the Parts of Fortune and Daemon. He often uses the word *locus* (place) rather than *pars* (part) or *sors* (lot) as well as the *locus Cupidinis et desideriorum* (place of Cupid and desires) and the *locus necessitatis* (place of necessity).[31] He does not call any of these 'planetary' lots. The formulae he uses for Cupid and Necessity are the reverse of Valens's formulae for Eros and Necessity.

G. Paulus of Alexandria (fl. 378 AD) and his commentator Olympiodorus (fl. 564 AD)

With his *Introduction to Astrology*, Paulus is the main expositor of what have become known as planetary lots. Olympiodorus the Younger, the Neoplatonist philosopher, wrote a Commentary on Paulus's *Introduction to Astrology* in 564 AD.[32] Paulus describes seven lots (Ch. 23), each associated with a planet,

27 See comparison in Greenbaum, *Daimon in Hellenistic Astrology*, p. 447.

28 This last possibility prompted by László, 'Rhetorius, Zeno's Astrologer', pp. 348–49.

29 Claudius Ptolemy, *Tetrabiblos*, ed. by Hübner, III, 11.5, p. 206.593 (= *Tetrabiblos*, ed. and trans. by Robbins, III, 10, pp. 276–77).

30 Claudius Ptolemy, *Tetrabiblos*, ed. by Hübner, III, 4.4, p. 177.197 (= *Tetrabiblos*, ed. and trans. by Robbins, III, 3, pp. 236–37); my translation.

31 Julius Firmicus Maternus, *Mathesis*, VI, 32.45–46, ed. by Kroll, Skutsch, and Ziegler, II, p. 187.3 and 13. In medieval astrology and later, the usual word for a lot is 'pars' in Latin and 'part' in English.

32 Though earlier editors, and the manuscripts, mention one 'Heliodorus' as the author, Olympiodorus has been established as the author of this astrological text: see Westerink,

though he does not specifically call the collection 'planetary lots'. Olympiodorus does say that there are seven lots to correspond with the number of the seven stars (i.e., planets, including the Sun and Moon), naming Hermes Trismegistus and a book called *Panaretos* ('All-virtuous') as their source.[33] Except for the Lots of Fortune and Daimon, which use the degree-arc between Sun and Moon, and whose positions are essential to the calculation of the other planetary lots,[34] each lot in this collection is calculated with the arc between a planet and either the Lot of Fortune or the Lot of Daimon: Daimon for the benefic planets, Venus and Jupiter; and Fortune for the malefics, Mars and Saturn, as well as for Mercury. The lots are listed in this order: Fortune, Daimon, Eros, Necessity, Courage, Victory, Nemesis. We may note here that rather than use Chaldean or other astronomical order, the first four lots are in the same order that Valens uses, with the lots of Mars, Jupiter and Saturn (which move from innermost to outermost planet) tacked on at the end.

H. Rhetorius (likely sixth century AD)[35]

Rhetorius was an Egyptian astrologer and compiler of works of earlier writers. In his *Compendium*[36] he gives considerations for judging a horoscope. The fifth consideration (in Ch. 54) addresses the interpretations of lots, specifically those of Fortune, Daimon, Basis, and Exaltation. No mention is made of planetary lots in this section. In his epitome of Antiochus's *Thesauri* (mentioned above in Section 3-D, Antiochus), Rhetorius cites the lots each associated with one of the seven classical planets in Ch. 47, though they are not specifically called 'planetary lots'. As we saw above in Section 3-D (Antiochus), there are similarities between this text and that of Paulus in his *Introduction*, and, as discussed above, this material may not be original to Antiochus. As a transmitter of this material, however, Rhetorius is a crucial source.

'Ein astrologisches Kolleg aus dem Jahre 564' and Warnon, 'Le commentaire attribué à Héliodore'.

33 Olympiodorus, Commentary on Paulus, *Introduction*, ed. by Boer, p. 42.6–7, 11–12: 'περὶ τούτων τῶν κλήρων γέγραπται Ἑρμῇ τῷ Τρισμεγίστῳ ἐν βίβλῳ λεγομένῃ Παναρέτῳ [...]. Οὗτος δὲ ὁ θεῖος ἀνὴρ ζ̅ ἔλεγεν εἶναι κλήρους πρὸς τὸν ἀριθμὸν τῶν ζ̅ ἀστέρων'. And see trans. by Greenbaum, p. 101. The reference to the *Panaretos* in Paulus is only in the title of Ch. 23, and may have been added later.

34 Olympiodorus, Commentary on Paulus, *Introduction*, ed. by Boer, p. 47.6–7 (trans. by Greenbaum, p. 104): 'τὸν τοῦ Δαίμονος καὶ τῆς Τύχης κλῆρον δεῖ προεκβάλλειν, ἐπειδὴ ἀπὸ τούτων καὶ τοὺς ἄλλους πάντας ἐκβάλλομεν·' (one must first cast out the Lot of Daimon and Fortune, since from these we cast out all the others).

35 See the persuasive argument of Holden in Rhetorius, *Astrological Compendium*, trans. by Holden pp. ix–x and 158; also László, 'Rhetorius, Zeno's Astrologer'.

36 Rhetorius, *Compendium astrologicum*, ed. by Pingree and Heilen. I am grateful to Stephan Heilen for help in providing some contents of this work prior to its publication.

I. Documentary and literary examples of charts (various dates)

'Documentary' are original charts on papyrus, ostraca or another medium; 'literary' are those in texts such as handbooks by the above authors. These examples contain groups of lots (the quartet Fortune/Daimon/Eros/Necessity, and 'planetary' lots as described above), up to the end of Late Antiquity.[37] In addition to the literary evidence already described, five extant charts calculate at least one and often all four of the lots of Fortune, Daimon, Eros, and Necessity, and one document calculates all seven 'planetary' lots:

1. P. Princeton 75 (138/161 AD), in Neugebauer and Van Hoesen, *Greek Horoscopes*, pp. 44–45, no. 138/161. The lots are mentioned and calculated in this order: Fortune, Daimon, Eros, Necessity. The Firmicus version of calculation is used.

2. P. Oxy. 4277 (probably second century AD), in Jones, *Astronomical Papyri from Oxyrhynchus*, I, pp. 284–86, II, pp. 420–27. In this order, the text calculates the Lots of Fortune, Daimon, Eros, and Necessity (some of the actual calculations are missing due to lacunae). The lots are designated as first to fourth in order. The Valens version of calculation is used.

3. PSI 23,a (338 AD), in Neugebauer and Van Hoesen, *Greek Horoscopes*, pp. 65–67, no. 338. The lots are mentioned and calculated in this order: Fortune, Daimon, Eros, Necessity. The Valens version of calculation is used.

4. P. Berlin 9825 (nativity of 319 AD), in Greenbaum and Jones, 'P. Berl. 9825'. The text contains all seven lots associated with a planet by Paulus, and all use his formulae. The table in which the planetary lots are presented, with their zodiacal positions, lists them in this order (as in Paulus): Fortune, Daimon, Eros, Necessity, Courage, Victory, Nemesis. The Paulus version of calculation is used.

5. Olympiodorus (fl. 564 AD), Commentary on Paulus, *Introduction to Astrology*, ed. by Boer, p. 59.8–15. A corrupted and tendentious passage calculates the Lot of Eros according to Paulus's formula.

6. In addition, the birth chart of Constantine VII Porphyrogenitus, though cast for a birth in 905 AD, closely follows the work of Dorotheus; its author quotes him frequently in the interpretation of the chart.[38] It calculates the

37 I am not including the documentary and literary charts that calculate only the Lot of Fortune and/or Lot of Daimon, but not any of the others, since these two lots have their own tradition separate from the other lots. See n. 8 above.

38 See Pingree, 'Horoscope of Constantine VII Porphyryogenitus'.

Lot of Eros, and its use of Hellenistic techniques permits its inclusion in this list of Hellenistic charts.[39] The Firmicus version of calculation is used.

Let us recap the salient points in the above material. The first appearance in extant sources of a group of seven lots each associated with a planet, and set apart as a group from other lots, occurs in the fourth century AD, specifically in the documentary chart cast for 319 AD, and in Paulus of Alexandria's *Introduction to Astrology* in 378 AD. Paulus's text shows similarities with the words in the *Thesauri* of Antiochus of Athens on these lots. An epitome of the *Thesauri* was later quoted by Rhetorius and ascribed to Antiochus, so possibly Paulus was either quoting (but not attributing) Antiochus, both Antiochus and Paulus were quoting from an earlier source no longer extant, or Rhetorius was actually quoting Paulus. Olympiodorus, Paulus's commentator, reiterates and adds to what Paulus has said, stating that these seven lots come from 'Hermes'. He specifically calls them the lots of the seven stars. A text called the *Panaretos* is said to be the Hermetic text in which these lots appear, but it is not mentioned in any extant writing until the work of Paulus and Olympiodorus.

We also know that there is another tradition that groups together the four Lots of Fortune, Daimon, Eros, and Necessity, does not associate them with planets, and uses different formulae than Paulus uses for the Lots of Eros and Necessity. These are used in original documentary horoscopes, discussed in detail by Vettius Valens as well as by Firmicus Maternus, and later mentioned by Hephaestio in connection with Dorotheus. This is the situation with the 'planetary' lots as we come to the work of Abū Maʿshar in the ninth century.

4. Abū Ma ʿshar and Planetary Lots

Abū Maʿshar extensively discusses lots in three extant works: *The Great Introduction to Astrology* (*Kitāb al-mudkhal al-kabīr ilā ʿilm aḥkām al-nujūm*), *The Abbreviation of the Introduction to Astrology* (*Kitāb al-mudkhal al-ṣaghīr*; *Ysagoga Minor*),[40] and *On the Revolutions of the Years of Nativities* (*Kitāb fī taḥāwil sinī al-mawālīd*; *Albumasaris de revolutionibus nativitatum*).[41]

---

39 For commentary and transcriptions of these charts, see Greenbaum, *Daimon in Hellenistic Astrology*, pp. 367–77.

40 Abū Maʿshar, *Abbreviation of the Introduction to Astrology*, ed. and trans. by Burnett, Yamamoto, and Yano (hereafter *Abbreviation*).

41 Abū Maʿshar, *On the Revolutions of the Years of Nativities*, trans. by Dykes; Greek edition, *De revolutionibus nativitatum*, ed. by Pingree. Other works of Abū Maʿshar, such as the *Book of Religions and Dynasties (On the Great Conjunctions)* mention lots from time to time, but not planetary lots as a rule and not as extensively as the three above-named treatises. The only 'planetary' lots mentioned in the *Book of Religions and Dynasties* (in Abū Maʿshar, *On Historical Astrology*, ed. and trans. by Yamamoto and Burnett, vol. I) are the Lot of Fortune (II, 4.4, 9, pp. 66–67; II, 7.9, pp. 74–75; and VIII, 1.11–12, pp. 480–81), the Lot of Absence and the Lot of Jupiter, called not that but the 'lot of elevation, victory, and prosperity' (II, 7.9,

The length of his expositions on lots in each of these texts demonstrates their importance in his estimation. *The Great Introduction* devotes an entire book to them; the *Abbreviation* an entire chapter; and they appear frequently in *On the Revolutions of the Years of Nativities*,[42] where they are considered interpretively important in nativities, profections, and revolutions. The planetary lots have a substantial presence in the overall lot discussions in each of these works. We shall consider the part they play in each of these works in turn. It is not easy to ascertain exactly where Abū Maʿshar obtained his information on the lots. In Book VIII of the *Great Introduction*, he mentions generically the ancients, Persians, Babylonians, Egyptians, and Greeks (VIII, 1.5; VIII, 2.3). His named sources are Ptolemy, Hermes, Valens, Zādān Farrūkh (al-Andarzaghar) and Tawfīl (Theophilus of Edessa) (VIII, 4.22, 24, 25). As authors for the formulae of the planetary lots he specifically cites 'Hermes and the astrologers preceding <us>' (VIII, 3.17).[43] More will be said about his sources below.

A. The Great Introduction

Abū Maʿshar divides the lots into three groups: first, the seven planetary lots; second, the eighty lots he associates with the astrological places; and third, the ten lots not falling into either of the previous categories — for a total of ninety-seven lots. These, in his view, are the 'authenticated lots' (VIII, 2.3), meaning those previously employed by prior writers. He says: وإنّما هي سهام يحتاج إليها في مواضع من المواليد وتحاويل السنين والمسائل والابتدائات (These are the lots that are necessary in <certain> places in nativities, revolutions of the years, questions, and beginnings) (VIII, 2.3).[44] For him, lots are an important tool in assessing the topics important to a client, such as wealth or marriage. Since a lot combines what he calls two 'natural indicators' of a topic, whose degree-arc is projected from a 'movable indicator', usually the Ascendant (VIII, 1.8), the astrological circumstances of where the lot falls can inform the astrologer about which indicator may be stronger, and therefore aid in accurate assessment.

---

pp. 116–17). A book on lots by Abū Maʿshar (*Kitab al-Sahmayn*), now lost, is mentioned in the *Book of Religions and Dynasties*, II, 5.17, in Abū Maʿshar, *On Historical Astrology*, ed. and trans. by Yamamoto and Burnett, I, p. 99; and also mentioned in the *Fihrist*: see Abū Maʿshar, *On Historical Astrology*, ed. and trans. by Yamamoto and Burnett, I, pp. 593–97.

42 See evidence of their frequency in Abū Maʿshar, *On the Revolutions of the Years of Nativities*, trans. by Dykes, index, p. 710. Forty-eight pages out of 239 in Pingree's Greek critical edition of *De revolutionibus nativitatum* (above, n. 41) mention the word κλῆρος (lot), often multiple times on the same page.

43 *Great Introduction*, I, pp. 844–45 (Arabic and English): هرمس والمتقدّمون من علماء النجوم; translation slightly modified by Liana Saif (private exchange).

44 *Great Introduction*, I, pp. 824–25 (Arabic and English).

The planetary lots are the first ones Abū Maʿshar enumerates, in VIII, 3.6–16,[45] a chapter entitled 'On the lots of the seven planets' (*fī ashām al-kawākib al-sabʿa; In partibus VII planetarum/De partibus stellarum*).[46] Each lot is associated with a planet, listed here with their day formula in parentheses (all project from the Ascendant and reverse for nocturnal births):

Table 1.2. Planetary Lots in the *Great Introduction*: Their Names and Day Formulae

Lot Name and its Associated Planet	**Day Formula for the Lot**
Lot of the Moon: Lot of Fortune	(from Sun to Moon)
Lot of the Sun: Lot of Absence/Hidden/Future	(from Moon to Sun)
Lot of Saturn: Lot of Bonds and Prison; the heavy lot	(from Saturn to Fortune)
Lot of Jupiter: Lot of Victory and Prosperity	(from Absence to Jupiter)
Lot of Mars: Lot of Courage	(from Mars to Fortune)
Lot of Venus: Lot of Love and Intimacy	(from Fortune to Absence)
Lot of Mercury: Lot of Poverty and Lack of Means	(from Absence to Fortune)

As is traditional in lot calculation in Hellenistic astrology, the indicator that is most strongly allied with the sect of the chart (diurnal or nocturnal) generally becomes the starting point from which the lot-arc is calculated (VIII, 3.3). For example, even though the Lot of Fortune is the lot of the Moon, the Lot of Fortune in a day chart is calculated from Sun to Moon, since the Sun is the luminary of the day (and from the Moon, the night luminary, to the Sun in a night chart). All the lots reverse the arc between the two fixed indicators for night charts.

Abū Maʿshar begins with the Lot of Fortune (VIII, 3.6–9), which he calls 'the lot of the Moon' and 'extracted from the Sun and Moon' (VIII, 3.6a).[47] He says: وهذا السهم مقدّم على السهام كما أنّ الشمس مقدّمة بالضياء على الكواكب وهو أعلى وأشرف السهام (This lot precedes the <other> lots, just as the Sun precedes the planets in brightness. It is the highest and most eminent of the lots) (VIII, 3.8).[48] Next (VIII, 3.10–11) is what is called the Lot of Daimon in Hellenistic astrology, but of course Abū Maʿshar does not call it that. He (or his Latin translator) calls it either the 'lot of absence' or 'hidden' (*sahm al-ghayb: absentiae* by Adelard of Bath,[49] *celati* by Hermann of Carinthia)[50] or the 'lot

45 *Great Introduction*, I, pp. 834–45 (Arabic and English).

46 *Great Introduction*, I, pp. 832–33 (Arabic and English); Abū Maʿshar, *Liber introductorii maioris*, ed. by Lemay, V, pp. 326–33, and VIII, pp. 152–54 (Latin translations by John of Seville and Hermann of Carinthia, respectively).

47 *Great Introduction*, I, pp. 834–35 (Arabic and English).

48 *Great Introduction*, I, pp. 836–37 (Arabic and English).

49 Abū Maʿshar, *Abbreviation*, Latin version of Adelard of Bath, *Sermo sextus*, 4, p. 128: 'Cehem vero abscentie'.

50 Abū Maʿshar, *Liber introductorii maioris*, ed. by Lemay, VIII, p. 153.162.

of future things' (*pars futurorum* by John of Seville)[51] (VIII, 3.10). The formula is the same as for the Lot of Daimon, from Moon to Sun by day; reverse at night. Abū Maʿshar says, وهذا السهم وسهم السعادة من أفضل السهام كلّها (This lot and the lot of fortune are the best of all the lots) (VIII, 3.10b).[52]

Having covered these two most important lots, he moves on to the planets, beginning with Saturn. Saturn's lot, whose two fixed indicators are Saturn and the Lot of Fortune, is 'heavy', indicating bonds and prison among other things (VIII, 3.12a).[53] The lot of Jupiter, 'possessor of property and assistance' (also power and victory), employs the Lot of Absence and Jupiter (VIII, 3.13).[54] The lot of Mars, 'the possessor of courage' (VIII, 3.14),[55] uses Mars and the Lot of Fortune. It also indicates bravery, boldness, harshness, murder, thievery and deception (VIII, 3.14) among other qualities.

He wraps up his introductory descriptions and formulae with the lots of Venus and Mercury. Venus, 'the possessor of love and intimacy' is reflected in the lot of love and intimacy, as well as desire, sexual intercourse, joy and pleasure (VIII, 3.15).[56] This lot by day takes the distance from the Lot of Fortune to the Lot of Absence (reverse at night). Finally, Mercury, 'the possessor of means' yields the 'lot of poverty and lack of means' (VIII, 3.16), indicating 'fear, hatred [...] enemies, anger [...] commerce, buying and selling [...] tricks, writing, calculation [...] and astronomy' (VIII, 3.16).[57] It takes the distance, by day, from the Lot of Absence to the Lot of Fortune (reverse at night). Thus we see that these two lots, unlike the others, do not use the position of the planet associated with them, and that their formulae reverse each other's.

B. Some Observations on Lot Order

Note that the planetary order in this list is Moon, Sun, Saturn, Jupiter, Mars, Venus, Mercury (thus luminaries, then planets from slowest to fastest). This translates to a lot order of Fortune, Absence, Bonds, Prosperity, Courage, Love, and Poverty. Interestingly, this is exactly the order used in Rhetorius's citation of *de facto* planetary lots in Ch. 47 (see above, Sections 3-D, Antiochus, and 3-H, Rhetorius). This suggests that Abū Maʿshar may have received it from a Persian or Arabic version of Rhetorius, or through an intermediary text utilising Rhetorius. We may also note that the two lots that do not employ a planetary position as one of the fixed indicators are given last in order, though this could merely be a function of the list moving, after the Moon and Sun, from outermost to innermost planets.

---

51 Abū Maʿshar, *Liber introductorii maioris*, ed. by Lemay, v, e.g. pp. 329.323, 330, 342 and 344.
52 *Great Introduction*, I, pp. 838–39 (Arabic and English).
53 *Great Introduction*, I pp. 840–41 (Arabic and English).
54 *Great Introduction*, I, pp. 842–43 (Arabic and English).
55 *Great Introduction*, I, pp. 842–43 (Arabic and English).
56 *Great Introduction*, I, pp. 844–45 (Arabic and English).
57 *Great Introduction*, I, pp. 844–45 (Arabic and English).

Some other observations can be made from looking at this list. First, the order of the lots and their planets is different than that used by Paulus and Olympiodorus.

Table 1.3. Orders of Lots and Planets, Abū Maʿshar and Paulus of Alexandria/Olympiodorus

Order of Lots/Planets in the *Great Introduction*		**Order of Lots/Planets in Paulus of Alexandria and Olympiodorus**	
Lot of Fortune	Moon	Lot of Fortune	Moon
Lot of Absence/Hidden/Future	Sun	Lot of Daimon	Sun
Lot of Bonds	Saturn	Lot of Eros	Venus
Lot of Prosperity	Jupiter	Lot of Necessity	Mercury
Lot of Courage	Mars	Lot of Courage	Mars
Lot of Love	Venus	Lot of Victory	Jupiter
Lot of Poverty/Lack of Means	Mercury	Lot of Nemesis	Saturn

Second, for the lots of the five non-luminaries, he uses the same formulae as Paulus for the lots of Saturn, Jupiter, and Mars, but for the lots of Venus and Mercury he follows the formulary tradition of Vettius Valens (see above, Table 1.1), using the degree-arc between the Lots of Fortune and Daimon (note that Abū Maʿshar uses the Valens formulae for these lots, not the version Firmicus employs). That he would use Valens rather than Firmicus makes sense because Valens ('Wālīs') is a commonly cited source for him, both in Book VIII and elsewhere in the *Great Introduction*, while Firmicus does not appear to be one of his sources.[58] But why switch from Paulus at all? Third, the descriptions of the lots of Venus and Mercury are interesting, especially the Lot of Mercury, whose qualities seem rather eclectic: poverty, lack of means, fighting, hatred, enemies, anger, buying and selling, tricks, writing, calculation, astronomy. We shall explore these odd switches and juxtapositions in more detail below.

C. *The Abbreviation of the Introduction*

In the *Abbreviation*, this group of lots is described in Ch. 6, 3–11.[59] The planetary lots are not identified as such, but they are listed in order at the beginning of the chapter, here shown with their names and day formula (all reverse for nocturnal births).

58 It is possible that Abū Maʿshar may have come across doctrines of Firmicus through an intermediary, but there is no evidence of him using Firmicus's formulae for the lots in his work. Arguably his most common source is Hermes, both generally in the *Great Introduction* and particularly in Book VIII.

59 Abū Maʿshar, *Abbreviation*, pp. 70–73 (Arabic and English).

Table 1.4. Lots in the *Abbreviation*: Name, Day Formula and Associated Planet

Name of the Lot and its Day Formula	**Planet Associated with the Lot**
'Lot of fortune [...] from the Sun to the Moon'	(Moon)
'Lot of the absent [...] from the Moon to the Sun'	(Sun)
'Lot of love and affection [...] from the lot of fortune to the lot of the absent'	(Venus)
Lot of firmness [...] basis of the ascendant [...] like the Lot of Venus [...] from the lot of fortune to the lot of the absent'[60]	(same formula as the 'Lot of Venus')
'Lot of poverty and lack of means [...] from the lot of the absent to the lot of fortune'	(Mercury)
'Lot of courage and boldness [...] from Mars to the degree of the lot of fortune'	(Mars)
'Lot of prosperity, overcoming and victory [...] from the lot of the absent to Jupiter'	(Jupiter)
'Lot of bonds, imprisonment [...] from Saturn to the degree of the lot of fortune'	(Saturn)

As in the *Great Introduction*, Abū Maʿshar uses the same formulae that Paulus does for the lots of Courage, Prosperity, and Bonds, and the formulae of Valens for the lots of Love, Firmness, and Poverty. However, his ordering of the lots here differs from that in the *Great Introduction* — and is, in fact, the same order that Paulus uses: Fortune, Daimon, Eros, Necessity, Courage, Victory, and Nemesis.

D. On the Revolutions of the Years of Nativities

Finally, in his book *On the Revolutions of the Years of Nativities*, Abū Maʿshar mentions three, what are *de facto* planetary, lots: Courage and Boldness (*al-shajāʿa wa-l-jurāʾ*), Prosperity and Victory (*al-falḥ wa-l-nuṣra*), and Reason and Logic (*al-ʿaql wa-l-manṭiq*).[61] These are calculated with exactly the Paulus

60 This lot, 'like the lot of Venus', is included in the series. It is mentioned as the Lot of Basis in: Rhetorius, *Compendium astrologicum*, ed. by Pingree and Heilen, Book V, Ch. 47, Sentence 3; Antiochus of Athens apud Rhetorius, *Thesauri*, Ch. 47, *Corpus Codicum Astrologorum Graecorum*, I, p. 160.16; Vettius Valens, *Anthologies*, II, 23.7–12, 18 and 22, and II, 27.2 and 6, ed. by Pingree, pp. 83.29–84.20, 85.1–5, 21–23 and 89.12, 21; and Olympiodorus in various lists of lots, sometimes with the usual formula, from Fortune to Daimon by day (Commentary on Paulus, *Introduction*, ed. by Boer, pp. 58.21 and 60.4–6; trans. by Greenbaum, p. 109), but also with the formula, by day, from Venus to Mercury (Commentary on Paulus, *Introduction*, ed. by Boer, pp. 54.5 and 58.21; trans. by Greenbaum, pp. 106 and 109).

61 Book III, Ch. 1. Transliterated and translated by Charles Burnett from BnF, MS ar. 2588, fol. 78^{r}, containing the 'Book of Abū Maʿshar On the Revolution of the years of the

formulae for the Mars lot, the Jupiter lot, and the Mercury lot. Since the chart is nocturnal, the nocturnal formulae are used for these lots.

Lot of Courage and Boldness: *sahm al-shajāʿa wa-l-jurāʾ* (Mars) (nocturnal formula: from Fortune to Mars)

Lot of Prosperity and Victory: *sahm al-falḥ wa-l-nuṣra* (Jupiter) (nocturnal formula, from Jupiter to Absence)

Lot of Reason and Logic: *sahm al-ʿaql wa-l-manṭiq* (Mercury) (nocturnal formula, from Fortune to Mercury)

These lots appear in the analysis of a birth chart often presumed to be Abū Maʿshar's own,[62] dated from the data to 10 August 787 at about 10:12 p. m. The position of the Lot of Courage and Boldness is given as Taurus 7°14', a position supported by its relationship to the given Ascendant mentioned in the text.[63] The position of the Lot of Prosperity is given as Taurus 12°06', and the Lot of Reason as Taurus 14°34'.[64] Their calculation is clearly based on the Paulus/Hermetic formulae.[65] These lots figure prominently in the subsequent interpretation associated with the chart. In this example none of the lots is specifically linked with a planet, nor are they called planetary lots,

Nativities' (fols 30ʳ–197ᵛ); sometimes these are abbreviated (e.g., fol. 77ᵛ, ll. 9–10) as 'Courage', 'Prosperity' and 'Reason'. See also Abū Maʿshar, *On the Revolutions of the Years of Nativities*, trans. by Dykes, pp. 291.26–292.39. This source helpfully lists the Arabic manuscripts of the text (p. 143), which led me to find BnF, MS ar. 2588. See also the list of manuscripts in Sezgin, *Geschichte des arabischen Schrifttums*, VII, p. 142. For the edition of the Greek text, see Abū Maʿshar, *De revolutionibus nativitatum*, III, 1, ed. by Pingree, pp. 130.1–131.14. The Greek names for the lots in question are κλῆρος τῆς ἀνδρείας καὶ τῆς τόλμης (lot of bravery and courage), κλῆρος τῆς προκοπῆς καὶ τῆς νίκης (lot of prosperity and victory), and κλῆρος τῆς φρονήσεως καὶ τῆς συνέσεως (lot of intellect and native wit).

62 See the bibliography pro and con in Abū Maʿshar, *Great Introduction*, I, p. 1 n. 2.

63 Abū Maʿshar, *On the Revolutions*, trans. by Dykes, pp. 290, 291, sentence 26 (4°20' between it and the Ascendant); this position is not included in the Greek version, but can be ascertained through the relationship to the Ascendant, which is supplied in the Greek, *De revolutionibus nativitatum*, III, 1, ed. by Pingree, p. 130.1–2.

64 Abū Maʿshar, *On the Revolutions*, trans. by Dykes, p. 290. There are mostly minor variations in the chart positions in the three manuscripts I have researched, BnF, MS ar. 2588, fol. 77ᵛ, ll. 3–6, 9–10 (*vidi*); Escorial MS Escorial ar. 917 (reproduced in chart form in *De revolutionibus nativitatum*, ed. by Pingree, p. 128); and Oxford, Bodl. Lib., MS Digby or. 5 (as translated by Dykes, pp. 289–90; *non vidi*). Note that the stated positions of the Lots of Courage and Prosperity are within a few minutes of their positions calculated using the given planetary data, but the Lot of Reason calculated using the planetary data among the three manuscripts is Taurus 18°52' ± 4', about 4 degrees from the stated position in the text. In the Greek, no exact positions for the three lots are given, but they are said to be in Taurus (their positions can be reconstructed from textual and planetary information given in the text); see Abū Maʿshar, *De revolutionibus nativitatum*, III, 1, ed. by Pingree, p. 129.3–8.

65 See the evidence for this laid out in Greenbaum, *Daimon in Hellenistic Astrology*, pp. 377–78 nn. 144–46.

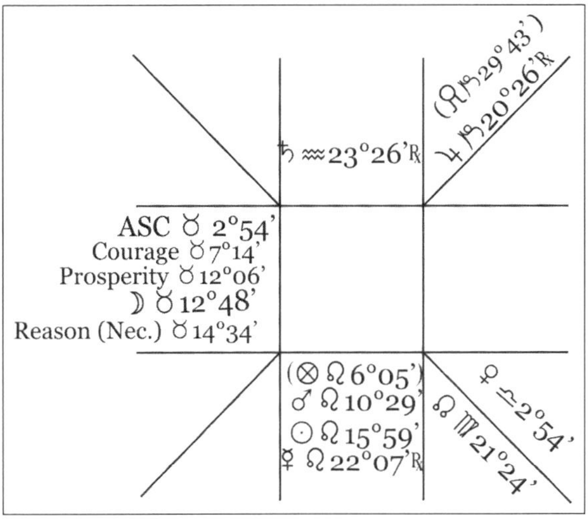

Figure 1.3. Birth chart of 10 August 787, 10:12 p. m. (diagram by the author). Positions inferred from text are in parentheses.

but the planetary lot formulae fit with the information given in the text, and we can connect them with the planets based on their names (e.g., 'Courage and Boldness' are qualities associated with Mars and with the name of his planetary lot; 'Prosperity and Victory' are associated with Jupiter and the name of the lot associated with him).[66]

The outlier in this example is the Lot of Reason and Logic, which was calculated with the same formula for Paulus's Mercury lot (i.e., his Lot of Necessity), but is not called, as in Paulus, the Lot of Necessity. And it certainly is not the Mercury lot of the *Great Introduction* (VIII, 3.16) — namely the lot of 'Poverty and Lack of Means' which uses by day the arc from the Lot of Absence to the Lot of Fortune (i.e., Daimon and Fortune). By calling it here the Lot of Reason and Logic, Abū Maʿshar is assigning to it something much closer to the usual qualities of Mercury, in that Mercury is typically associated with intellect, reason, speech, writing and calculation.[67] One might wonder whether this text, which illustrates an astrological practice, requires a different version of a 'Mercury lot', with a name reflecting its purpose. It also confirms that Abū Maʿshar was aware of the Mercury lot formula used by Paulus (it seems unlikely that he made up the formula himself, given its

66 Note that these lots, of Prosperity and Victory and of Courage and Boldness, are exactly the same names given for the planetary lots of Jupiter and Mars in the *Great Introduction*, VIII, 3.13 and 3.14 respectively, vol. I, pp. 842–43 (Arabic and English), but the Lot of Reason and Logic is not what the planetary lot of Mercury is called. For the Greek text, English translation and analysis, see Greenbaum, *Daimon in Hellenistic Astrology*, pp. 376–78 and 480–82.

67 In his description of the qualities of Mercury in the *Great Introduction*, VII, 9.8a, vol. I, pp. 812–13 (Arabic and English), Abū Maʿshar even uses, consecutively, the exact words, *ʿaql* and *manṭiq*, that are used in the name of this lot.

association in the text with two other known planetary lots), and that not using the formula he employs for the planetary lot of Mercury in the *Great Introduction* and *Abbreviation* (as he did for the other two 'planetary' lots here) was a deliberate choice.

5. Conflating and Creating Planetary Lot Traditions

We have now seen how Abū Maʿshar has presented his version of planetary lots, as well as a *de facto* use of planetary lot formulae in practice. What can this tell us about the transmission of the doctrine of planetary lots to the Arabic-speaking astrological world? What sources could Abū Maʿshar have drawn on, and what changes did he make to the doctrine of planetary lots? The following discussion will address these concerns.

A. Abū Maʿshar's Changes in Planetary Lot Doctrine

Since he has provided us with both descriptions and formulae for these lots, we can see that Abū Maʿshar has accepted the nomenclature of 'planetary lots' but he has not always used the formulae that go with this designation. In the *Great Introduction*, Abū Maʿshar designates a specific category of lots each associated with one of the seven planets. As we have seen, five of these lots do indeed use a planet or planets in their calculation, aligning with the practice of Paulus of Alexandria and Olympiodorus.[68] His Venus and Mercury lots, however, do not use the Paulus formulae that he employs for the other lots. He inserts the Valens formulae for these instead. Thus he is combining the traditions exemplified by two different authors in his descriptions and formulae. He still calls *all* of these lots 'planetary' (*ashām al-kawākib al-sabʿa*), though he does not address the fact that the lots of Venus and Mercury contain no planet in their formulae. Furthermore, he inserts qualities of Mercury into his description of the Lot of Mercury, and the Lot of Venus descriptions emphasise qualities of Venus. A closer look at these descriptions, and comparing them with the descriptions of earlier authors, will allow us to see what is similar and what is different in the way he treats these lots.

When Abū Maʿshar discusses 'The lot of Venus' (VIII, 3.15), he breaks the formulary pattern he has previously used for the three superior planets, Saturn, Jupiter, and Mars, whose lots use the arc between them and either the Lot of Fortune (for Saturn and Mars) or Absence (for Jupiter). Instead, as we have seen, he uses the arc between the lots of Fortune and Daimon. This is how he describes the Lot of Love:

---

68 The formulae for the lots of Fortune and Daimon remain stable across all ancient astrological authors except for Ptolemy, who only accepts the day formula for the Lot of Fortune because he can provide a physical basis for it, in its designation as a lunar ascendant (that is, there is the same proportion between the Moon and the Lot of Fortune as there is between the Sun and the Ascendant).

سهم الزهرة ذو الحبّ والألفة لمّا كانت موافقة الناس بعضهم لبعض وممازجتهم إنّما تكون بالحبّ والمقة والهوى والمزاوجة وكان جميع الأزواج والنكاح والمجامعة والتزويج والموافقات واللهو والطرب منسوبا إلى الزهرة وهي الدالّة عليه والمودّة والحبّ وسائر ما ذكرنا من دلالات الزهرة من السرور والفرح هي كلّها سعادات حسبوا سهم الزهرة من سهم النيّرين الدالّين على السعادة وقالوا سهم الزهرة ذو الحبّ والألفة يؤخذ بالنهار من سهم السعادة إلى سهم الغيب وبالليل مقلوبا ويزاد عليه ما طلع من أوّل برج الطالع ويلقى من أوّل الطالع فحيث ينتهي الحساب فهناك السهم وهو يدلّ على الشهوة والرغبة في الجماع والمودّة والطلب لما هويت النفس وسرّت وقويت به والحبّ وجميع أمور النكاح والأزواج والألفة واللهو واللذّة والطيب.

(Since the agreement of people with each other and their mixture is with love, tender love, affection, and pairing, and all couples, love-making, sexual intercourse, marrying, agreements, amusement, and joy are related to Venus who indicates this; and love, and affection and the other indications of Venus we mentioned among delights and joy are all good fortunes, they counted the lot of Venus from the lot<s> of the luminaries, indicating good fortune, and said that the lot of Venus, the possessor of love and intimacy, is taken by day from the lot of fortune to the lot of absence, and by night the opposite. [...] It indicates desire and wish for sexual intercourse, love and the pursuit of what the soul likes and delights in, and by which it is strengthened, love and all matters of love-making; and couples, intimacy, amusement, joy and pleasure.) (VIII, 3.15)[69]

This begs the question: why, if Venus bestows such good things, is she not included as a component in the formula for the lot? We know that Abū Maʿshar has used the Valens formula for the Lot of Eros to calculate this lot. How does Valens describe the Lot of Eros, which for him is fundamentally concerned with matters of love, desire and intimacy? He says: 'ἔρως περὶ ἐπιθυμίας' ((*The Lot of*) Eros (*gives signs*) about desire);[70] and, in interpretation:

Ὁ ἔρως παραδιδοὺς ἢ παραλαμβάνων ἐν χρηματιστικοῖς τόποις, καὶ ἀγαθοποιῶν ἐπόντων ἢ μαρτυρούντων, εὐπροαιρέτους ἐπιθυμίας κατασκευάζει καὶ καλῶν ἐραστάς· οἱ μὲν γὰρ περὶ παιδείαν καὶ ἄσκησιν σωματικὴν ἢ μουσικὴν τρέπονται καὶ μεθ' ἡδονῆς κολακευόμενοι τῇ μελλούσῃ ἐλπίδι ἀκοπίαστον ἡγοῦνται τὴν πρόνοιαν, οἱ δὲ ἀφροδισίοις καὶ συνηθείαις θελχθέντες θηλυκῶν τε καὶ ἀρρενικῶν ἀγαθὸν ἡγοῦνται.

(The (*Lot of*) Eros handing or taking over in profitable places, and with benefics in them or witnessing, furnishes desires with a good moral purpose and lovers of beautiful things. For some are turned toward education and bodily or musical training and, being softened up with pleasure in future hope, they believe in foresight (*pronoia*) as

69 *Great Introduction*, I, pp. 844–45 (Arabic and English).
70 Vettius Valens, *Anthologies*, II, 16.1, ed. by Pingree, p. 67.8–9.

> untiring; but some, being beguiled by sexual pleasures and intimacies both with women and men, believe it good.)[71]

When the lot with this name appeared in Paulus's *Introduction*, those descriptions did not change much, but Paulus now specifically linked the lot with Venus.[72] He says: '῾Ο δὲ ῎Ερως σημαίνει τὰς ὀρέξεις καὶ τὰς ἐπιθυμίας τὰς κατὰ προαίρεσιν γινομένας, φιλίας τε καὶ χάριτος παραίτιος καθέστηκεν' (Eros signifies appetites and desires occurring by choice, and it becomes responsible for friendship and favour).[73] Olympiodorus concurs (he often quotes Paulus *verbatim*).[74] The descriptor 'friendship' appears in documentary chart evidence as well: '[ο κληρος ε]ρωτος [...] [σημαιν]ει τον περι φιλιας και συστασεως' ((*the Lot of*) Eros [...] (*signifies*) friendship and association).[75] In the medieval *Liber Aristotilis*, the name of the lot uses similar wording: 'pars <appetitus> vel desiderii de die a parte fortune ad partem spiritalem' (part of appetite or desire by day from the Part of fortune to the Spiritual part).[76]

When Abū Maʿshar moves to the 'Lot of Mercury', again there is a break with the formulary pattern:

> سهم عطارد ذو الحيلة سهم عطارد وهو سهم الفقر وقلّة الحيلة يؤخذ بالنهار من سهم الغيب إلى سهم السعادة وبالليل مخالفا ويزاد عليه ما طلع من أوّل برج الطالع ويلقى من الطالع فحيث ينتهي الحساب فهناك هذا السهم وهو يدلّ على الفقر والقتال والخوف والبغض وكثرة الخصومة والأعداء والغضب والخصومة في وقت الغضب والتجارة والشرى والبيع والمكر والحيل والكتابة والحساب وطلب العلوم المختلفة والنجوم.

> (The lot of Mercury, the possessor of means. The lot of Mercury, which is the lot of poverty and lack of means (*sahm al-faqr wa-qilla al-ḥīla*), is taken by day from the lot of absence to the lot of fortune and by night the opposite. [...] It indicates poverty, fighting, fear, hatred, frequent quarrelling, enemies, anger, quarrels in the moment of anger, commerce, buying and selling, cunning, tricks, writing, calculation, seeking out various sciences and astronomy.) (VIII, 3.16)[77]

This is a rather odd list, even if we put aside, for the moment, the obvious question of what Mercury has to do with poverty and lack of means. The first half, up to 'quarrels in the moment of anger' seems to describe indications opposite to those of the lot of Venus — love and affection versus hatred and anger, agreements versus enmities, etc. But the second half of the list

71 Vettius Valens, *Anthologies*, IV, 25.5, ed. by Pingree, p. 191.27–192.1, my translation.
72 Paulus of Alexandria, *Introduction to Astrology*, Ch. 23, ed. by Boer, p. 49.13.
73 Paulus of Alexandria, *Introduction to Astrology*, Ch. 23, ed. by Boer, p. 50.1–3.
74 Olympiodorus, Commentary on Paulus, *Introduction*, Ch. 22, ed. by Boer, p. 57.5–7.
75 Neugebauer and Van Hoesen, *Greek Horoscopes*, p. 44, no. 138/161, ll. 6–7.
76 Hugo of Santalla, *Liber Aristotilis*, III, xii, 3.3, ed. by Burnett and Pingree, p. 97.
77 *Great Introduction*, I, pp. 844–45 (Arabic and English).

clearly contains Mercury-related qualities. It seems that two different sets of indications comprise the description of this lot.[78]

For this lot Abū Maʿshar uses the same formula that Valens does for his Lot of Necessity. Valens tells us that the Lot of Necessity, which is the inverse of the Lot of Eros and is its mirror image in the chart, gives signs 'περὶ ἐχθρῶν' (about enemies)[79] and produces 'ἐχθρῶν καθαιρέσεις ἢ θανάτους [...] ἀντιδικίας καὶ κρίσεις' (subjugations or deaths of enemies [...] litigations and trials).[80] For Paulus, the Lot of Necessity (using the arc between Mercury and Fortune) signifies: 'συνοχὰς καὶ ὑποταγὰς καὶ μάχας καὶ πολέμους, ἔχθρας τε καὶ μῖσος καὶ καταδίκας καὶ τὰ ἄλλα πάντα τὰ τοῖς ἀνθρώποις συμβαίνοντα βίαια πράγματα ἐν γέννᾳ ποιεῖ' (confinements, subjections, battles and wars, and it makes enmities, hatred, condemnations and all the other constraining circumstances that happen to humans as their lot in life).[81]

Though the delineations of Valens and Paulus are similar, they apply to lots with different formulae. In these descriptions we see that Eros and Necessity form an oppositional pair — love-hate, friendship-enmity. This accords with their earlier, Egyptian, formulae, where they form a mirror image that connects them but also allows for some contrasting qualities. The same documentary chart that tells us that the Lot of Eros 'concerns friendship and association' says that the Lot of Necessity 'περι εκχθρων και παντ̣ο̣δα̣π̣ου [δυςτυχημα]τοc' (concerns enemies and every kind of [misfortune]).[82]

It is also important to note here that love (Eros) and necessity (Ananke) have a long cultural history as opposing pairs, or opposite sides of the same coin, in the Graeco-Roman world.[83] The concept of necessity includes the idea of constraint, confinement, compulsion and subjugation. Possibly the idea of constraint allows a link to the idea of poverty or lack of means that we see in Abū Maʿshar's description for the Lot of Mercury. But here Abū Maʿshar does not use what is certainly the cognate to ἀνάγκη in Arabic, *iḍṭirār*,[84] the word

78 We should note that in *Great Introduction*, VII, 9.8c, vol. I, pp. 812–13 (Arabic and English) some qualities of Mercury include 'fighters, enmity, heavy loss from enemies and much fear from them'. But whether these are listed as a result of Abū Maʿshar's understanding of the lot of Mercury, or instead they were there beforehand as Mercurial qualities, cannot be determined. Also, Mercury is not associated with poverty in this section.

79 Vettius Valens, *Anthologies*, II, 16.1, ed. by Pingree, p. 67.9.

80 Vettius Valens, *Anthologies*, IV, 25.14, ed. by Pingree, p. 192.27–28.

81 Paulus of Alexandria, *Introduction to Astrology*, Ch. 23, ed. by Boer, p. 50.4–7; Olympiodorus repeats the list: Olympiodorus, Commentary on Paulus, *Introduction*, ed. by Boer, p. 57.8–10.

82 Neugebauer and Van Hoesen, *Greek Horoscopes*, p. 44, no. 138/161, ll. 9–10.

83 See Greenbaum, *Daimon in Hellenistic Astrology*, pp. 339–56.

84 He does use this word for another lot, that he calls the lot of necessity, the 8th lot of the 11th place, formula from the lot of absence to Mercury, in both day and night (VIII, 2.5l, vol. I, p. 828.14 Arabic, p. 829 English; also, with its formula, VIII, 4.78, p. 904.16 Arabic, p. 905 English, and VIII, 6.14b, p. 930.10 Arabic, p. 931 English); but a full examination of this issue is beyond the scope of this paper.

used by at least one Arabic author for the Lot of Necessity in this formula.[85] Instead, he connects the lot with poverty (*faqr*), but also chooses to use a word with Mercurial connotations: *ḥīla*, a word that, in addition to 'means', can mean trick, ruse, device, artifice, or strategy.[86] The first sentence of VIII, 3.16, *al-sahm al-ʿuṭārid dhū al-ḥīla* (The lot of Mercury, possessor of means),[87] seems worded (like the other planets in this section) to emphasise this quality of the planet, though connected with a lot that does not use Mercury as one of its indicators. But unlike the other planetary lots, where the description of both planet and lot match, the next sentence calls it not the 'lot of means' but the 'lot of *poverty and lack* of means'.[88]

So because he calls it the Lot of Mercury and uses words with Mercury-like meanings to describe it, clearly qualities of Mercury are important to Abū Maʿshar for this lot, even though he uses Valens's formula for the Lot of Necessity for his Lot of Poverty. We do not see either 'poverty' or Mercurial qualities associated with the lot in the earlier Hellenistic tradition: the best that Olympiodorus can do in linking Necessity with Mercury is to say that Mercury 'κύριος ἐστι λόγου, ὁ δὲ λόγος ἀναγκαστικὸς ὑπάρχει' (is lord of reason, and reason is actually necessary).[89] So it is clear that Abū Maʿshar is innovating here in conflating the two traditions both through the formulae he chooses and in descriptions. He appears to be the first to do this. In addition, he is the first Arabic writer to mention a group of 'lots of the seven planets'. What sources could have been available to him from previous planetary lot traditions?

---

85 Sahl ibn Bishr quoting al-Andarzaghar, in *Kitāb al-mawālīd* (*Book of Nativities*), Escorial, MS Escorial ar. 1636, fol. 92v, l. 7. See also Sahl ibn Bishr, *Book of Nativities*, trans. by Dykes, p. 727. See the text accompanying n. 96 for the quotation and further discussion.

86 'Arabic, English, Greek and Latin Glossary', in *Great Introduction*, II, p. 185. It can also connote the idea of engineering or using a technique. I thank Charles Burnett for discussing this word and its Mercurial connotations with me.

87 *Great Introduction*, I, pp. 844.9 (Arabic) and 845 (English).

88 Using *faqr* to name the lot (a word that can mean lack in general as well as poverty), followed by *qilla* (also 'lack' in the sense of smallness or lesser amount), reminds us of a surely coincidental, but intriguing, parallel using the vocabulary of love and poverty in Plato, *Symposium*, 202d: Eros, Diotima's 'great daimon', is the offspring of Poros (Provision or Resources) and Penia (Poverty). The Greek translation of this passage even uses πενία to translate *faqr* (*Great Introduction*, VIII, 3.16, vol. I, p. 844.11 (Arabic); for the Greek, Ch. 61, VIII, 3.16[2], vol. II, p. 101.

89 Olympiodorus, Commentary on Paulus, *Introduction*, ed. by Boer, p. 42.23–43.1.

B. Possible Sources for Abū Maʿshar for the Lots of 'Love' and 'Poverty' and the Other Planetary Lots

As general sources of astrological doctrines, we know that Abū Maʿshar had access to Ptolemy, Hermes, Dorotheus and Vettius Valens. Sources cited for specific lots in Book viii of the *Great Introduction* include Hermes, the ancients, Valens, Zādān Farrūkh (al-Andarzaghar) and Tawfīl (Theophilus of Edessa).

Let us first consider the order in which Abū Maʿshar lists the planetary lots in the *Great Introduction*. We have seen above in Section 4-B that this order strongly suggests his source was Rhetorius, the only one who uses exactly that order in his quotation from Antiochus's *Thesauri*, Ch. 47: Fortune/Moon, Daimon/Sun, Nemesis/Saturn, Victory/Jupiter, Courage/Mars, Love/Venus, and Necessity/Mercury.[90] We know that material traceable to Rhetorius was available to Arabic astrologers (more on this below in this section).

The sources for the formulae must also be taken into account. At the end of his enumerations of the planetary lots, Abū Maʿshar says: فهكذا يستخرج سهام الكواكب السبعة على ما ذكره هرمس والمتقدّمون من علماء النجوم (Thus the lots of the seven planets are extracted according to what Hermes *and the astrologers preceding* <*us*> related) (viii, 3.17).[91] Such a statement permits room for sources for the lot formulae other than Hermes (whose lot system is earliest named and exemplified in Paulus), by offering alternative origins for some of them (presumably those of Venus and Mercury, since these are the only formulae different from the Hermetic ones).

And here the use of the Valens formulae (perhaps originally derived from Dorotheus) for the Lot of Love and Lot of Poverty suggest as well a source in Valens, who is one of the authors specifically mentioned for lots in Book viii. In addition, an excerpt of Dorotheus in Greek also contains the same formula for the Lot of Eros:

90 The list is in: Rhetorius, *Compendium astrologicum*, ed. by Pingree and Heilen, Book V, Ch. 47.1–8, in preparation; Antiochus of Athens apud Rhetorius, *Thesauri*, Ch. 47, ed. by Franz Boll, *Catalogus Codicum Astrologorum Graecorum*, i, p. 160.11–28; and see Greenbaum, *Daimon in Hellenistic Astrology*, pp. 447–48 for the CCAG text and translation. Between the lots of the Sun and Saturn, Rhetorius (in fact, quoting Paulus) gives a 'Lot of the Ascendant', i.e. the Lot of Basis or Foundation, responsible for 'life and breath': Rhetorius, *Compendium astrologicum*, ed. by Pingree and Heilen, Book V, Ch. 47.3, in preparation; Antiochus of Athens, *Thesauri*, Ch. 47, ed. by Franz Boll, *Catalogus Codicum Astrologorum Graecorum*, i, p. 160.16. The order is otherwise identical to Abū Maʿshar's. It makes sense that Abū Maʿshar would have left that lot out of his list because he was specifying only 'lots of the seven planets' (the title of Ch. 3 of Book viii). Ch. 4 covers the lots of the twelve places, of which the Ascendant is first, and specifically what corresponds to the Greek Lot of Basis appears in the *Great Introduction*, viii, 4.6a, vol. i, pp. 852–53 (Arabic and English): 'the lot of firmness and survival [...] the lot of the support of the ascendant'.

91 *Great Introduction*, i, pp. 844–45 (Arabic and English). My italics; translation slightly modified by Liana Saif.

τηρείτω δὲ καὶ [...] τὸν κλῆρον ἔρωτος (ὅς ἐστιν ἡμέρα<ς> ἀπὸ κλήρου τύχης εἰς τὸν κλῆρον δαίμονος καὶ τὰ ἴσα ἀπὸ ὡροσκόπου, νυκτὸς δὲ τὸ ἀνάπαλιν)

> (Let attention be paid to [...] the Lot of Eros (which is by day from [the] Lot of Fortune to the Lot of Daimon and the equal amount from the Hour-marker [Ascendant], but by night the reverse.)[92]

Since it is doubtful that Abū Maʿshar had any knowledge of Greek,[93] such an excerpt had to have been available to him in Arabic or Persian, but no such passage directly ascribed to Dorotheus exists to provide definite proof of this. However, it seems certain that material on the lots in question, whether ascribed to Valens, Dorotheus, and/or Rhetorius (or all three), or transmitted, whether ascribed or not, through an Arabic writer, was available in Arabic (or Persian). Evidence in the much later *Liber Aristotilis* may point to such a source used by Sahl ibn Bishr (*c.* 786–*c.* 845), a contemporary of Abū Maʿshar, as a transmitter of some of this doctrine.

The *Liber Aristotilis* (at III, xii, 1.2) lists the four traditional lots of Fortune, Daimon, Eros and Necessity in a passage on friendship:[94] 'Prima quidem pars fortune, secunda pars spiritalis quam absentis partem nuncupant, tercia pars voluptatis sive appetitus, quarta namque pars necessitatis dicitur' (First the Part of Fortune, second the spiritual part which they call the Part of Absence, third the Part of Passion or Appetite, and the fourth is said to be the Part of Necessity).[95]

A similar passage on friendship in Sahl ibn Bishr's *Kitāb al-mawālīd* (*Book of Nativities*) cites al-Andarzaghar and mentions the same lots, preceded by the lot of friends:

الاندوزغر انظر الى هذه الخمسة ... الثاني سهم السعادة الثالث سهم الدين الرابع سهم الشهو الخامس سهم الاضطرار.

> (Andūzaghar: look at these five [lots] [...]. The second is the lot of fortune (*saʿāda*). The third is the lot of religion (*dīn*) [i.e., absence]. The fourth is the lot of desire (*shahwa*). The fifth is the lot of necessity (*iḍṭirār*).)[96]

---

92 Dorotheus of Sidon, Excerpt XVI, 6; my translation. This is one of 69 Dorothean excerpts from BAV, MS Vat. gr. 1056, fols 238^{r}–241^{v} (this passage fol. 239^{r}.8–10). The excerpts first appeared, transcribed by Pingree, in Appendix II of Hugo of Santalla, *Liber Aristotilis*, pp. 204–14, at 206.

93 Charles Burnett considers it unlikely that Abū Maʿshar knew Greek (email communication 19 Sept 2018).

94 Similar passages on friendship in Hephaestio and *De triginta sex decanis* (*Liber Hermetis*) are associated with Dorotheus. See Section 3-B above.

95 Hugo of Santalla, *Liber Aristotilis*, ed. by Burnett and Pingree, p. 96; my translation.

96 Escorial, MS Escorial arab. 1636, fol. 92^{v}, 4–7; Tehran, Majlis, MS Majlis 6484, fol. 134^{v}, 11–13, transcription and translation by Charles Burnett; cited and translated by David Pingree in Hugo of Santalla, *Liber Aristotilis*, p. 192 (manuscripts listed on p. 11). I warmly thank Rosa Comes for kindly obtaining for me a copy of fol. 92^{v} of Escorial, MS Escorial arab. 1636, and

The *Liber Aristotilis* further mentions the Lot of Appetite (Passion) or Desire (*pars <appetitus> vel desiderii*) with its formula at Book III, xii, 3.3; and the entire corresponding chapter of Book III, xii, 3.1–10 appears in Sahl's *Book of Nativities*.[97] In addition, the Lot of Necessity is discussed without mention of its formula in *Liber Aristotilis*, Book III, xii, 5, with another corresponding passage preserved in Sahl's *Book of Nativities*.[98] This text of Sahl, in fact, is full of material traceable to both Dorotheus and Rhetorius.[99] Abū Maʿshar could, therefore, have drawn on sources similar to Sahl's, such as al-Andarzaghar.[100] The information on lots in these texts and their sources, therefore, provides (at the least) evidence of parallel Arabic interests and connections to the Lots of Love and Poverty discussed by Abū Maʿshar, as well as other lots (i.e., Fortune and Absence) connected to what becomes his planetary lot scheme. Certain proof of the use of a specific Arabic text by Abū Maʿshar, though, remains elusive. The circumstantial evidence, however, involves sources we already know Abū Maʿshar used in other contexts, namely Dorotheus, Valens, Rhetorius, Zādān Farrūkh al-Andarzaghar, Māshāʾallāh, and Theophilus of Edessa.

Charles Burnett and Juan Acevedo for transliterating and translating lines 4–7 of the Arabic. This research was done prior to the 2019 publication of an English translation of Sahl's *Kitāb al-mawālīd*, in *The Astrology of Sahl B. Bishr Volume I: Principles, Elections, Questions, Nativities*, trans. by Dykes; the same passage is at Ch. 11, sentence 5, p. 727. This text has supplied more material supporting the involvement of al-Andarzaghar in Sahl's work: on pp. 30–32, Dykes listed 11 passages explicitly attributed to him by Sahl, as well as other material traceable to him. These conclusions are in line with my own. Thus Sahl seems to be an important transmitter of al-Andarzaghar's work (who himself is a transmitter of the work of Dorotheus and Rhetorius).

97 Escorial, MS Escorial arab. 1636, fols 94^{r}.16–94^{v}.12; Tehran, Majlis, MS Majlis 6484, fols 137^{r}.6–137^{v}.4; cited and translated by Pingree in Hugo of Santalla, *Liber Aristotilis*, pp. 193–94. The corresponding passage in Sahl, *Nativities*, is Ch. 11.2, sentence 5, in *Astrology of Sahl B. Bishr Volume 1*, trans. by Dykes, p. 733.

98 Escorial, MS Escorial arab. 1636, fol. 96^{r}.11–15; Tehran, Majlis, MS Majlis 6484, fol. 139^{v}.4–8; cited and translated by Pingree in Hugo of Santalla, *Liber Aristotilis*, p. 195; Sahl, *Nativities*, Ch. 11.4, sentence 18, in *Astrology of Sahl B. Bishr Volume 1*, trans. by Dykes, p. 740. Given the order of the lots at III xii, 1.2, it seems highly likely that the formula for the Lot of Necessity is the one that reverses the indicators for the Lot of Love; Pingree was also of this opinion, p. 195; Dykes's comment, p. 740 n. 62, assumes it.

99 Hugo of Santalla, *Liber Aristotilis*, ed. by Burnett and Pingree, p. 8. The notes in Dykes's translation of Sahl's *Nativities* also cite these authors.

100 One could plausibly suggest that the source used by both Sahl and Abū Maʿshar is al-Andarzaghar's *Book of Nativities*, which would necessarily have included more than the material on anniversary horoscopes (as published in Burnett and al-Hamdi, 'Zādānfarrūkh al-Andarzaghar on Anniversary Horoscopes'). We find other citations of al-Andarzaghar's *Nativities* by al-Qabīṣī (*Introduction to Astrology*, ed. and trans. by Burnett, Yamamoto, and Yano, Ch. 1.57, pp. 51–55; see also p. 9), which show his organization of topics by place; and his Ch. v, on lots, is organised by place as well; interestingly, there is no section specifically for planetary lots (see below, Section 6-A). Abū Maʿshar's listings of the place lots in the *Great Introduction* show some similarity in order (though not quantity) with Sahl's inclusion of lots in *Nativities* within his chapters on the places and with the *Liber Aristotilis*. I am preparing a study of these place lots.

For the formulae of the other planetary lots of Courage, Prosperity and Bonds, we must further infer that he had some kind of access to the information about lots in Paulus or Olympiodorus or both, through a source in Persian or Arabic.[101] In writing about sources for the *Liber Aristotilis*, Pingree states Māshāʾallāh 'certainly had an Arabic copy of Rhetorius' work' through Theophilus of Edessa.[102] The *Liber Aristotilis* also has connections to Buzurjmihr and, as we saw above, al-Andarzaghar.[103] So such texts were undoubtedly circulating among Arabic astrological writers. But it was not until Abū Maʿshar made his detailed investigation of lots (and for our purposes here, specifically planetary lots) that a systematic lot doctrine came into medieval Arabic astrology.[104]

As we shall now see, his authority was such that his innovation became the standard for other Arabic astrologers, as well as medieval astrologers writing in other languages like Abraham Ibn Ezra.[105] Latin translators, as well, chose him as the representative of Arabic astrology;[106] and Guido Bonatti quoted Abū Maʿshar extensively, in the translation of the *Great Introduction* by John of Seville.[107] The following will show by brief examples the transmission of this 'new' tradition of planetary lots.

---

101 In the case of Olympiodorus, not only for the planetary lots: many of the other lots in his lists and discussion of 97 lots in the *Great Introduction* appear in Olympiodorus's commentary. Paulus's Ch. 23 on lots is contained in an abridgment of his *Introduction* that appears in Rhetorius, *Compendium astrologicum*, Book VI, as Ch. 29. Material on lots from Olympiodorus appears as well in Book VI, Ch. 5.103–22, including some lots later mentioned by Abū Maʿshar.

102 Hugo of Santalla, *Liber Aristotilis*, ed. by Burnett and Pingree, p. 140. This claim has recently been contested by Benjamin Dykes (ed.), *Astrological Works of Theophilus of Edessa*, pp. 36–42, who suggests that al-Andarzaghar was more likely the transmitter of Rhetorius into Arabic astrology, through authorship of the *Liber Aristotilis* (with an Arabic translation by Māshāʾallāh). He expands on this position, with more evidence and detail, in *Astrology of Sahl B. Bishr*, I, pp. 27–32.

103 Hugo of Santalla, *Liber Artistotilis*, ed. by Burnett and Pingree, p. 140.

104 While the systems of earlier astrologers may have included lots within their discussions of astrological places (here I am thinking of al-Andarzaghar), and this may have influenced Abū Maʿshar's second group of lots based on places, it is clear that it was Abū Maʿshar's innovative and specific system of lots in three groups of 7, 80 and 10 that gained currency in the medieval tradition. (Note that no less an authority than Fuat Sezgin, *Geschichte des arabischen Schrifttums*, VII, p. 141, described the *Great Introduction* as 'anscheinend das populärste astrologische Buch im Abendland').

105 For Ibn Ezra's use of Abū Maʿshar, and as a translator from Arabic to Hebrew, see Ibn Ezra, *Beginning of Wisdom*, ed. and trans. by Sela, pp. 10–11.

106 'Introduction', in *Great Introduction*, I, p. 5.

107 This Latin version is in Abū Maʿshar, *Liber introductorii maioris*, ed. by Lemay, V.

6. The Afterlife of Abū Maʿshar's System of Planetary Lots

As we examine the systems of astrologers who followed Abū Maʿshar on the planetary lots, it is important to keep in mind that all of them use the same formulae as Abū Maʿshar does, and in many cases Abū Maʿshar's influence is obvious.

A. al-Qabīṣī (fl. 945–967)

In al-Qabīṣī's *Introduction*, the lots associated with the seven planets first appear in Ch. 2 (on the natures and conditions of the seven planets). Each planet's lot is given except for the Moon's lot (i.e., Lot of Fortune) — perhaps it was missed out? No formulae are given and the descriptions are brief, though typical. In order, Saturn's lot (2.7) is called the lot of power and firmness; Jupiter's the lot of prosperity (2.12); Mars's the lot of boldness and courage (2.17); the Sun's the lot of the absent (2.24); Venus's the lot of greed, friendliness and desire (2.30); and Mercury's the lot of commerce (2.35).[108] The planets are listed in Chaldean order (from slowest to fastest in movement). Though the descriptions of what the lots indicate cover roughly similar topics in comparison to those in Abū Maʿshar's *Great Introduction* (some using the same word or synonym), these descriptions do not appear to depend on the *Great Introduction*; nor, for that matter, on the *Abbreviation*.

Chapter 5, however, is entirely devoted to lots and is obviously based on Abū Maʿshar, specifically his *Abbreviation*, as Burnett, Yamamoto and Yano have pointed out.[109] Sometimes the quotations are more or less *verbatim*, though he mixes up the categories of planetary and place lots. The formulae are included. He begins with the Lot of Fortune (5.3; = *Abbrev.* 6.3), and moves on in 5.4 to 'place' lots which include planetary lots. So under the first place lots we find the lot of the absent (= *Abbrev.* 6.4), the lot of love and affection (= *Abbrev.* 6.6), the lot of firmness and survival that is equivalent to the lot of love (= *Abbrev.* 6.7) and the lot of courage and boldness (= *Abbrev.* 6.9). Under second place lots (5.5) we find the lot of poverty and lack of means (= *Abbrev.* 6.8) and the lot of prosperity, overcoming and victory (= *Abbrev.* 6.10). The eighth place lots (5.11) include the lot of bonds and imprisonment (= *Abbrev.* 6.11).

The planetary/lot order in which these are listed are Moon/fortune, Sun/absence, Venus/love, Mars/courage, Mercury/poverty, Jupiter/prosperity, Saturn/bonds. This order does not follow any traditional planetary order, nor

108 al-Qabīṣī, *Introduction to Astrology*, ed. and trans. by Burnett, Yamamoto, and Yano, pp. 64–65, 68–69, 70–71, 74–75, 80–81 respectively. My thanks to Charles Burnett for pointing out this mention of the planetary lots in Ch. 2.

109 al-Qabīṣī, *Introduction to Astrology*, ed. and trans. by Burnett, Yamamoto, and Yano, pp. 7–8.

does it correspond to any usual lot order. It does not appear that al-Qabīṣī has noticed how Abū Maʿshar has ordered the planetary lots, or their segregation from place lots. This supports the premise that he primarily consulted the *Abbreviation* and not the *Great Introduction*.[110]

B. al-Bīrūnī (973–1048)

In Section 476 of the *Book of Instruction in the Elements of the Art of Astrology*,[111] al-Bīrūnī says:

> امّا بطلميوس فلم يتجاوزه وامّا غيره فقد افرطوا في المواليد ونحن نورد ما ذكره ابو معشر في جداول.

> (Ptolemy recognized only one Part of Fortune, but others have introduced an excessive number of methods of casting lots at nativities. We reproduce in tables those which Abū Maʿshar has mentioned.)[112]

Bīrūnī then remarks that he will list 97 different lots, 7 of which belong to the planets, 80 to the houses, and 10 to neither. Naturally this is exactly the way in which Abū Maʿshar categorises the lots in the *Great Introduction*, with the same number of lots in each category. Al-Bīrūnī's table following this passage lists the planetary lots first:

Lots (*Sahām*) of the Seven Planets

1. Lot of Fortune or Lunar Horoscope
2. Lot of the unseen and religion (*sahm al-ghayb wa-l-dīn*)
3. Of friendship and love
4. Of despair and penury and fraud
5. Of captivity, prisons and escape therefrom
6. Of victory, triumph and aid
7. Of valour and bravery[113]

The order here is interesting because it combines the orders given in the *Great Introduction* and the *Abbreviation*. Al-Bīrūnī begins as if quoting from the *Abbreviation*, with Fortune, Absence, Love, and Despair (this is the traditional Hellenistic order of Fortune, Daimon, Eros, and Necessity); but

110 See al-Qabīṣī, *Introduction to Astrology*, ed. and trans. by Burnett, Yamamoto, and Yano, p. 8. However, although the influence of the *Abbreviation* is clear, this could also reflect some dependence on al-Andarzaghar's *Nativities* and its organisation of lots by astrological place.

111 al-Bīrūnī, *Book of Instruction*, trans. by Wright.

112 al-Bīrūnī, *Book of Instruction*, ed. and trans. by Wright, pp. 282–83 (English and Arabic).

113 al-Bīrūnī, *Book of Instruction*, § 476, trans. by Wright, p. 283, translation modified. He translates *Sahām*, 'lots', as 'Fortunes', and *sahm al-ghayb wa-l-dīn* as 'Part of Daemon and religion' (his n. 2 modifies this to 'lot of the unseen and religion').

then he switches to the order of the *Great Introduction*: Captivity/Saturn, Victory/Jupiter, and Valour/Mars. So possibly he consulted both of Abū Maʿshar's texts in his work.

C. Abraham Ibn Ezra (c. 1089–c. 1161)

The dissemination of these lots into the wider medieval astrological world shows their continuing importance and popularity. For example, we find them in *The Beginning of Wisdom*, written in Hebrew by Abraham Ibn Ezra, for whom Abū Maʿshar was an important named source. In Ch. 9, Ibn Ezra lists the planetary lots, using Abū Maʿshar's formulae and order in the *Great Introduction*:[114]

Lot of the Moon (Lot of Good Fortune)

Lot of the Sun (Lot of Mystery)[115]

Lot of Saturn, from Saturn to Good Fortune (depth of thought, working the soil, loss, theft, poverty, prison, captivity, death)

Lot of Jupiter, from the Lot of Mystery to Jupiter (truth, benevolence, wisdom, honour, good reputation, money)

Lot of Mars, from Mars to Fortune by day; reverse at night (force, might, anger, swiftness, treachery)

Lot of Venus, from Fortune to the Lot of Mystery by day, reverse at night (love, joy, pleasure, food, drink, lust and intercourse)

Lot of Mercury, from the Lot of Mystery to Good Fortune by day, reverse at night (poverty, enmity, hostility, negotiations, mathematics, knowledge)[116]

Not only does Ibn Ezra use the *Great Introduction* order and formulae, the lots also have similar qualities. The Lot of Mercury includes both the traditional Lot of Necessity characteristics and those aligned to Mercury, as in the *Great Introduction*. And, notably, the assignment of poverty to the lot of Mercury reiterates Abū Maʿshar's characterisation of it.

---

114 Ibn Ezra, *Beginning of Wisdom*, ed. and trans. by Sela, pp. 234–37; Sela notes the correspondences with the *Great Introduction*, pp. 467–69. See also Ibn Ezra, *Beginning of Wisdom*, trans. by Epstein, pp. 139–41.

115 Sela translates 'the absent' in Ibn Ezra, *Beginning of Wisdom*, ed. and trans. by Sela, p. 235.

116 Translation by Meira Epstein, in Ibn Ezra, *Beginning of Wisdom*, pp. 139–41.

D. Guido Bonatti (1210–c. 1296)

Finally, Guido Bonatti's *De astronomia tractatus X*,[117] Tractate 8, Part 2, Ch. 2, 'Capitulum secundum in partibus septem planetarum et earum significationibus particularibus' (Ch. 2 on the parts of the seven planets and their particular significations) relies heavily on the *Great Introduction*, VIII, 3. In fact, Bonatti quotes Abū Maʿshar by name (in the translation of John of Seville) in numerous chapters in this section.

Part of Fortune 'pars fortunae' from Sun to Moon (day)

Part of Future Things 'pars futurorum' from Moon to Sun (day)

Part of Saturn 'pars ponderosa' from Saturn to Fortune (day)

Part of Jupiter 'pars beatitudinis' from Future to Jupiter (day)

Part of Mars 'pars audaciae' from Mars to Fortune (day)

Part of Venus 'pars amoris et concordiae' from Fortune to Future (day)

Part of Mercury 'pars paupertatis et modici intellectus' from Future to Fortune (day)

All reverse at night.[118]

The order in which the planetary lots are listed is identical to Abū Maʿshar's in the *Great Introduction*. In following John of Seville's translation, the wording of the names of the lots are close to the Arabic original, e.g. the Part of Venus as *pars amoris et concordiae* and the Part of Saturn as 'the heavy lot' (Bonatti's *pars ponderosa*). He deviates slightly from John of Seville with the *pars paupertatis et modici intellectus* (John uses *ingenii* instead of *intellectus*). The qualities associated with each lot follow Abū Maʿshar's descriptions. It appears that at least one other astrologer writing in Latin used these locutions and order as well: Leopold of Austria (contemporary with Bonatti) wrote a late thirteenth-century *Compilatio de astrorum scientia* that, in its section on lots, clearly relies on the same source as Bonatti, namely John of Seville's Latin version of Abū Maʿshar's *Liber introductorii maioris*, whether word for word or paraphrased.[119]

---

117 Also known as the *Liber astronomiae*.

118 The names and formulae are in Guido Bonatti, *De astronomia*, cols 627–32.

119 For an example of this reliance, see the section on the Part of Mercury in: Leopold of Austria, *Compilatio de astrorum scientia*, Tractate 4, Part 5, 'Pars que est ingenii [...] et astronomie', ll. 21–26; compared to Abū Maʿshar, *Liber introductorii maioris*, ed. by Lemay, V, p. 332.438–49. Note that Leopold does not directly cite Abū Maʿshar in his sections on lots, but the reliance on him is obvious.

7. Conclusion

While at times the doctrine of planetary lots becomes garbled in transmission, it seems undeniable that its later development was heavily dependent on Abū Maʿshar. Abū Maʿshar, as we have shown, is clearly following two strands of lot traditions in Hellenistic astrology: one (Egyptian) which groups the Lots of Fortune, Daimon, Eros and Necessity; and one (Hermetic) which comes to be known as planetary lots, associated with Hermes Trismegistus and using each of the seven planets with the Lot of Fortune or Lot of Daimon in its calculation.

One question to be considered is why Abū Maʿshar breaks with the Paulus tradition and inserts Valens's versions (possibly from a Dorothean source) of the lots of Venus (Eros) and Mercury (Necessity) into his system. Although a definitive answer is elusive, it seems clear that he was aware of, and understood that there were, two traditions for the lots of Eros and Necessity, and significantly, that he recognised the importance of the quartet of the lots of Fortune, Daimon, Eros, and Necessity.[120] Although the planetary lot order used in the *Great Introduction* seems likely to have depended on a Rhetorius source, the other common order of Arabic planetary lots begins with Fortune, Absence, Love, and Poverty (the 'Fortune/Daimon/Eros/Necessity' order of Hellenistic astrology), followed by superior planets either starting with Saturn and moving down, or with Mars and moving up. This suggests that the four lots linked by their formulae (and, in fact, continuing to be so linked in Abū Maʿshar) are ingrained in a lot doctrine that persists in keeping them together in spite of doctrinal change: they remain first and foremost.

When these lots are mentioned in medieval texts in a context unrelated to planetary lots, this order is always used whether the text is in Arabic, Latin or Greek. The Fortune/Daimon/Eros/Necessity order is consistently used, for example, by Abū Maʿshar's predecessor al-Andarzaghar (cited by Sahl ibn Bishr), his contemporary Sahl, and the later *Liber Aristotilis* (from intermediary Arabic sources), showing the staying power and continuity of this tradition. The prominence of this order and the antiquity of its lots, therefore, may have contributed to Abū Maʿshar's decision to use the earlier formulae for the lots of Venus and Mercury; equally, the authority of Dorotheus or Valens or both, however transmitted, could have been influential.

By his innovation of combining two earlier lot traditions into one, and revising the interpretations for these lots, Abū Maʿshar set the stage and named the players for subsequent material on planetary lots. Furthermore, his division of the lots into three categories, beginning with a chapter specifically devoted to the planetary lots, may have contributed to the development of a

120 I discuss this quartet extensively in Greenbaum, *Daimon in Hellenistic Astrology*, Chs 9–10. As we have seen above, the Egyptian tradition is evident in the medieval *Liber Aristotilis* as well as in Sahl's *Book of Nativities*.

full-fledged doctrine of lots. A case could even be made that his sharp distinction of planetary lots from other types of lots influenced modern scholarship on these lots, to the point that many scholars assumed that the Hermetic tradition was authoritative and neglected the earlier lot tradition of the four lots of Fortune, Daimon, Eros, and Necessity, or assumed its incorporation into planetary lot schemes while ignoring the different formulae for Eros and Necessity. From Bouché-Leclercq[121] to Neugebauer and Van Hoesen,[122] Pingree,[123] Holden,[124] and Barton,[125] the Hermetic planetary lots have been emphasised and privileged over the earlier Egyptian tradition, and differences between the two traditions have rarely been noticed or considered.[126] As a result, Abū Maʿshar's insertion of lot formulae from the earlier tradition, and the rationale of his revision of meanings of these lots, have gone unnoticed as the innovations they certainly are. By close examination of this material and its history, I hope to have shed some light on lot doctrine in general, and on the part played by Abū Maʿshar in the historical and cultural trajectory of the doctrine of planetary lots.

121 Bouché-Leclercq, *L'astrologie grecque*, pp. 306–08.

122 Neugebauer and Van Hoesen, *Greek Horoscopes*, p. 9.

123 For example, assuming the spuriousness of their formulae in his critical edition of Valens's *Anthologies*; see above, n. 9.

124 Holden, *History of Horoscopic Astrology*, pp. 71–78.

125 Barton, *Ancient Astrology*, p. 81.

126 The only scholar, as far as I know, to take a serious look at lot doctrine prior to my interest in it was Giuseppe Bezza, *Arcana Mundi*, II, pp. 963–1012. I have mentioned my examinations of this doctrine in *Daimon in Hellenistic Astrology* above: for this topic specifically, see p. 361 and n. 105. I also mentioned the conflation of traditions by Abū Maʿshar in my 2009 PhD thesis, 'The Daimon in Hellenistic Astrology: Origins and Influence', p. 250 and n. 178.

Bibliography

Manuscripts

Città del Vaticano, Biblioteca Apostolica Vaticana, MS Vat. gr. 1056
Berlin, Staatsbibliothek zu Berlin, MS or. oct. 2663
Istanbul, Süleymaniye Library, MS Yeni Cami 784
Oxford, Bodleian Library, MS Digby or. 5
Paris, Bibliothèque nationale de France, MS fonds arabe 2588
San Lorenzo de El Escorial, Real Biblioteca, MS Escorial arab. 917
San Lorenzo de El Escorial, Real Biblioteca, MS Escorial arab. 1636
Tehran, Majlis Library, MS Majlis 6484

Primary Sources

Antiochus of Athens, *Thesauri*, ed. by Franz Boll, *Corpus Codicum Astrologorum Graecorum*, I, pp. 140–64, and *Corpus Codicum Astrologorum Graecorum*, VII, pp. 107–28 (Brussels: Henri Lamertin, 1898, 1908)

Abū Maʿshar al-Balkhī, *Albumasaris De revolutionibus nativitatum*, Greek ed. by David Pingree (Leipzig: Teubner, 1968)

——, *The Abbreviation of the Introduction to Astrology, together with the Medieval Latin Translation of Adelard of Bath*, Arabic and Latin ed. and trans. by Charles Burnett, Keiji Yamamoto, and Michio Yano, Islamic Philosophy, Theology and Science, 15 (Leiden: Brill, 1994)

——, *Liber introductorii maioris ad scientiam judiciorum astrorum*, Arabic and Latin ed. by Richard Lemay, 9 vols (Naples: Istituto Universitario Orientale, 1995–1996)

——, *On Historical Astrology: The Book of Religions and Dynasties (On the Great Conjunctions)*, Arabic and Latin ed. and trans. by Keiji Yamamoto and Charles Burnett, Islamic Philosophy, Theology and Science, 33, 2 vols (Leiden: Brill, 2000)

——, *The Great Introduction to Astrology*, Arabic ed. and trans. by Keiji Yamamoto and Charles Burnett, with an edition of the Greek version by David Pingree, Islamic Philosophy, Theology and Science, 106, 2 vols (Leiden: Brill, 2019)

——, *On the Revolutions of the Years of Nativities: Profections, Distributions, Fardārs, Transits, Solar Revolutions & more*, trans. by Benjamin N. Dykes (Minneapolis: The Cazimi Press, 2019)

Astronomical Papyri from Oxyrhynchus, ed. and trans. by Alexander Jones, 2 vols (Philadelphia: American Philosophical Society, 1999)

al-Bīrūnī, *The Book of Instruction in the Elements of the Art of Astrology*, facsimile of British Museum MS Or. 8349, and trans. by R. Ramsay Wright (London: Luzac & Co., 1934)

Bonatti, Guido, *De astronomia tractatus X* (Basel: Nicolaus Prückner, 1550)

Burnett, Charles, and Ahmed al-Hamdi, 'Zādānfarrūkh al-Andarzaghar on Anniversary Horoscopes', *Zeitschrift für Geschichte der arabisch-islamischen Wissenschaften*, 7 (1991–1992), 294–398

Dorotheus of Sidon, *Carmen astrologicum*, Arabic ed. and trans. by David Pingree (Leipzig: Teubner, 1976)

——, *Carmen astrologicum: The 'Umar al-Tabarī Translation*, trans. by Benjamin N. Dykes (Minneapolis, MN: The Cazimi Press, 2017)

Firmicus Maternus, Julius, *Matheseos libri VIII*, ed. by Wilhelm Kroll, Franz Skutsch, and Konrat Ziegler, 2 vols (Leipzig: Teubner, 1897–1913)

Greenbaum, Dorian, and Alexander Jones, 'P.Berl. 9825: An Elaborate Horoscope for 319 CE and its Significance for Greek Astronomical and Astrological Practice', *Institute for the Study of the Ancient World (ISAW) Papers*, 12 (2017) <http://dlib.nyu.edu/awdl/isaw/isaw-papers/12/> [accessed 19 April 2023]

Hephaestio of Thebes, *Apotelesmatica*, ed. by David Pingree, 2 vols (Leipzig: Teubner, 1973)

Hermes Trismegistus, *De Triginta Sex Decanis*, ed. by Simonetta Feraboli, Corpus Christianorum Continuatio Mediaeualis, 144, Hermes Latinus, IV, Part 1 (Turnhout: Brepols, 1994)

Hugo of Santalla, *The Liber Aristotilis of Hugo of Santalla*, ed. by Charles Burnett and David Pingree, Warburg Institute Surveys and Texts, 26 (London: The Warburg Institute, 1997)

Ibn Ezra, Abraham, *The Beginning of Wisdom (Reshit Hochma)*, ed. by Robert Hand, trans. by Meira B. Epstein (Reston, VA: ARHAT, 1998)

——, *Abraham Ibn Ezra's Introductions to Astrology: A Parallel Hebrew-English Critical Edition of the Book of the Beginning of Wisdom and the Book of the Judgments of the Zodiacal Signs*, ed. and trans. by Shlomo Sela, Études sur le Judäisme Médiéval, 69, Abraham Ibn Ezra's Astrological Writings, V (Leiden: Brill, 2017)

Leopold of Austria, *Compilatio de astrorum scientia decem continens tractatus* (Augsburg: Erhard Ratdolt, 1489)

Nechepso, and Petosiris, *Fragmenta magica*, ed. by Ernst Riess, *Philologus*, Suppl. Bd. 6, Part 1 (1892), 325–94

Neugebauer, Otto, and H. B. Van Hoesen, *Greek Horoscopes* (Philadelphia: The American Philosophical Society, 1959; repr. 1987)

Olympiodorus, *Eis ton Paulon <Heliodorou>. Heliodori, ut dicitur, in Paulum Alexandrinum Commentarium*, ed. by Emilie Boer (Leipzig: Teubner, 1962)

Olympiodorus, *Commentary on Paulus of Alexandria's 'Introduction to Astrology'* (trans. by Greenbaum): see Paulus of Alexandria, *Introduction to Astrology*

Paulus of Alexandria, *Eisagogika. Pauli Alexandrini Elementa Apotelesmatica*, ed. by Emilie Boer (Leipzig: Teubner, 1958)

Paulus of Alexandria, *Introduction to Astrology*, with Olympiodorus' commentary, in *Late Classical Astrology: Paulus Alexandrinus and Olympiodorus with the Scholia from Later Commentators*, trans. by Dorian Gieseler Greenbaum (Reston, VA: ARHAT, 2001)

Pingree, David, 'The Horoscope of Constantine VII Porphyrogenitus', *Dumbarton Oaks Papers*, 27 (1973), 217 and 219–31

Ptolemy, *Tetrabiblos*, ed. and trans. by Frank Egleston Robbins, Loeb Classical Library, 435 (Cambridge, MA: Harvard University Press, 1940; repr. 1994)

——, Claudius, *Ἀποτελεσματικά*, ed. by Wolfgang Hübner, Opera quae exstant omnia, III, 1 (Leipzig: Teubner, 1998)
al-Qabīṣī, *Al-Qabīṣī (Alcabitius): The Introduction to Astrology*, ed. and trans. by Charles Burnett, Keiji Yamamoto, and Michio Yano, Warburg Institute Studies and Texts, 2 (London: The Warburg Institute, 2004)
Rhetorius, *Rhetorius the Egyptian: Astrological Compendium Containing his Explanation and Narration of the Whole Art of Astrology*, trans. by James H. Holden (Tempe, AZ: American Federation of Astrologers, 2009)
——, *Compendium astrologicum secundum epitomen in cod. Paris. gr. 2425 servatam*, ed. by David Pingree and Stephan Heilen (Berlin: De Gruyter, in preparation)
Sahl ibn Bishr, *Book of Nativities*, trans. by Benjamin N. Dykes, *The Astrology of Sahl B. Bishr*, I (*Principles, Elections, Questions, Nativities*) (Minneapolis, MN: The Cazimi Press, 2019)
Theophilus of Edessa, *Astrological Works of Theophilus of Edessa*, intro. by Benjamin N. Dykes, trans. by Eduardo J. Gramaglia (Minneapolis, MN: The Cazimi Press, 2017)
Vettius Valens, *Anthologiarum libri novem*, ed. by David Pingree (Leipzig: Teubner, 1986)

Secondary Studies

Barton, Tamsyn, *Ancient Astrology* (London: Routledge, 1994)
Beck, Roger. *The Religion of the Mithras Cult in the Roman Empire: Mysteries of the Unconquered Sun* (Oxford: Oxford University Press, 2006)
Bezza, Giuseppe, *Arcana Mundi: Antologia del pensiero astrologico antico*, 2 vols (Milan: Rizzoli, 1995)
Bouché-Leclercq, Auguste, *L'astrologie grecque* (Paris: E. Leroux, 1899)
Bull, Christian H., *The Tradition of Hermes Trismegistus: The Egyptian Priestly Figure as a Teacher of Hellenized Wisdom*, Religions in the Graeco-Roman World, 186 (Leiden: Brill, 2018)
Greenbaum, Dorian Gieseler, 'The Lots of Fortune and Daemon in Extant Charts from Antiquity (First Century BCE to Seventh Century CE)', *Mēnē* 8 (2008), 173–90
——, 'The Daimon in Hellenistic Astrology: Origins and Influence' (unpublished doctoral thesis, The Warburg Institute, University of London, 2009)
——, *The Daimon in Hellenistic Astrology: Origins and Influence*, Ancient Magic and Divination, 11 (Leiden: Brill, 2016)
——, 'The Hellenistic Horoscope', in *Hellenistic Astronomy: The Science in Its Contexts*, ed. by Alan C. Bowen and Francesca Rochberg (Leiden: Brill, 2020), pp. 443–71
——, 'Divination and Decumbiture: Katarchic Astrology and Greek Medicine', in *Divination and Knowledge in Greco-Roman Antiquity*, ed. by Crystal Addey (New York: Routledge, 2021), pp. 109–37

Heilen, Stephan, 'Some Metrical Fragments from Nechepsos and Petosiris', in *La poésie astrologique dans l'Antiquité*, ed. by Isabelle Boehm and Wolfgang Hübner (Paris: De Boccard, 2011), pp. 23–93

——, *Hadriani genitura. Die astrologischen Fragmente des Antigonos von Nikaia. Edition, Übersetzung und Kommentar*, Texte und Kommentare, 43, 2 vols (Berlin: De Gruyter, 2015)

Holden, James Herschel, *A History of Horoscopic Astrology from the Babylonian Period to the Modern Age* (Tempe, AZ: American Federation of Astrologers, 1996)

László, Levente. 'Rhetorius, Zeno's Astrologer, and a Sixth-Century Astrological Compendium', *Dumbarton Oaks Papers*, 74 (2020), 329–50

Pingree, David. 'Antiochus and Rhetorius', *Classical Philology*, 72 (1977), 203–23

——, *From Astral Omens to Astrology: From Babylon to Bīkāner* (Rome: Istituto Italiano per l'Africa e l'Oriente, 1997)

Sezgin, Fuat, *Geschichte des arabischen Schrifttums*, VII: *Astrologie – Meteorologie und Verwandtes bis ca. 430 H* (Leiden: Brill, 1979)

Warnon, Jean, 'Le commentaire attribué à Héliodore sur les εἰσαγωγικά de Paul d'Alexandrie', *Recherches de philologie et de linguistique* (1967), 197–217

Westerink, Leendert G., 'Ein astrologisches Kolleg aus dem Jahre 564', *Byzantinische Zeitschrift*, 64 (1971), 6–21

GODEFROID DE CALLATAŸ

The Ikhwān al-Ṣafāʾ on the *Ṣūrat al-Arḍ*

A Geography in Motion

We moderns tend to believe that Muslim geographers of the Middle Ages conceived of the Earth as a reality that was profoundly immutable. The displacement of mountains on the surface of the terrestrial globe is an image which the Qurʾān reserves for the end of times, along with the wrapping of the Sun in darkness, the vanishing of the star-lights and other similar events of catastrophic dimensions.[1] In contrast to future apocalyptic times, the mountains as humans see them now are repeatedly presumed in medieval texts to act as anchors — the proper meaning of the Qurʾānic *rawāsin* ('anchored mountains')[2] — and this is precisely to prevent the Earth's surface from pitching. We have no reason to suspect that the geographers of medieval Islam held a different view than that exposed in the Qurʾān. The tables of coordinates they left may be at odds with one another on the location of a given landmark, but this disagreement likely results from a difference of measuring, or from the contingency of the manuscript transmission. Deep in his heart, every medieval geographer must have been thoroughly convinced that the positions he marked on the terrestrial map were valid once and for all.

* This article was written with the support of the ERC project 'The origin and early development of philosophy in tenth-century al-Andalus: the impact of ill-defined materials and channels of transmission' (ERC 2016, AdG 740618) which I am currently conducting at the University of Louvain (Université catholique de Louvain). It was partly elaborated from a paper on 'The Seven Planets and the Seven Climes' presented on 22 June 2017 in the conference 'Climates and Elements: Man and his Environment from Antiquity until the Renaissance' (John of Seville and Lima Conference' series) organized at the Warburg Institute by Charles Burnett and Pedro Mantas. It also includes elements from an article previously published as 'Kishwār-s, planètes et rois du monde'. I thank the editors of that volume for allowing me to reproduce, in English, some parts of my former study. Unless otherwise specified, all translations from the Arabic appearing in the present contribution are mine.

1 See for instance: Qurʾān 81:1–13.

2 See: Qurʾān 15:19; 21:31; 31:10; and 79:32.

Godefroid de Callataÿ (Louvain-la-Neuve, Belgium) is a historian of medieval Arabic science and philosophy and their role in transmitting knowledge from Greek Antiquity to the Latin West.

Mastering Nature in the Medieval Arabic and Latin Worlds: Studies in Heritage and Transfer of Arabic Science in Honour of Charles Burnett, ed. by Ann Giletti and Dag Nikolaus Hasse, CAT 4 (Turnhout: Brepols, 2023), pp. 57–81 BREPOLS PUBLISHERS 10.1484/.CAT-EB.5.134026

This article examines the geographical conceptions of the Ikhwān al-Ṣafāʾ (the 'Brethren of Purity'), the anonymous authors of a major and influential encyclopaedia most probably written in Iraq between the third and the fourth centuries of Islam (ninth/tenth centuries).[3] Their system, remarkably, does not rest on the presupposition that a given point on the surface of the Earth is fixed forever with respect to the others. On the contrary, it postulates that the different parts of the Earth's surface never cease to shift from one place to another in accordance with, and resulting from, the cyclical motions of the heavenly bodies. As shall be seen, the *Rasāʾil Ikhwān al-Ṣafāʾ* ('Epistles of the Brethren of Purity') include numerous passages of a geographical nature that make sense only when one is willing to accept that for the authors the three dimensions of space cannot be separated from the dimension of time. In the following pages, we will review a number of these passages, taken in particular from Epistle 36 ('On Cycles and Revolutions'), the astrological epistle par excellence of the corpus, Epistle 19 ('On Minerals'), where we find original and quite remarkable views on geological transformations, and Epistles 38 ('On Rebirth and Resurrection') and 48 ('On the Call to God'), both of which open up interesting perspectives on eschatology that are also critical to understand the Ikhwān's approach. However, before discussing these points, we must begin by presenting, in more detail, Epistle 4 ('On Geography'), which is the one the authors have devoted specifically to the description of the surface of the Earth, its divisions, and its inhabitants. Unsurprisingly, the Ikhwān base their conception on the classical theory of climes and their system follows essentially the same lines as those of other scholars in Antiquity and the Middle Ages. But even so, the epistle concludes with views that widely depart from the rest of the treatise and that, as we shall see, introduce an ideologically-orientated notion of astrological cycles into the very heart of this descriptive geography.

The first critical edition of the Epistle of Geography has recently been established by James Montgomery and Ignacio Sánchez as part the *Epistles of the Brethren of Purity* series published by Oxford University Press in association with the Institute of Ismaili Studies.[4] In the canonical list of *Rasāʾil* ('Epistles') as we have been accustomed to using it through the testimonies of the previous — and uncritical — editions of Bombay, Cairo and Beirut,[5] the one on geography occupies the fourth position, which means that it appears to break the classical order of the quadrivium of mathematical sciences, with arithmetic in the first position, geometry in the second, astronomy in the third, and music relegated to the fifth. We can explain this order. Presumably the epistle on geography immediately follows that of astronomy because, using essentially the same system of coordinates as astronomy, geography

3 For a recent overview, see de Callataÿ, 'Brethren of Purity'.

4 Ikhwān al-Ṣafāʾ, Epistle 4, ed. by Sanchéz and Montgomery.

5 On these uncritical editions, see: Poonawala, 'Why We Need an Arabic Critical Edition'.

was regarded already from Antiquity as an appendix or ancillary to the prime and nobler science of astronomy. The new edition does not confirm this order. Rather, it shows that in the great majority of the 12 manuscripts used for the edition, the epistle on geography occupies not the fourth, but the fifth position.[6] We also have several manuscripts that put music in the fourth position, by which the order of the quadrivium is fully re-established.[7] This kind of discrepancy is very common in the Ikhwānian corpus, and it should be noted that nothing is known for certain regarding the authors, the period of redaction, and the process by which the *Rasāʾil* came to be organized in the way we read them today.

The epistle on geography is not particularly long. The English translation provided by Montgomery and Sánchez consists of about 30 pages in all, to which some extra pages are appended in the form of tables to indicate the variants of the manuscripts concerning the coordinates of the cities for each of the seven climes. The contents of the epistle were summarized by Montgomery and Sánchez in the following manner: (1) Proemium: geography as one of the noble sciences (*al-ʿulūm al-sharīfa*); (2) Description of the Earth, the equator and the cardinal points (with diagrams of the Earth); (3) Discussion of common beliefs regarding the highest and the lowest points; (4) Discussion of the sphericity of the Earth; (5) Discussion of the four theories explaining the immobility of the Earth; (6) Description of the shape of the Earth; (7) Description of the inhabited quarter of the Earth; (8) Boundaries of the seven climes and the origin of these divisions; (9) Allegory of the city of the wise king and the value of scientific knowledge; (10) Geographical features, lands, and cities of the seven climes (with cartographical tables); and (11) Section on the influence of the stars, cosmic dualism, and the opposition between the virtuous people (*ahl al-khayr*) and the evil-doers (*ahl al-sharr*).[8]

A remarkable characteristic of this epistle is that it was translated into Latin during the Middle Ages. This translation, which seems to have been realised in Italy in the thirteenth or the fourteenth century, is surprisingly literal, if we except the omission of a few passages (such as the allegory of the city of the wise king) which were probably deemed too 'Islamic' to be reproduced by the translator. This Latin text, under the appellation *Epistola fratrum sincerorum in cosmographia*, was discovered by Patrick Gautier Dalché in a manuscript of the Library of Palermo.[9] It is exceptional in two respects. First, in the overall context of transmission of scientific disciplines in the Middle Ages, translations

6 Ikhwān al-Ṣafāʾ, Epistle 4, ed. by Sanchéz and Montgomery, pp. 17–21.

7 On the quadrivium (and trivium) in Islam, see: de Callataÿ, 'Trivium et quadrivium en Islam'; Cottrell, 'Trivium and Quadrivium'.

8 Ikhwān al-Ṣafāʾ, Epistle 4, ed. by Sanchéz and Montgomery, p. 30.

9 Gautier Dalché, '*Epistola fratrum sincerorum in Cosmographia*'. See also: Jules Janssens, 'The Latin Translation of the Epistle on Geography of the Ikhwān aṣ-Ṣafāʾ: A Few Preliminary Remarks in View of a Critical Edition', *Studi Magrebini*, 12 (2014), pp. 367–80; Hasse and Büttner, 'Notes on Anonymous Twelfth-Century Translations'.

of Arabic geographical works were extremely rare.[10] Second, with respect to the rest of the Ikhwānian corpus, this translation appears almost as a unicum, the only other Latin translation of an individual epistle known thus far to us being that of the *Liber introductorius in artem logicae demonstrationis*, whose text closely corresponds to that of Epistle 14, *Fī l-āfūdiqṭīqā / Fī l-ānālūṭīkā* ('On the Posterior Analytics'), published by Carmela Baffioni as part of the same editorial project.[11]

The general framework, the scientific principles and most of the technical notions of the epistle of geography are of Greek provenance. The overall configuration of the world (*ṣūrat al-arḍ*), based on a network of coordinates longitude (*ṭūl*) and latitude (*arḍ*), the conception of the *ecumene* or 'inhabited quarter' (*rubʿ al-maskūn*), and the theory of the seven climes (*al-aqālīm al-sabʿa*), inevitably point towards Ptolemy, a fact that the Ikhwān duly acknowledge.

1. Climes

The epistle preserves various traces of elements originating from Iran, as I have shown elsewhere.[12] The most important of these remnants from Sassanid times is found in the amalgamation of two ancient theories, a feature which the Ikhwān share with many other Muslim scholars of their age. The first one is inherited from Greek science. It is the classical theory of climes (*aqālim*, from the Greek *klimata*) in the form of horizontal bands or carpets ranked from 1 (near the equator) to 7 (near the pole) in the northern hemisphere. The frontiers between one band and the next are determined by the position of the Sun at noon on the day of the solstice. The second theory also divides the ecumene into seven zones or regions, but these regions are now distributed in such a way that six of them circumscribe the seventh, which occupies the centre of the representation. It is the ancient Sassanid conception of the seven *kishwār*-s (from the Persian '*kēshvar*'), which still finds an echo in both the *Avesta* and the *Bundahishn*, and where Iran — or, more precisely, Īrānshahr, the Iranian-Iraqi region around Babylon — serves as the omphalos at the centre of the inhabited world. The delimitation of the *kishwār*-s was not the result of scientific measurements. According to the extant sources, mainly

10 Gautier Dalché explains ('*Epistola fratrum sincerorum in Cosmographia*', pp. 137–38) that the reason for this is to be looked for in the low esteem in which geographical studies were generally held in the Latin West: 'la géographie ne fut jamais considérée comme constituant une discipline en soi'. Amongst medieval translations other than in Latin, one should also mention here the Arabic treatise formerly known as the 'Anonymous of Almería'; see: Mahammad Hadj-Sadok, 'The *Kitāb al-djaʿrāfiyya*'; Bramon, *El Mundo en el siglo XII*.

11 See: Farmer, 'Who was the author of the *Liber Introductorius in artem logicae*?'; Baffioni, 'Il *Liber introductorius in artem logicae demonstrationis*'. For Baffioni's edition of Epistle 14, see: Ikhwān al-Ṣafāʾ, Epistles 10–14, *On Logic*, ed. by Baffioni, pp. 129–200 of the Arabic.

12 de Callataÿ, 'Kishwār-s, planètes et rois du monde'.

found in Arabic literature, the seven regions correspond to the territories allegedly conquered at the dawn of history by 'the ancient kings' (*al-mulūk al-awwalūn*). In their epistle, the Ikhwān mention amongst these ancient kings the names of Alexander the Greek, Tubbaʿ al-Ḥimyarī, Sulayman b. Dāwūd the Israelite, Afrīdūn the Nabatean and Ardāshir b. Bābakān the Persian.[13]

While this ancient Iranian division of the *ecumene*, with its geopolitical and, as it were, mythological resonance, has little to do with the scientific considerations establishing the Greek system, two common features allow us to understand why these two conceptions came to be amalgamated, and why this amalgamation was more or less consciously propagated by the Arab-Muslim authors themselves.[14] In the first place, the two systems contain the same number of 'climes' or 'regions'. In the second place — and this is more important — they both reserve the pride of place to the same part of the world, since Ptolemy already situated Babylon, Mesopotamia, and Iran around the centre of the fourth clime, which is the pivotal clime of his representation, at an equal distance from the two extreme climes. The Muslim geographers of the Middle Ages — many of them Iranians themselves, like Bīrūnī (d. 1111) — were all the more inclined to adopt the '*kishwār* model' because this system enabled them to make Bagdad, the glorious capital of the 'Abbasid empire, the new omphalos of the representation, in the same way as Babylon was in Sassanid times.

Let us now briefly consider the systematic description of the 'geographical features, lands, and cities of the seven climes' as provided by the Ikhwān in the epistle. For the measurements of the frontiers (inferior, central, and superior) of each clime, their indications are comparable to those of other scientists of their age. These indications coincide almost perfectly with those mentioned by Farghānī in his *Jawāmiʿ ʿilm al-nujūm* ('Elements of Astronomy').[15] For the mention of the regions making up each clime, the Ikhwān also largely agree with these scholars. Like many of their contemporaries, the Ikhwān highlight the excellency of the fourth clime, that of Baghdad, the centre of Iraq, and Fārs. While nothing similar is found for any of the other six climes, the Brethren's description of the fourth clime takes the form of a genuine *laudatio*:

> This clime is the clime of the prophets and the sages (*wa-hādha al-iqlīm huwa iqlīm al-anbiyāʾ wa-l-ḥukamāʾ*), because it is at the centre of the climes (*li-anna-hū wasaṭ al-aqālīm*), with three climes to the south and three to the north, and also because it is in the compartment of the Sun (*wa-huwa ayḍan fī qism al-shams*), the most brilliant luminary. The people of this clime have the most equilibrated natures and equable morals. After them come the people of the two adjacent climes, I mean the third and the fifth. The inhabitants of the remaining climes, such as the Zanj, the

---

13 Ikhwān al-Ṣafāʾ, Epistle 4, ed. by Sanchéz and Montgomery, p. 21.

14 On this amalgamation, see: Miquel, *La géographie humaine du monde musulman*, I, pp. 56–68.

15 Honigmann, *Sieben Klimata*, p. 163.

> Abyssinians, and most of the communities which inhabit the first and the second climes, lack the nature of the most excellent [clime], because their appearance is hideous and their behaviour savage. This is also the case for the communities in the sixth and seventh clime, such as Gog and Magog, the Bulgars, the Slavs, and similar people.[16]

The exaltation of the fourth clime — that of Baghdad and ancient Babylon — is a *topos* of Arabic geographical literature. The Ikhwān are thus not innovative here, except when they specify that this clime is that of the prophets and the wise men. In the overall perspective adopted by the Brethren of Purity, this specification is far from being trivial. It brings us to the two great paths of knowledge, the prophetic and the philosophical, and towards comprehension as postulated by the Ikhwān and manifested through their classification of the sciences. In a study specifically focused on the fourth clime in the *Rasāʾil Ikhwān al-Ṣafāʾ*, Carmela Baffioni has conjectured that the indications provided by the Brethren in this epistle for the fourth clime should be put into relation with the emergence and the expansion of the Ismāʿīlī movement from the ninth to eleventh centuries.[17]

Typical of the Ikhwān's proclivity for analogies of all sorts are the two networks of associations which they establish in this passage among the climes, planets, and colours of skin of the inhabitants, as in Table 2.1.

Table 2.1. Networks of Associations among Climes, Planets, and Skin Colour

Clime	Planet	Colour of Skin
1	Saturn	Black
2	Jupiter	Black/Brown
3	Mars	Brown
4	Sun	Brown/White
5	Venus	White
6	Mercury	White/Rosy
7	Moon	Red

The spectrum of colours goes from black (clime 1) to red (clime 7), via brown/white for the inhabitants of clime 4, the clime of temperance and moderation *par excellence*. For the planets, the sequence has Saturn in clime 1, and then the other planets in decreasing order of distance from the Earth, up to the Moon in the clime 7. This is the order traditionally referred to as 'Chaldean', in which the Sun is placed, like a king, at the centre of the system.

16 Ikhwān al-Ṣafāʾ, Epistle 4, ed. by Sanchéz and Montgomery, pp. 43–44, and the discussion provided on 'The Fourth Clime: Geography and Prophetism' on pp. 36–42 of the introduction to the edition.

17 Baffioni, 'Il "quarto clima" nell'*Epistola sulla Geografia*', pp. 45–60.

Ernst Honigmann has pointed out that the same list of correspondences between planets and climes can also be found in Abū Maʿshar and Qummī.[18] In his *Kitāb al-milal wa-l-duwal* ('The Book of Religions and Dynasties'), also known as 'The Book of Great Conjunctions', Abū Maʿshar (d. 886) lists several systems of correspondences that are clearly divergent from one another and which seem to have been inherited from various traditions of a different provenance. The most coherent of these systems from a geographical point of view is the one found in the work known as 'On the cities, their climes, signs and planets' that was appended to the manuscript tradition of the *Kitāb al-milal wa-l-duwal*, and which was published by Keiji Yamamoto and Charles Burnett as appendices to the Arabic and Latin editions of Abū Maʿshar's treatise.[19] What follows is a recapitulation chart of the principal regions mentioned by the Ikhwān under each clime of their description, listed from east to west as they appear in the text.

Table 2.2. Correspondences between Climes and Regions

Clime	Main regions
1	Sea of China, Southern China, Central India, Persian Gulf, Yemen, Abyssinia, Nile, Nubia, Mauritania, Ocean
2	Central China, Northern India, Persian Gulf, Central Arabia, Nile, Berber Regions, Mauritania, Ocean
3	Northern China, Kabul, Kirmān, Fars, Southern Iraq, Northern Arabia, Syria, Egypt, Qayrawān, Tanger, Ocean
4	Northern China, Balkh, Fars, Central Iraq, Syria, Southern Mediterranean, Cyprus, Sicily, African Coast, Ocean
5	People of Gog and Magog, Farghāna, Khurasān, Persia, Northern Iraq, Anatolia, Constantinople, Northern Mediterranean, Rome, South from *Haykal al-Zuhra*, Pyreneans, Andalus, Ocean
6	People of Gog and Magog, Sogdiana, Khwārizm, Daylam, Caspian, Armenia, Macedonia, Central Europe, Baltic, North from *Haykal al-Zuhra*, Ocean
7	People of Gog and Magog, Sijistān, Northern Caspian, Caucasus, Baltic, Ocean

18 Honigmann, *Sieben Klimata*, pp. 141–42. For another mention of the same list, see: Masʿūdī, *Kitāb al-tanbīh wa-l-ishrāf*, ed. by De Goeje, pp. 33–34.

19 Yamamoto and Burnett appendices in Abū Maʿshar, *On Historical Astrology*, I, pp. 514–19 (Arabic original), and II, pp. 141–47 (Latin version). The correspondences as listed by Abū Maʿshar are: (1) India = Saturn; (2) Ḥijāz and Ethiopia = Jupiter; (3) Egypt = Mars; (4) Babylon and Iraq = Sun; (5) Rūm = Venus; (6) Gog and Magog = Mercury; (7) Fārs and China = Moon.

2. *Ahl al-Khayr* vs *Ahl al-Sharr*

For the present considerations, however, the most significant section of the geographical epistle remains the final one, which Montgomery and Sánchez entitle 'The influence of the stars, cosmic dualism, and the opposition between the virtuous people and the evil-doers'. Having just reported that one of the ancient kings 'found out that there were a little over 17,000 cities, not counting the villages (*fa-wajada sab'a 'ashar alf madīna wa-kasr siwā al-qurā*)' — a statement unmatched in medieval geographical literature, as far as I know — the Ikhwān embark on a prolonged discussion of the heavenly cycles which they hold responsible for a variety of changes in this world, including the number of cities of the *ecumene*:

> You must know that there are times when the number of the cities of the Earth increases and [there are] times when this number decreases. This depends on the rules of the conjunctions and the revolutions of the spheres (*bi-ḥasab mawjabāt aḥkām al-qirānāt wa-adwār al-aflāk*) in terms of: the power of felicity, the evenness of the period and the equilibrium of the natures of the elements, the advent of the prophets, the perpetuation of revelation, the abundance of scholars, the justice of kings, the welfare of human condition, and the descent of blessings from heavens in the form of rain, in such a way that land and plants becomes fruitful, that animals multiply, that countries get civilized, and that buildings and cities develop. [And this also depends on the rules of] the conjunctions that indicate (*fa-l-qirānāt al-dālla*) the power of misfortune, the corruption of the period, the rupture of the equilibrium of the mixture, the interruption of revelation, the paucity of scholars, the cessation of good deeds and the liberality of kings, the corruption of morals in people and the wickedness of their actions, the divergences of their opinions, and the withholding of blessings descended from the heavens in the form of rain, in such a way that lands do not become fruitful, that plants wither, that animals perish, and that cities in the countries fall into ruin.[20]

Elaborating from these highly deterministic views, the Ikhwān next insist that every cycle in this world of coming to be and passing away, whether long or short, is made of two phases, one ascending from its starting point to a point they refer to as 'the extreme limit' (*aqṣā ghāyati-hā*) of the circle, and the other descending from this apogee to the end of the circle. Their aim in doing this is to deliver a religious and political message directly linked with their inner convictions, as is clear from the following statement:

> Such is also the order of time for the cycle (*dawla*) of the excellent people (*ahl al-khayr*) and the cycle of the wicked people (*ahl al-sharr*).

20 Ikhwān al-Ṣafā', Epistle 4, ed. by Sanchéz and Montgomery, pp. 56–57.

> At one time the cycle, the power, and the manifestation of actions in the world are for the excellent people; at other times the cycle, the power, and the manifestation of actions in the world are for the wicked people, as God — How powerful and lofty He is — said: 'And these days [of varying conditions] We alternate among the people' (Q. 3:140) 'but none will understand them except those of knowledge' (Q. 29:43). Dear, pious and merciful Brother, may God stand by you and us with a spirit coming from Him, we observe that the cycle of the wicked people has already reached its peak, that their power has become manifest and their actions have multiplied in the world (*anna-hā qad tanāhat dawlat ahl al-sharr wa-ẓaharat quwwatu-hum wa-kathurat afhālu-hum fī l-ʿālam*). But after reaching the limit of increasing there is only decline and diminution (*wa-laysa baʿd al-tanāhī fī l-ziyāda illā al-inḥiṭāṭ wa-l-nuqṣān*) You must know that power and kingship are transferred in every era, period, cycle or conjunction from one community to the next, from one dynasty to the next, and from one country to the next (*wa-ʾlam bi-anna al-dawla wa-l-mulk yantaqilān fī kull dahr wa-zamān wa-dawr wa-qirān wa-min ummat ilā umma wa-min ahl bayt ilā ahl al-bayt wa-min balad ilā balad*). And you must know, my Brother, that the beginning of the cycle of the excellent people is now appearing with a group of wise, outstanding and virtuous scholars who agree with one another in having one doctrine and in adhering to one religion (*wa-ʾlam yā akhī bi-anna dawlat al-khayr yabdū awwalu-hā min qawm ʿulamāʾ ḥukamāʾ akhiyār fuḍalāʾ yajtamiʿūn ʿalā raʾy waḥid wa-yattafiqūn ʿalā dīn wāḥid*).[21]

Sánchez and Montgomery refer to this passage as 'one of the most — if not the most — relevant political statements to be found in the *Rasāʾil*.'[22] I cannot say if they are right, but what is certain is that it is most unusual to find a theory of cycles of excellent people versus cycles of wicked people in the conclusion of a work on the *ṣūrat al-arḍ*. In a treatise otherwise exclusively concerned with mathematical and physical data, the intrusion of a reflection such as this one only makes sense when the corpus of *Rasāʾil* as a whole is considered together with the ideological commitment of its authors.

3. Cycles

The Ikhwān have devoted one epistle to the cycles of the supralunar world and their influences upon the world of coming to be and passing away. This is Epistle 36, *Fī l-akwār wa-l-adwār* ('On Cycles and Revolutions'), which is found in the third section of the corpus, on the sciences of the soul and intellect.

21 Ikhwān al-Ṣafāʾ, Epistle 4, ed. by Sanchéz and Montgomery, pp. 59–60.

22 Ikhwān al-Ṣafāʾ, Epistle 4, ed. by Sanchéz and Montgomery, p. 45.

In the introductory part of this epistle, the authors identify the four 'extreme' celestial periods, two of which are referred to as 'revolutions' (*adwār*), in the sense of the circular movement accomplished by any given celestial body on its sphere, and the other two as 'conjunctions' (*qirānāt*), in the sense of the alignment of two or more stars on the same degree of the zodiac with respect to the Earth. The 'shortest revolution' is, for them, the diurnal movement of the entire vault of the heavens, from east to west, in a day of 24 hours. The 'shortest conjunction' is the alignment of the Moon, every month, with each of the other planets. The 'longest revolution' is that of the starry sphere on the zodiac, once in every 36,000 years, corresponding to the movement of equinoctial precession with the value traditionally assigned to it since Ptolemy in the second century AD. The 'longest conjunction' is the 'Great Year' by which all the seven planets (Moon, Mercury, Venus, Sun, Mars, Jupiter, and Saturn) are assumed to come back together again at the vernal point (1° of Aries) once every 360,000 years, according to Indian speculations.[23] In the same place of the introduction is also found the affirmation, with no further definition or justification, that 'there are four species of the "thousand" revolutions (*adwār al-ulūf*), namely: 7000 years; 12,000 years; 51,000 years; 360,000 years'. Having already discussed each one of the above figures in my edition and translation of Epistle 36,[24] I shall limit myself here to the two periods that have the greatest implications for the Ikhwānian conception of the *ṣūrat al-arḍ*, namely the 36,000-year and the 7000-year cycles.

4. Geological Changes

As previously mentioned, 36,000 years is the value traditionally ascribed to what is technically referred to as the movement of equinoctial precession. By this appellation scientists designate the slow motion which the vernal point (i.e., the point of intersection between the celestial equator and the ecliptic) completes with respect to the sphere of the fixed stars. Hipparchus (second century BC) is generally credited with the discovery of this important motion affecting the coordinates of all the points of the heavenly vault, but it is Ptolemy who was responsible for the value which became canonical for centuries.[25] As for the Ikhwān, what is worth recalling about their views on this period is not so much the astronomical definition of the cycle as the kind of influences it was supposed to exert upon the sublunary world, and which they describe as follows:

---

23 Ikhwān al-Ṣafāʾ, Epistle 36, ed. by de Callataÿ, pp. 135–36.

24 Ikhwān al-Ṣafāʾ, Epistle 36, ed. by de Callataÿ, pp. 197–201.

25 For the history of the precession theory, see: Duhem, *Le Système du Monde*, II, pp. 180–266; Neugebauer, 'The Alleged Babylonian Discovery'; Mercier, 'Studies in the Medieval Conception of Precession'.

> Among the motions [that are] slow, of long period, and of infrequent recommencement, there are the motions of the fixed stars on the sphere of the zodiac, once every 36,000 years, together with [those] of the apogees, perigees, and nodes of the planets. What results from this motion in the world of coming-to-be and passing-away, and during this time span, is the transfer of civilization on the face of the Earth, from one quarter to the next (*tanaqqul al-ʿimāra ʿalā saṭḥ al-arḍ min rubʿ ilā rubʿ*), and the fact that the regions of deserts become seas, the regions of seas [become] deserts, the regions of mountains become seas, and the regions of seas become mountains (*wa-an taṣīru mawāḍīʿ al-barārī biḥāran wa-mawāḍiʿ al-biḥār barāriyan wa-mawāḍiʿ al-jibāl biḥāran wa-mawāḍiʿ al-biḥār jibālan*), as we have explained, [regarding] the modalities of this, in the epistle 'On Minerals'.[26]

Indeed, in Epistle 19, *Fī l-maʿādin* ('On Minerals'), is found a more detailed account of the phaenomenon. The passage is quoted here in full:

> Now, we say that the Earth in its whole is divided into two halves, one southern and one northern, and that the external shape (*ẓāhir*) of each of these halves is divided into two halves, so that its whole consists of four quarters. Each of these quarters is characterized by four species [of regions]: regions such as deserts, wastelands, waterless areas, and ruins (*mawāḍiʿ wa-hiya barārī wa-qifār wa-falwāt wa-kharāb*); regions of seas, reedy areas, and ponds (*mawāḍiʿ al-biḥār wa-l-ājām wa-l-ghudrān*); regions of mountains, hills, crests and valleys (*mawāḍiʿ al-jibāl wa-l-tilāl wa-l-irtifāʿ wa-l-inkhifādh*); and regions of pastures, villages, cities, and civilized spots (*mawāḍiʿ al-marāʿī wa-l-qurā wa-l-mudun wa-l-ʿumrān*). Know that these regions are changed and transformed in the course of ages and eras (*iʿlam bi-anna hādhihi al-mawāḍiʿ tataghayyaru wa-tatabaddalu ʿalā ṭūl al-duhūr wa-l-azmān*), the regions of mountains becoming steppes and deserts, the regions of steppes becoming seas, ponds, and rivers, the regions of seas becoming mountains, hills, marshes, reedy places, and sands, the regions of civilization becoming desolate places, and the desolate places becoming regions of civilizations. It is suitable that we mention some of these particularities, since this kind of science is a weird and far-reaching field for most trained scientists, not to speak of the other people. You must know, my Brother — may God stand by you and by us with a spirit coming from Him — that every 3,000 years the fixed stars, as well as the apogees and the nodes of the planets, are carried along the degrees (of one) of the zodiacal signs. Every 9,000 years they shift from one quarter of the sphere to the next. Every 36,000 years they perform one revolution through the twelve signs. For this reason, the azimuths of the stars as well as the projections of their rays upon the

26 Ikhwān al-Ṣafāʾ, Epistle 36, ed. by de Callataÿ, p. 188.

> spots of the Earth and the atmospheres of countries vary (*fa-bi-hādha al-sabab yakhtalifu musāmat al-kawākib wa-maṭāriḥ shuʿāʿāti-hā ʿalā biqāʾ al-arḍ wa-ahwiyat al-bilād*), as vary also the succession of night and day and that of winter and summer, whether it be from a state of equilibrium and equality or from the increasing or diminishing in the excess of heat or coldness. These are the causes and the reasons why the quarters of the Earth differ in condition from one another, why the atmosphere of countries and spots changes and gets transformed from one condition to another. Those who study the science of the *Almagest* and the natural sciences (*al-nāẓirūn fī ʿilm al-majisṭī wa-ʿulūm al-ṭabīʿiyyāt*) know the reality of what we say. For these reasons and causes, civilized regions get transformed into ruins, and ruins into civilized regions, and deserts into seas, and seas into deserts and mountains. Those who study the natural and the divine sciences (*al-nāẓirūn fī l-ʿulūm al-ṭabīʿiyyāt wa-l-ilāhiyyāt*) know the reality of what we say, as well as those who seek for the causes of the things that come-to-be and pass-away below the sphere of the Moon, and the way these things change.[27]

Let us try to schematize the theory that underlies the above description. The Earth's surface is divided into four quarters, two of which are in the northern hemisphere and two in the southern hemisphere. Each one of these quarters is made of four different types of region, which can grossly be characterized as: (1) deserts, (2) seas, (3) mountains, and (4) civilized areas. In virtue of the precessional movement, all these regions slowly shift with respect to one another, but here we are already faced with a question. Do the interchanges of regions only take place within the borders of a given quarter of the Earth; or should we also consider the possibility that these regions trespass these frontiers by shifting, after 9000 years, from one quarter to the next? This passage from the epistle on minerals does not allow one to answer the question, but the passage quoted above from Epistle 36 does, since there the Ikhwān unambiguously refer to 'the transfer of civilization on the face of the Earth, from one quarter to the next'. Incidentally, this is also what is explicitly affirmed by Masʿūdī (d. 956), who deals with the same doctrine while ascribing it to the Indians:

> He [the first king Brahman, i.e. the astronomer Brahmagupta] was the first to define the apogee of the Sun, asserting that the Sun stays in each sign for 3,000 years and that it accomplishes one revolution around the sphere in 36,000 years. At the present time, i.e. 332 AH [= 943 AD], the apogee is, according to the Brahmans, in the sign of Gemini. But when it will be transferred to the Southern signs, the inhabited region will be transferred: the inhabited region will be covered with sea and the region

27 Ikhwān al-Ṣafāʾ, Epistle 19, ed. by Baffioni, pp. 257–60.

> covered with sea will be inhabited; the North will become the South and the South will become the North.[28]

Unfortunately, Masʿūdī's report is not free of errors, betraying his limited familiarity with astronomy. Like many scholars of the age, he confuses the 360,000 years of the conjunctional Great Year (which the Ikhwān indeed derive from India), and the 36,000 years of the equinoctial precession. These two periods have nothing to do with one another, but since both involve the starry sphere and both could somehow be regarded as 'the larger cycle of the universe', it is not surprising to find that some scientists confused them with one another during the Middle Ages, and even later. As I have shown in my *Annus Platonicus*, a survey of conjunctional world cycles from Antiquity to the Renaissance, the history of the Great Year doctrine is one of multiple distortions and amalgamations, and this certainly holds true also for the Islamic period.[29] In the current state of knowledge, it is not possible to establish who was first responsible for connecting the 36,000-year period with the theory of interchanges between regions of the Earth's surface. What is clear, however, is that the Ikhwān strove a good deal harder than Masʿūdī to incorporate this idea into a global and coherent system to account for changes in this sublunar world. In a study concerned with our authors' underestimated achievements in a field that can be defined as pre-modern geology, I have aimed to draw from ancient and medieval sources the principal elements that enable us to better appreciate the most remarkable originality and coherence of the Brethren's views in that respect.[30] The most relevant passage is also found in the epistle on minerals, in the immediate continuation of the section already quoted from. It reads as follows:

> You must know, my Brother — may God stand by you and by us with a spirit coming from Him — that seas are like swamps on the surface of the Earth, and that mountains are like dams and barriers for them, so as to separate the seas from one another, and so that the surface of the Earth be not entirely covered with water. For, if mountains were at the level of the surface of the Earth, and if their surface were rounded and smooth, the waters from the seas would have spread over their surface, wrapped them from all sides, and surrounded them — in the same way as the circle of the atmosphere entirely surrounds the Earth, the whole surface of the Earth becoming one single sea. But Divine Providence and Divine Wisdom have made it so that a part of the surface of the Earth be uncovered, so as to be a residence for mainland animals, and so that a part of it be a place for the cultures of grass, trees, and crops, since these

28 Masʿūdī, *Murūj al-dhahab*, I, Ch. 7, pp. 150–51. See also: Masʿūdī, *Kitāb al-tanbīh wa-l-ishrāf*, ed. by De Goeje, p. 222.

29 de Callataÿ, *Annus Platonicus*. See also: de Callataÿ, 'Eternity and World Cycles'.

30 de Callataÿ, 'World Cycles and Geological Changes'.

> are a food for the animals and a substance for their bodies — for all this [results] from the judgement of [God] the Mighty, the Knowing. Know, my Brother, that *wadī*-s and rivers all take their origin from mountains and hills, that they all run their courses to seas, reedy places, and ponds. [Know also] that mountains, having their wetness dried up in the course of ages and eras by the intensity of the rays from the Sun, the Moon, and the stars, increase in dryness and desiccation, break apart, and weather, especially under the assaults of thunders, and that they become stones, rocks, pebbles, and sands. Then, rains and flows lay these rocks, stones and sands down to the beds of *wadī*-s and rivers (*thumma inna al-amṭār wa-suyūla-hā taḥuṭṭu tilka al-ṣukhūr wa-l-aḥjār wa-l-rimāl ilā buṭūn al-awdiya wa-l-anhār*), and the strength of the stream carries this to seas, ponds, and reedy places. Then, owing to the strength of their waves and the intensity of their turmoil and bubbling, seas spread these sands, clay, and pebbles onto their beds, layer after layer, in the course of ages and eras (*tanbasiṭu tilka al-rimāl wa-l-ṭīn wa-l-ḥaṣā fī qaʿri-hā sāf ʿalā sāf bi-ṭūl al-zamān wa-l-duhūr*). These [layers] stick together, one upon another, coagulate, and build mountains and hills on the seabeds (*tatalabbadu baʿḍa-hā fawqa baʿḍ wa-tanʿaqidu wa-tanbutu fī quʿūr al-biḥār jibālan wa-tilālan*), in the same way as dunes and sands stick together in deserts and wastelands by the blowing of the winds.[31]

From a modern scientific standpoint, the theory explained in this passage is not free of weaknesses. We can hardly agree with the Ikhwān, for instance, when they claim that mountains are formed in the seas by the coagulation of sands, clay and pebbles, and even less so when they make the waves of the sea responsible for this coagulation. Neither could one admit that 'the intensity of the rays of the Sun, the Moon, and the stars' play a role in desiccation and, consequently, the breaking apart of these mountains. At a fundamental level, we would call archaism the idea of connecting these geological changes to an astronomical cycle such as the equinoctial precession. Still, as was convincingly argued by the modern geologist François Ellenberger in his *Histoire de la géologie*, the vision here presented by the Ikhwān sharply contrasts for its intrinsic coherence with all the rest of ancient and medieval sources, including Aristotle and Avicenna. As Ellenberger himself put it:

> We have at last discovered the founding text we had in vain been looking for in the Greek-Roman world — a text where all the geodynamic processes active on the surface of the Earth are logically related to one another in a never-ending circle; secular erosion, stratified layers of marine sediments, uplifting of these layers into new mountains while the sea covers ancient lands again. It is already the grand vision of

31 Ikhwān al-Ṣafāʾ, Epistle 19, ed. by Baffioni, pp. 261–62.

James Hutton, including its finality (although it lacks, as with Buffon), a plausible orogeny [i.e. theory of the formation of mountains].[32]

5. Prophets and Millenaries

In the Ikhwān's own words, the theory of interchanges on the surface of the Earth is a far-reaching science, accessible only to some experts in the *Almagest*, the natural sciences and the divine sciences. How are we to understand this reference to the divine sciences in this context? In Epistle 7, *Fī l-ṣanāʿiʾ al-ʿilmiyya* ('On the Scientific Arts'), where a purposefully-designed classification of human knowledge has been designed, the Ikhwān list as 'divine sciences' (*al-ʿulūm al-ilāhiyya*) the five following subjects: (1) the knowledge of the creator (*maʿrifat al-bārī*); (2) the science of spiritual beings (*ʿilm al-ruḥāniyyāt*); (3) the science of psychic beings (*ʿilm al-nafsiyyāt*); (4) the science of governance (*ʿilm al-siyāsa*), itself divided into five sub-disciplines; and (5) the science of [the soul's] return (*ʿilm al-maʿād*).[33] I would assume that, by referring to the divine sciences in this passage of the epistle on minerals, the Ikhwān mean in particular the science of the *ruḥāniyyāt*, since this is the science that accounts for the powers exerted by the souls attached to the celestial bodies upon the realities of this terrestrial world. The Ikhwānian corpus as we read it today includes one *risāla*, that is Epistle 49, *Fī l-ruḥāniyyāt* ('On the Spiritual Beings'), specifically devoted to these angelic entities, but their presence permeates, as it were, the whole work and is particularly noticeable in those parts of the encyclopaedia that are concerned with disciplines of an esoteric nature such as astrology and magic.[34]

Amongst the millenary cycles mentioned in Epistle 36, the Brethren appear to have attached an unparalleled significance to the 7000-year period. This measure does not correspond to the revolution of any celestial body as such, but is arrived at, with some approximation, from a calculation based upon the returns into conjunction of Saturn and Jupiter, which already played a role of paramount importance in Sassanid astrology.[35] The theory is briefly described or alluded to in various passages of the *Rasāʾil*, such as this one at the beginning of Epistle 36:

---

32 Ellenberger, *Histoire de la géologie*, I (*Des anciens à la première moitié du xvii*^e^ *siècle*), pp. 78 and 80.

33 Ikhwān al-Ṣafāʾ, Epistle 7, ed. by de Callataÿ, pp. 88–91. On the Ikhwānian classification of sciences, see also: de Callataÿ, 'The Classification of Knowledge in the *Rasāʾil*'. See also: de Callataÿ, 'Encyclopaedism on the Fringe of Islamic Orthodoxy'.

34 On the angelology of the Ikhwān al-Ṣafāʾ, see: de Callataÿ, 'The Ikhwān al-Ṣafāʾ on Angels and Spiritual Beings'.

35 On this, see for instance: Panaino, 'The Cardinal Asterisms in the Sasanian Urography'; Panaino, 'Saturn, the Lord of the Seventh Millennium'; Raffaelli, *L'oroscopo del mondo*; Pingree, 'Sasanian Astrology in Byzantium'.

> There is also one conjunction which takes place once every 20 years approximately, namely, the conjunction of Jupiter and Saturn. Among the long-period conjunctions, there is the revolution which recommences once every 240 years, namely, when Saturn and Jupiter complete twelve conjunctions in one and the same triplicity [*fī al-muthallathāt al-wāḥida*]. There are also conjunctions which take place once every 960 years, namely, when Saturn and Jupiter complete 48 conjunctions in the four triplicities.[36]

There is no need here to go into the astronomical and astrological details of these conjunctions, nor to delve into the reasons the theory acquired such importance in the Middle Ages in Islam, Byzantium and the Latin West.[37] Suffice it to recall that these types of Saturn/Jupiter conjunction were traditionally regarded as having a major influence on the destinies of entire human communities. Although not mentioning there the two planets in an explicit manner, it is evidently to these three types of conjunction that the Ikhwān, near the end of Epistle 36, allude when they retain as the subjects of maximal concern to astrologers:

1. the religions and empires (*al-milal wa-l-duwwal*), about which one seeks indications from the grand conjunctions (*al-qirānāt al-kibār*) that take place once every 1000 years, approximately (*bi-l-taqrīb*);
2. the transfer of royalty (*tanaqqul al-mamlaka*) from one nation to the next, or from one land to the next, or from one family to the next — things that are generated and about whose existence one seeks indications from the conjunctions that take place once every 240 years;
3. the replacement of individuals on the royal throne (*tabaddul al-ashkhāṣ ʿalā sarīr al-malik*) and the wars and dissensions (*al-ḥurūb wa-l-fitan*) that occur as a result of this — things about which one seeks indications from the conjunctions that take place once every 20 years.[38]

6. Eschatological Prospects

Of these three types of conjunction — the major (after 960 years, here given as 'approximately 1000 years'), the medium (after 240 years) and the minor (after 20 years) — it is undoubtedly to the one concerned with the changes of religions and empires that the Ikhwān have attached the greatest significance. Taking for granted that the 960 solar years of the grand Saturn/Jupiter conjunction corresponds almost exactly to 1000 lunar years, the *Rasāʾil* contain frequent allusions to a larger period made of seven such 'millenaries'

36 Ikhwān al-Ṣafāʾ, Epistle 36, ed. by de Callataÿ, p. 141.

37 On this, see for instance: Kennedy and Pingree, *The Astrological History of Māshāʾallāh*; North, 'Astrology and the Fortune of Churches'.

38 Ikhwān al-Ṣafāʾ, Epistle 36, ed. by de Callataÿ, pp. 189–90.

and which the Ikhwān regarded as a complete cycle of prophethood. According to this doctrine, each one of these seven 'millenaries' is heralded by a prophet, namely Adam, Noah, Abraham, Moses, Jesus, Muḥammad, and the *Qāʾim* of Resurrection. This is closely reminiscent of eschatological speculations found in various Ismāʿīlī works of the age,[39] but the Brethren's originality is to have given this scheme of prophetic history an astronomical/astrological interpretation or, better said, justification. At the end of Epistle 38, *Fī l-baʿth wa-l-qiyāma* ('On Rebirth and Resurrection'), the Ikhwān even incorporate this sequence of prophets into a narrative elaborated, in a very personal and decidedly unorthodox manner, from the Qurʾānic story of the Seven Sleepers.[40] There, the authors, who do not hesitate to identify themselves with the Sleepers, make various transparent allusions that allow one to draw lines of correspondences between the seven prophets of the cycle and the seven planets, in the following order: (1) Adam/Sun, (2) Noah/Saturn, (3) Abraham/Jupiter, (4) Moses/Mars, (5) Jesus/Venus, (6) Muḥammad/Mercury, and (7) *Qāʾim*/Moon. The text also includes a few cryptic numerical indications which, it would seem, can only be interpreted at the light of other passages from the *Rasāʾil*, such as these lines from Epistle 48, *Fī l-daʿwa ilā Allāh* ('On the Call to God'), in which the return of what appears to be a most favourable Saturn/Jupiter conjunction of the long type is announced by the Brethren to their partisans:

> Among the features of our Brethren is that they are learned in the field of religion, that they know the secrets of prophecies and that they are well-trained in the philosophical disciplines. When you meet one of them and seem to note integrity in him, tell him something that will please him and remind him of the recommencement of the revolution of revealing and awakening, as well as of the dissipation of worries for mankind, from the transfer of the conjunction, from the sign of fiery triplicities to the sign of vegetal and animal triplicities, in the tenth circle which corresponds to the house of power and the appearance of the eminent people.[41]

The elliptic character of the Ikhwānian prose and the sophistication of the astrological art at stake here make it questionable that a fully satisfactorily interpretation of these passages can ever be provided. Yet modern scholarship devoted thus far to the subject suggests that what the Ikhwān have in mind is a millenary period which started with a 'major Saturn/Jupiter conjunction' at the time of Muḥammad's birth in 571, and which would end in about 1524 with a conjunction of the same type. The midpoint of this millenary was expected to be reached 480 lunar years (i.e., two 'medium conjunctions')

39 On this, see: Madelung, 'Das Imamat in der frühen ismailitischen Lehre'; Halm, *Kosmologie und Heilslehere der frühen Ismāʿīlīya*; De Smet, 'La *taqiyya* et le jeûne du Ramadan', pp. 363–65.

40 Ikhwān al-Ṣafāʾ, Epistle 38, in *Rasāʾil Ikhwān al-Ṣafāʾ*, ed. by Buṭrus al-Bustānī, III, pp. 315–20.

41 Ikhwān al-Ṣafāʾ, Epistle 48, ed. by Hamdani and Soufan, p. 155.

after the starting point, in the year 1047, which is the only date in which a Saturn/Jupiter conjunction such as the one described in Epistle 48 actually took place.[42] In fact, this scheme exactly corresponds to the one used by the Byzantine astrologer Theophilus of Edessa (d. 785) for his prediction about the end of Islam. The passage, which I am quoting here from Rosenthal's translation, reads:

> Theophilus, the Byzantine astrologer of the Umayyad period, said the Muslim dynasty would have the duration of the great conjunction, that is, 960 years. When the conjunction occurs again in the sign of Scorpio, as it had at the beginning of Islam, and when the position of the stars in the conjunction that dominates Islam has changed, it will be less effective, or there will be new judgements that will make a change of opinion necessary.[43]

The Ikhwānian re-elaboration of the Seven Sleepers story leaves one to understand that this millenary is the sixth of the prophetic cycle, the last before the one to be heralded by the *Qāʾim* of Resurrection. In agreement with the authors' Shīʿī and Imsāʿīlī-like inclinations, the Cave in which the seven prophets-brothers converge to sleep is a symbol of the obligation for a group of eminent people to preserve the purity of a message. In other words, we are dealing here with a good example of a Shīʿī *taqiyya* with its typical function of protecting a community of believers in a double way, as was pointed out by Michael Ebstein:

> *Taqiyya* (prudence, the concealment of true beliefs, or dissimulation) plays a double role in the Shīʿī tradition. To begin with, *taqiyya* serves the Shīʿī believers as a means of self-protection against their Sunnī rivals [...]. At the same time, *taqiyya* entails an important esoteric aspect: it is designed to safeguard the secrets of the Shīʿī faith and to hide them from the uninitiated, be they Sunnī Muslims or the common Shīʿī believers, who are unable to understand the mysteries of their own religion.[44]

According to these chosen few of the Cave simile, amongst whom the Ikhwān al-Ṣafāʾ evidently counted themselves, the genuine sense of revelation became corrupted in the aftermath of the Karbalāʾ disaster, not long after the beginning of the sixth millennium, by unjust and mundane tyrants who never ceased to be in power since. In this context, the mid-millenary conjunction of Saturn and Jupiter is meant to represent the moment from which, after a period of

---

42 See: Casanova, 'Une date astronomique dans les Épîtres des Ikhwān al-Ṣafāʾ'; de Callataÿ, 'Astrology and Prophecy'; de Callataÿ, *Brotherhood of Idealists*, pp. 35–58 ('Millenarianism'). For a different approach, see: Widengren, 'La légende des Sept Dormants'.

43 Ibn Khaldūn, *Muqaddimah*, trans. by Rosenthal, II, Ch. 3, Section 52, p. 216. On this passage, see: Hartner, 'Quand et comment s'est arrêté l'essor de la culture scientifique dans l'Islam', pp. 319–20. For speculations in the same vein about the duration of Islam, see: *Abū Maʿshar. On Historical Astrology*, ed. by Yamamoto and Burnett, I, pp. 525–43.

44 Ebstein, 'Absent yet All Times Present', pp. 387–88.

forced concealment, the original message will begin to emerge anew, its full disclosure being expected to coincide with the appearance of the *Qāʾim*, after another two Saturn/Jupiter conjunctions of the medium type.

In the concluding section of their geographical epistle, as we have seen, the Ikhwān remind their readers that the cycles of the excellent people alternate with the cycles of the wicked people, and that we now find ourselves in an epoch when the cycle of the wicked people has reached its peak, coinciding with the first glimmers of the cycle of the excellent people. As mentioned above, these considerations may look strange at the end of an epistle of descriptive geography, but they acquire a good deal of sense as soon as they are put in connection with the Ikhwān's convictions and political/religious aspirations. There is little doubt that the Saturn/Jupiter conjunction of Epistle 48 is also what the Brethren had in mind when they concluded their geographical epistle. Far from being an anecdotal complement to the rest of the *risāla*, this final meditation on alternating cycles of excellent and wicked people is, as I see it, the genuine purpose for which the whole epistle was written.

7. The Harmonization of Data and the Limits of a Model

I hope to have succeeded in showing how misleading it is to interpret the Ikhwānian representation of the *ṣūrat al-arḍ* by conceiving it from a purely static viewpoint. Instead of a system where every component is fixed in the same place for eternity, we are invited to imagine the surface of the Earth as an entity in constant reformation under the action of various astronomical cycles operating simultaneously. The different types of Saturn/Jupiter conjunction affect, politically and religiously, the destinies of large communities of humans. These conjunctions determine periods of 20, 240 and 960 years, these latter two being characterized by a transfer of power from one place to another. In the Ikhwānian system, the 960-year periods are especially important because together they define a larger cycle of 6720 years (or seven lunar millenaries) corresponding to the transfer of prophethood from one community of believers to another. Conversely, the Ikhwān also ascribe to the 36,000-year precessional cycle the power of affecting the geological conditions of the Earth's surface, via a mechanism of transfer of the regions themselves. Although accounting for a constant process of reformation over the whole period, the theory implies that, after the completion of 9000 years, seas, mountains, deserts and even civilized areas are shifted from one quarter of the Earth's surface to the next. On a still larger chronological scale, the Ikhwān also postulate the existence of Great Year of 360,000 years identified as the greatest cycle of the universe, although they do not indicate the implication of this cycle in terms of geography or geology.

Did the Ikhwān succeed in bringing all these elements together into the coherent and fully-satisfactorily system they claim in so many places to have provided? Not really. The overall impression is rather that they prudently abstained from going into the details of these explications because they felt that at some stage it would be impossible to harmonize the data. Let us take a few examples. In their account of the Sleepers, where the Ikhwān associate each prophet with a planet, the two luminaries are placed at the ends of the sequence, as follows: (1) Sun, (2) Saturn, (3) Jupiter, (4) Mars, (5) Venus, (6) Mercury, and (7) Moon. In their geographical epistle, where planets are associated with climes, the Ikhwān use the 'Chaldean order' instead: (1) Saturn, (2) Jupiter, (3) Mars, (4) Sun, (5) Venus, (6) Mercury, and (7) Moon. The place on Earth offering the best conditions is another vexing issue for which we note a serious discrepancy between the geographical epistle and other parts of the corpus. In the epistle, as we have seen, the Brethren are unequivocal in asserting that the fourth clime is the most harmoniously balanced and, therefore the best of all (although, remarkably, they do not identify any place or city in particular as being the omphalos of the world). In the famous animal fable occupying most of Epistle 22, *Fī aṣnāf al-ḥayawānāt* ('On the Species of Animals'), the place offering the most favourable conditions — and where the Ikhwān situate the spontaneous generation of Adam, the father of mankind — is located on the island of Sarandīb, 'under the equatorial line (*min taḥta khaṭṭ al-istiwāʾ*) which is where night and day are equivalent, and where the climate is always temperate between hot and cold, and where matter, ready to receive the forms, is always present (*wa-l-mawādd al-mutahayyiʾa li-qubūl al-ṣuwar mawjūda dāʾiman*)'.[45] What can we make of these diverging data? How can we account for the fact that the Ikhwān do not relate Adam to the fourth clime, since they purposefully associate him with the Sun in the allegory of the Sleepers? Likewise, and still more challenging, how can the Brethren identify the fourth clime with 'the clime of the prophets' when they are perfectly aware of the fact that Muḥammad was born in Mecca, a city pertaining to the second clime? As was rightly observed by Sánchez and Montgomery, the identification of the fourth clime with the clime of the prophets:

> is in keeping generally with how the fourth clime is treated by the Ikhwān in other epistles, but it is puzzling. How can it be reconciled with the location of the sacred city of Islam? In all extant versions of this epistle, Mecca belongs to the second clime and, consequently, and by implication, Muḥammad would be excluded from the land of the prophets. By linking geographical position and prophecy, the Ikhwān seem to preclude any possibility of harmonizing ideology and religion by means of a double omphalic system such as that of Ibn Khurradādhbih. But they do not make any attempt to explain this oddity either.[46]

45 Ikhwān al-Ṣafāʾ, Epistle 22, ed. by Goodman and McGregor, pp. 7–8 of the Arabic. See also: de Callataÿ, 'For Those with Eyes to See'.

46 Ikhwān al-Ṣafāʾ, Epistle 4, ed. by Sanchéz and Montgomery, pp. 37–38.

In fact, it appears that the theory of climes as described in the epistle of geography — and which the Ikhwān must have derived without change from an older source — is not compatible with their scheme of prophetic history. In the Brethren's own words, every 9000-year period sees the transfer of the four types of region — deserts, seas, mountains, and civilized areas — from one quarter of the Earth's surface to the next. How can one account for this theory, which is so much at odds with the classical postulate of the fixity of the inhabited quarter in the northern hemisphere? What are we to make of the theory of climes itself in these conditions? The Ikhwānian corpus does not offer any insight into these intricacies. Nor do the *Rasāʾil* provide any clue as to the manner to understand how the 9000-year seasons of the precessional cycle can interact with the 7000-year prophetical cycles. To put it briefly, one may even doubt whether the authors of the epistle of geography considered the full implications of their doctrine of cyclical transfers on the *ṣurat al-arḍ*. Reconciling their esoteric and astrologically-orientated views on cycles with the positive data inherited from past geographers appears to have been an insurmountable task for our encyclopaedists.

Bibliography

Primary Sources

Abū Maʿshar, *On Historical Astrology: The Book of Religions and Dynasties (On the Great Conjunctions)*, ed. by Keiji Yamamoto, and Charles Burnett, 2 vols, Islamic Philosophy, Theology and Science, 33–34 (Leiden: Brill, 2000)

Ikhwān al-Ṣafāʾ, *Rasāʾil Ikhwān al-Ṣafāʾ*, ed. by Buṭrus al-Bustānī, 4 vols (Beirut: Dār Ṣādir – Dār Bayrūt, 1957)

——, *Epistles of the Brethren of Purity. The Case of the Animals versus Man Before the King of the Jinn. An Arabic Critical Edition and English Translation of Epistle 22*, ed. by Lenn Evan Goodman, and Richard McGregor (Oxford: Oxford University Press in association with the Institute of Ismaili Studies, 2009)

——, *Epistles of the Brethren of Purity. On Logic. An Arabic Critical Edition and English Translation of Epistles 10–14*, ed. by Carmela Baffioni (Oxford: Oxford University Press in association with The Institute of Ismaili Studies, 2010)

——, *Epistles of the Brethren of Purity. On the Natural Sciences. An Arabic Critical Edition and English Translation of Epistles 15–21*, ed. by Carmela Baffioni (Oxford: Oxford University Press in association with The Institute of Ismaili Studies, 2013)

——, *Epistles of the Brethren of Purity. On Geography. An Arabic Critical Edition and English Translation of Epistle 4*, ed. and trans. by Ignacio Sanchéz, and James Montgomery (Oxford: Oxford University Press in association with The Institute of Ismaili Studies, 2014)

——, *Epistles of the Brethren of Purity. Sciences of the Soul and Intellect, Part I, An Arabic Critical Edition and English Translation of Epistles 32–36*, ed. by Paul Ernest Walker, Ismail K. Poonawala, David Simonowitz, and Godefroid de Callataÿ (Oxford: Oxford University Press in association with The Institute of Ismaili Studies, 2015)

——, *Epistles of the Brethren of Purity. On Composition and the Arts. An Arabic Critical Edition and English Translation of Epistles 6–8*, ed. by Nader El-Bizri, and Godefroid de Callataÿ (Oxford: Oxford University Press in association with the Institute of Ismaili Studies, 2018)

——, *Epistles of the Brethren of Purity. The Call to God. An Arabic Critical Edition and English Translation of Epistle 48*, ed. by Abbas Hamdani, and Abdallah Soufan (Oxford: Oxford University Press in association with The Institute of Ismaili Studies, 2019)

Ibn Khaldūn, *The Muqaddimah: An Introduction to History*, trans. by Franz Rosenthal, 3 vols (Princeton: Princeton University Press, 1967)

Masʿūdī, *Kitāb al-tanbīh wa-l-ishrāf*, ed. by Michael Jan De Goeje, Bibiotheca Geographorum Arabicorum, 8 (Leiden: Brill, 1967)

Masʿūdī, *Murūj al-dhahab*, in *Les prairies d'or*, ed. and French trans. by Charles Barbier de Meynard and Abel Pavet de Courteille, 9 vols (Paris: Imprimerie impériale – Imprimerie nationale, 1861–1876)

Secondary Studies

Baffioni, Carmela, 'Il "quarto clima" nell'*Epistola sulla Geografia* degli Ikhwān al-Ṣafāʾ', in *Oriente Occidente. Scritti in memoria di Vittorina Langella*, ed. by Filippo Bencardino (Naples: Istituto Universitario Orientale, 1993), pp. 45–60

——, 'Il *Liber introductorius in artem logicae demonstrationis*: Probemi storici e filogici', *Studi Filosofici*, 17 (1994), 69–90

Bramon, Dolors, *El Mundo en el siglo XII. Estudio de la version castellana y del 'original' árabe de una geografia universal: 'El tratado de al-Zuhrī'* (Sabadell: Editorial Ausa, 1991)

Casanova, P., 'Une date astronomique dans les Épîtres des Ikhwān al-Ṣafāʾ', *Journal Asiatique*, series 11.5 (1915), 5–17

Cottrell, Emily, 'Trivium and Quadrivium: East of Baghdad', in *Universality of Reason: Plurality of Philosophies in the Middle Ages*, XII Congresso Internazionale di Filosofia Medievale de la SIEPM, Palermo, 17–22 settembre 2007, ed. by Alessandro Musco, 3 vols (Palermo: Officina di Studi Medievali, 2012), III: 11–26

de Callataÿ, Godefroid, *Annus Platonicus. A Study of World Cycles in Greek, Latin and Arabic Sources*, Publications de l'Institut Orientaliste de Louvain, 47 (Louvain: Peeters, 1996)

——, 'Astrology and Prophecy, The Ikhwān al-Ṣafāʾ and the Legend of the Seven Sleepers', in *Studies in the History of the Exact Sciences in Honour of David Pingree*, ed. by Charles Burnett, Jan P. Hogendijk, Kim Plofker, and Michio Yano (Leiden: Brill, 2004), pp. 758–85

——, *Ikhwān al-Ṣafāʾ. A Brotherhood of Idealists on the Fringe of Orthodox Islam*, Makers of the Muslim World (Oxford: Oneworld, 2005)

——, 'The Classification of Knowledge in the *Rasāʾil*', in *Epistles of the Brethren of Purity. The Ikhwān al-Ṣafāʾ and their Rasāʾil. An Introduction*, ed. by Nader El-Bizri (Oxford: Oxford University Press in association with The Institute of Ismaili Studies, 2008), pp. 58–82

——, 'Trivium et quadrivium en Islam: Des trajectoires contrastées', in *Une lumière venue d'ailleurs: Héritages et ouvertures dans les encyclopédies d'Orient et d'Occident au Moyen Age. Actes du colloque international tenu à Louvain-la-Neuve du 19 au 21 mai 2005*, ed. by Godefroid de Callataÿ and B. Van den Abeele (Turnhout: Brepols, 2008), pp. 1–30

——, 'World Cycles and Geological Changes According to the Ikhwān al-Ṣafāʾ', in *In the Age of al-Fārābī: Arabic Philosophy in the Fourth/Tenth Century: Proceedings of the Conference held at the Institute of Classical Studies and the Warburg Institute (London, 19–21 June 2006)*, ed. by Peter Adamson, Warburg Institute Colloquia, 12 (London: The Warburg Institute, 2008), pp. 179–93

——, 'Brethren of Purity (Ikhwān al-Ṣafāʾ)', in *Encyclopaedia of Islam*, 3rd edn, Part 2013–2014, ed. by Kate Fleet, Gudrun Krämer, Denis Matringe, John Nawas, and Everett Rowson (Leiden: Brill, 2013), pp. 84–90

——, 'Kishwār-s, planètes et rois du monde: Le substrat iranien de la géographie arabe, à travers l'exemple des Ikhwān al-Ṣafāʾ', in *Perspectives on Islamic Culture: Essays in Honour of Emilio G. Platti*, ed. by Bert Broeckaert, Stef Van den

Branden, and Jean-Jacques Pérennès, Les Cahiers du MIDEO, 6 (Leuven: Peeters, 2013), pp. 53–71

——, 'Eternity and World Cycles', in *Eternity: A History*, ed. by Yitzhak Y. Melamed (Oxford: Oxford University Press, 2016), pp. 64–69

——, 'Encyclopaedism on the Fringe of Islamic Orthodoxy: The *Rasāʾil Ikhwān al-Ṣafāʾ*, the *Rutbat al-ḥakīm* and the *Ghāyat al-ḥakīm* on the Division of Science', *Asiatische Studien*, 71.3 (2017), 857–77

——, 'For Those with Eyes to See: On the Hidden Meaning of the Animal Fable in the *Rasāʾil Ikhwān al-Ṣafāʾ*', *Journal of Islamic Studies*, 29.3 (2018), 357–91

——, 'The Ikhwān al-Ṣafāʾ on Angels and Spiritual Beings', in *The Intermediate World of Angels: Islamic Representations of Celestial Beings in Transcultural Contexts*, ed. by Sara Kühn, Stefan Leder, and Hans-Peter Pökel, Beiruter Texte und Studien, 114 (Beirut: Orient-Institut, 2019), pp. 347–64

De Smet, Daniel, 'La *taqiyya* et le jeûne du Ramadan: Quelques réflexions ismaéliennes', *Al-Qantara*, 34.2 (2013), 357–86

Duhem, Pierre. *Le Système du Monde. Histoire des doctrines cosmologiques de Platon à Copernic*, 10 vols (Paris: Hermann, 1913–1959)

Ebstein, Michael, 'Absent yet All Times Present: Further Thoughts on Secrecy in the Shīʿī Tradition and in Sunni Mysticism', *Al-Qantara*, 34.2 (2013), 387–413

Ellenberger, François, *Histoire de la géologie*, 2 vols (Paris: Technique et Documentation–Lavoisier, 1988–1994)

Farmer, Henry George, 'Who was the Author of the *Liber Introductorius in artem logicae*?', *Journal of the Royal Asiatic Society of Great Britain and Ireland*, 66 (1934), 553–56

Gautier Dalché, Patrick, '*Epistola fratrum sincerorum in Cosmographia*: Une traduction latine inédite de la quatrième *Risāla* des Ikhwān al-Ṣafāʾ', *Revue d'histoire des textes*, 18 (1988), 137–67

Hadj-Sadok, Mahammad, 'The *Kitāb al-djaʿrāfiyya*, Mappemonde du calife al-Maʾmūn reproduite par Fazārī (iiie/ixe s.) rééditée et commentée par Zuhrī (vi[e]/xiie s.)', *Bulletin d'études orientales*, 21 (1968), 1–312

Halm, Heinz, *Kosmologie und Heilslehere der frühen Ismāʿīlīya: Eine Studie zur Islamischen Gnosis*, Abhandlungen fûr die Kunde des Morgenlandes, 44.1 (Wiesbaden: Franz Steiner GmbH, 1978)

Hartner, Willy, 'Quand et comment s'est arrêté l'essor de la culture scientifique dans l'Islam', in *Classicisme et déclin culturel de l'Islam. Actes du symposium international d'histoire de la civilisation musulmane (Bordeaux, 25–29 juin 1956)*, ed. by Robert Brunschvig, and Gustave E. von Grunebaum (Paris: Editions Besson Chantemerle, 1957), pp. 319–37

Hasse, Dag Nikolaus, and Andreas Büttner, 'Notes on Anonymous Twelfth-Century Translations of Philosophical Texts from Arabic into Latin on the Iberian Peninsula', in *The Arabic, Hebrew and Latin Reception of Avicenna's Physics and Cosmology*, ed. by Dag Hasse and Amos Bertolacci, Scientia Graeco-Arabica, 23 (Berlin: De Gruyter, 2018), pp. 313–70

Honigmann, Ernst, *Die Sieben Klimata und die Poleis Episemoi. Eine Untersuchung zur Geschichte der Geographie und Astrologie im Altertum und Mittelalter* (Heidelberg: Winter, 1929)

Janssens, Jules, 'The Latin Translation of the Epistle on Geography of the Ikhwān aṣ-Ṣafāʾ: A Few Preliminary Remarks in View of a Critical Edition', *Studi Magrebini*, 12 (2014), 367–80

Kennedy, Edward Stewart, and David Pingree, *The Astrological History of Māshāʾallāh* (Cambridge MA: Harvard University Press, 1971)

Madelung, Wilferd, 'Das Imamat in der frühen ismailitischen Lehre', *Der Islam*, 37 (1961), 43–135

Mercier, Raymond, 'Studies in the Medieval Conception of Precession', *Archives Internationales d'Histoire des Sciences*, 26 (1976), 197–220 (Part I), and 27 (1977), 33–71 (Part II)

Miquel, André, *La géographie humaine du monde musulman jusqu'au milieu du 11ème siècle*, 2 vols (Paris: Mouton, 1975)

Neugebauer, Otto, 'The Alleged Babylonian Discovery of the Precession of the Equinoxes', *Journal of the American Oriental Society*, 70 (1950), 1–8

North, J. D., 'Astrology and the Fortune of Churches', *Centaurus*, 24 (1980), 181–211

Panaino, Antonio, 'Saturn, the Lord of the Seventh Millennium', *East and West*, 46 (1996), 235–50

——, 'The Cardinal Asterisms in the Sasanian Urography', in *La Science des cieux. Sages, mages, astrologues*, ed. by Rika Gyselen, and Anna Caiozzo (Bures-sur-Yvette: Groupe pour l'étude de la civilisation du Moyen Orient, 1999), pp. 183–90

Pingree, David, 'Sasanian Astrology in Byzantium', in *La Persia e Bisanzio: Convegno internazionale (Roma, 14–18 ottobre 2002)* (Rome: Accademia Nazionale dei Lincei, 2004), pp. 539–54

Poonawala, Ismail K., 'Why We Need an Arabic Critical Edition with Annotated English Translation of the Rasāʾil Ikhwān al-Ṣafāʾ', in *Epistles of the Brethren of Purity. The Ikhwān al-Ṣafāʾ and their Rasāʾil. An Introduction*, ed. by Nader El-Bizri (Oxford: Oxford University Press in association with The Institute of Ismaili Studies, 2008), pp. 33–57

Raffaelli, Enrico G., *L'oroscopo del mondo. Il tema di nascita del mondo e del primo uomo secondo l'astrologia zoroastriana* (Milan: Mimesis, 2001)

Widengren, G., 'La légende des Sept Dormants dans les écrits des Frères Purs', in *Démythisation et idéologie. Actes du colloque organisé par le Centre International d'Études Humanistes et par l'Institut d'Études Philosophiques de Rome, 4–9 janvier 1973*, ed. by E. Castelli (Paris: Aubier-Montaigne, 1973), pp. 509–26

PEDRO MANTAS-ESPAÑA

Adelard of Bath on Climates and the Elements

An Adaptive View on Nature

This article explores what, in his *Questions on Natural Science* and occasionally in *De opere astrolapsus*, Adelard of Bath (d. 1152) claimed to have learned about climates and the elements from Arab scientific masters after his long journey through the Mediterranean basin — knowledge which was also consistent with the classical tradition.[1] In comparing Adelard's conceptions and explanations with those of the Spanish *converso* Petrus Alfonsi (d. 1140) in his *Dialogue against the Jews*, we discover in Adelard a view more adaptive than Petrus's on the relationship between nature as provided by climate and human physical and intellectual prosperity.

The term 'climate' today refers to a region of the earth having specified climatic conditions (e.g., an expert can teach a seminar on 'warm climate regions in Europe'), or to the general weather conditions of a geographic area or over a long time period (e.g., a physicist can write about the rainfall conditions of southern Spain during the Little Ice Age). The climatic conditions and the consequences derived from it have a determining influence on our way of life, individual character, and social behaviour. This study examines the understandings of climate and its determining factor in medieval perspectives, according to Adelard and Petrus Alfonsi. As Charles Burnett has noted, the concern of human beings' relationship with our natural environment is not new: we are all part of a common nature, and have long held that heavenly

* This study is the outcome of reflections and conclusions from the conference 'Climates and Elements'. *Climates and Elements: Man and his Environment from Antiquity until the Renaissance. 2nd John of Seville Conference*, The Warburg Institute (London), 26–27 June 2017. Focusing on Adelard of Bath is both a scholarly and grateful acknowledgement to the dedicatee of this volume.

1 See Adelard of Bath, *De opere astrolapsus*, ed. by Dickey, pp. 45–46; Adelard of Bath, *Questions on Natural Science*, ed. and trans. by Burnett, pp. 83, 95 and 101.

Pedro Mantas-España (Córdoba, Spain) is a historian of medieval philosophy and the transmission of scientific and philosophical knowledge between Arab and Latin thinkers.

Mastering Nature in the Medieval Arabic and Latin Worlds: Studies in Heritage and Transfer of Arabic Science in Honour of Charles Burnett, ed. by Ann Giletti and Dag Nikolaus Hasse, CAT 4 (Turnhout: Brepols, 2023), pp. 83–105 BREPOLS PUBLISHERS 10.1484/.CAT-EB.5.134027

bodies and phenomena have an effect on human health, living conditions and destiny. While today we think about how individuals can reduce bad effects of the environment through appropriate social practices, historically astrological and cosmological traditions have linked different climates to certain human characteristics, and have evaluated how living in a particular environment and region of the planet could be more determinant of health and cleverness than any other physical condition.

1. Adelard of Bath and Petrus Alfonsi on Climate and Human Character

In his *Traces from the Rhodian Shore*, written some decades ago yet still a reference work, Clarence Glacken pointed to human beings' persistent questioning about the habitable earth and our relationship with it:

> Is the earth, which is obviously a fit environment for man and other organic life, a purposefully made creation? Have its climates, its relief, the configuration of its continents influenced the moral and social nature of individuals, and have they had an influence in molding the character and nature of human culture? In his long tenure of the earth, in what manner has man changed it from its hypothetical pristine condition?[2]

Such concerns are clearly present in three twelfth-century treatises, Adelard's *De opere astrolapsus* ('Treatise on the Astrolabe') and *Quaestiones naturales* ('Questions on Natural Science'), and Petrus's *Dialogus contra iudaeos* ('Dialogue against the Jews'). These prominent medieval intellectuals were part of a rising tendency to observe and investigate nature and its structure, composition, and behaviour. As we shall see, while Adelard's account of the climate proper to excellent intellective life in *De opere astrolapsus* might seem to agree with Petrus's *Dialogus contra iudaeos* (though put in different terms), his *Quaestiones* show he had a more nuanced view.

Taking *De opere astrolapsus* first, we read about the 'first clime',[3] as the Greeks had identified the zone of the Mediterranean and Mesopotamia, characterised by mild temperature conditions:

> Unde fit ut naturali sedis positione domus philosophica esse perhibeatur. Illic enim et omnia semina sponte proueniunt et indigene tam morum

---

2 Glacken, *Traces from the Rhodian Shore*, p. vii. As Ravi Rajan has written, '*Traces* remains in print well over four decades after its initial publication, and regarded as one of the foundational classics in the field of environmental history': Rajan, 'Clarence Glacken', p. 1.

3 As we read below (see n. 13), in early Hellenistic times *klíma* (terrestrial latitude) did not refer to a coordinate as such but to a specific 'band' of the earth where the same phenomena (e.g., length of longest daylight) are found. It was in this conceptual context that arose the ancient notion of the division of the known world into seven standard *climata*.

honestatem quam uerborum ueritatem modis omnibus illesam custodiunt, solique deo principaliter, stellarum uero numinibus secundario obnoxii, in communi omnia ponentes feliciter degunt solamque nature et rationis uiam sequentes. Cum aliquem cuiuslibet legis uirum in continentiis suis uident, prouerbio utuntur tali, 'aiekaeleb', id est 'caue bestiam'. Quibus, si arabes sequimur, eam patriam habitare datum est in qua primus homo, omnibus planetis preter mercurium in regnis suis existentibus, creatore uolente statuque celi ad generationem applicante, exortus est.

(Hence it comes about that in the first clime, they say, the home of philosophers has its natural position. For there all seeds spring up spontaneously and the inhabitants always do the right thing and speak the truth. Obeying only God first, and the spirits of the planets second, and sharing everything in common, they live happily. Following the way of nature and reason only, when they meet anyone of any religion (*lex*) in their everyday life they greet him with this motto: '*Iyyāka <wa> dābba*', which means 'beware of the Beast'. According to the Arabs, this is the fatherland that the philosophers were granted. It is here that when all the planets except Mercury were in their exaltations, when the Creator willed and the condition of the heavens was encouraging generation, the first man was born.)[4]

Aside from certain astrological connotations to which these lines belong, and the interesting attitude towards religious tolerance that Adelard displays ('Following the way of nature and reason only, when they meet anyone of any religion'), the paragraph clearly mentions a Hellenistic geographical notion as the division of the habitable regions of the earth into seven climates. We will focus on this topic below.

In his *Dialogus contra iudaeos*, when trying to explain graphically to his interlocutor how climates operate, Petrus Alfonsi wrote this description of the city of Aren:[5]

Visu enim probamus Aren in medio terrae sitam, et initium arietis et libra: super eam recta progredi linea, aeremque ibi temperatissimum esse, adeo, ut veris, aestatis, autumni et hiemis semper ibi fere tempus sit aequale. Ibi aromatizae species pulchri coloris et melliflui nascuntur saporis. Corpora quoque hominum non macilenta ibi sunt nimis aut pinguia, sed mediocris succi discretione decora. Temporum quoque temperies hominum corpora

4 Adelard of Bath, *De opere astrolapsus*, ed. by Dickey, pp. 169–70; trans. by Burnett, *Introduction of Arabic Learning*, pp. 44–45.

5 Aren is modern Ujjain, a city in the Ujjain district of the Indian state of Madhya Pradesh. The city holds an ancient tradition as astronomical centre, yet Petrus's information was incorrect because Ujjain is far from the equator, in Central-West India. Through his explanation, Petrus highlights the habitability of the first clime and the temperate nature of the city of Aren (on the equator, mid-way between east and west).

> sibi consona reddit et pectora, quia ineffabili pollent sapientia et materiali iusticia. [...] Potius totum terrae habitabile spatium existit continuum a predicto loco usque ad septemtrioanlem globum, quod antiqui in septem diviserunt partes, quas septem clymata vocaverunt, secundum numerum septem planetarum. Primum exhibet media linea, ubi Aren civitas est condita, septimum autem septemtrionalis orbis tenet extremum, reliquia vero medium continent spacium. Nullusque est inhabitabilis locus, nisi ubi vel multarum harenarum siccitas cum aqua modica vel montium asperitas aratri vel rastri non patitur opera. Haec autem supradicta omnia oculis subiecta figura demonstrat.
>
> > (From observation (*visus*) we prove that Aren is situated in the middle of the earth, and a straight line drawn from the beginning of Libra passes over the city, and that the air is very temperate, so that the temperature is almost always the same in spring, summer, autumn, and winter. There all sorts of things grow that are fragrant, beautifully colored, and sweet to taste. There men's bodies are neither fat nor thin, but well-fitted with modest vigour. The temperate nature of the seasons makes men's bodies and hearts harmonious, since they reign with ineffable wisdom and material justice. [...] Rather, the entire inhabitable area of the earth extends from this place to the north of the globe, an area that the ancients divided into seven parts, which they called the seven climates, according to the number of the seven planets. The first begin on the median line where the city of Aren was founded, the seventh is at the extreme north of the world, and the remaining climates occupy the space in between. And no place is uninhabitable except where either the dryness of many deserts with their paucity of water or the mountains' ruggedness does not permit the work of plowing or reaping. The following figure [Fig. 3.1] demonstrates to the eye all the things already described.)[6]

The map in Figure 3.1 reproduces the location of Aren (*Aren civitas*), the first habitable climate (*primum clyma habitabile*), over the rest of six habitable climates (from *secundum* to *septimum*). At the bottom of the representation (medieval maps are oriented with north at the bottom) is located the coldest habitable area (due to freezing, *extremitas septemtrionis videtur habitabilis frigore*).[7]

6 Petrus Alfonsi, *Dialogus contra iudaeos*, ed. by Mieth, p. 21; trans. by Irven M. Resnick, *Dialogue against the Jews*, pp. 60–61.

7 Cambridge, St John's College, MS E.4 (12th c.) fol. 122[r]. We find a slight yet important variation in the text describing the most extreme zones of the earth: 'Medietas terre inhabitabilibis pro nimio calore and extremitas septemtrionalis pro frigore inhabitable'. For another similar map, see: Cambridge, St John's College, MS D.11 (early 13th c.), fols 6[r] and 38[v].

Figure 3.1. Petrus Alfonsi, *Dialogus contra iudaeos*, Oxford, Bodl. Lib., MS Laud Misc. 356 (14th c.), fol. 120r. Photo: © Bodleian Libraries, University of Oxford. Terms of use: CC-BY-NC 4.0.

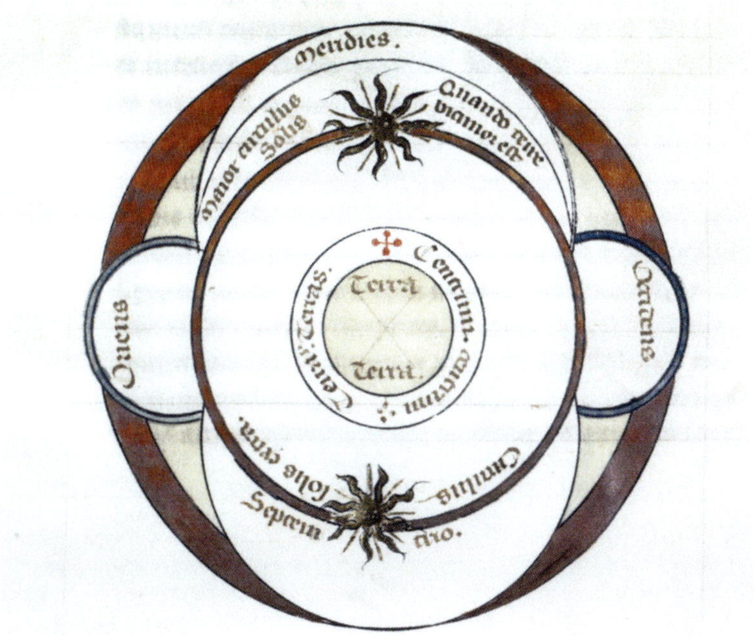

Figure 3.2. Petrus Alfonsi, *Dialogus contra iudaeos*, Oxford, Bodl. Lib., MS Laud Misc. 356 (14th c.), fol. 120r. The map shows the centre of the earth, graphical compass rose and zenithal/nadir position of the Sun. Photo: © Bodleian Libraries, University of Oxford. Terms of use: CC-BY-NC 4.0.

While Charles Burnett has shown the similarities shared by Adelard's and Petrus's descriptions,[8] we also find notable differences which are important to elucidate. However, before entering into the divergences and how they affect these thinkers' views on nature, we should contextualise the meaning of the passages quoted from their works. In this regard, we should first address some issues related to the Hellenistic and Arab/Persian roots of the division of the earth into climates.

2. Hellenistic and Arab/Persian Roots

In the creation of ancient maps or drawings like the one displayed in Figure 3.3, we find a particular worldview.[9] As Adam Silverstein has pointed out,[10] although the idea of a 'worldview' may seem somewhat incongruous for pre-modern societies, it is nevertheless meaningful in some historical and cultural contexts, particularly in the case of the medieval Islamic world and its immense geographical knowledge and interest: 'Medieval Islamic civilization produced a body of geographical literature that is unparalleled in its scope and size among pre-modern cultures: of the numerous works composed in Arabic and Persian that are directly concerned with geography, over two dozen survive'.[11]

This literature drew on great ancient sources. Indeed, Arabic geographies can be traced to traditions in three ancient civilizations: Hellenistic, Iranian and Mesopotamian, with the Hellenistic contributing to the latter two. Some Muslim geographers mention having consulted the works of Ptolemy

8 Irven Resnick has written: 'My thanks to Charles Burnett, for drawing my attention to Adelard of Bath's similar description of the first clime in his *De opere astrolapsus*. Adelard identifies this climate as the "home of philosophers"': Petrus Alfonsi, *Dialogus contra iudaeos*, trans. by Resnick, 1st *Titulus*, p. 33 n. 60. On interesting parallels between Adelard and Petrus, and a reformulation of the connection between them, see Burnett, 'Petrus Alfonsi and Adelard of Bath'. The paper cites a selection of Charles Burnett's papers in which he has addressed those parallels.

9 In these pages, the use of the concept 'worldview' is restricted to a context and meaning consistent with Silverstein's reflection on the topic. While a general meaning of the complex German *Weltanschauung* is present, we are alluding to a general apprehension of the world from a geographical standpoint. However, if we consider the conception of the world derived from Petrus's presentation of Aren, we understand that it has connotations linked not just to geographical and astronomical issues but also to doctrinal concerns: locating Aren mid-way between east and west, Petrus dissociates himself from the Jewish tradition that claimed 'Cum omnes Stellas in occidentem decidunt [...] Et quoniam haec stellarum supplicatio fit in occidente, ideo deum in occidente assenant esse': Petrus Alfonsi, *Dialogus contra iudaeos*, ed. by Mieth, p. 16. This is thus the 'worldview' of a *converso* who enters the polemist argumentations against Jewish cosmology.

10 Silverstein, 'Medieval Islamic Worldview', p. 273.

11 Silverstein, 'Medieval Islamic Worldview', p. 273. Most of those maps were done over a period of one and a half centuries (850–1000).

(d. 168 AD) and Marinus of Tyre (d. 130 AD), translated from Greek into Arabic by the ninth century. The influence of Hellenistic geography in the Near East is also evident in the work of the seventh-century Armenian writer Ananias of Shirak,[12] who refers to authors such as Ptolemy and Pappus of Alexandria; and his geography is also indebted to Iranian predecessors. That early Muslim geographers produced their works under the influence of Hellenistic geography was due at least partly to the large-scale translation of Greek scientific works into Arabic in ninth-century Iraq.

Among the Greek geographical notions which are present in many Arabic geographical works, we find: a division of the habitable regions of the earth into seven climates (Greek *klíma*),[13] and the existence of three continents (Asia, Libya, and Europe); the idea that the habitable earth is encircled by a surrounding body of water (Greek *okeanos*); and the practice of locating places by using mathematical coordinates of longitude and latitude. Although Iranian and Mesopotamian contributions to the medieval Islamic worldview have often been ignored, they played a role as important as the Greek in the formation of Islamic geography.[14] An Iranian geographical notion that is

12 The Armenian Ananias of Shirak wrote during the Late Sasanian period. Robert Hewsen has studied and translated Ananias's *Long and Short Recensions*, in Ananias of Shirak, *Ašxarhac'oyc'*, *The Long and the Short Recensions*, trans. by Hewsen. As Adam Silverstein has pointed out, the Armenian Ananias of Shirak 'whose geography is heavily indebted to both Hellenistic and Iranian predecessors but who, strikingly, refers only to Greek writers such as Ptolemy and Pappus of Alexandria when enumerating his sources': Silverstein, 'Medieval Islamic Worldview', p. 274. Yet describing the Sasanid provinces, he abandons his Greek sources and distinctly turns to Persian archival: 'Other Iranian geographical ideas permeate Arabic geographical works from this period. For instance, the notion that the habitable world consisted of seven *kishwars* or "regions" [...] competed with the Hellenistic theory of climates for primacy in Islamic geographies [...]. Although echoes of the Hellenistic climate system are detectable in this later phase of the theory's development (particularly the idea that a region's geographical position affects its fortunes), there is no doubt that the general idea of the *kishwars* is an Iranian one that persisted long into the Islamic period': Silverstein, 'Medieval Islamic Worldview', p. 276.

13 As G. J. Toomer reminds us, in the context of latitude, Ptolemy uses *plátos* to refer to the celestial coordinate and *klíma* ('inclination') to the unrelated terrestrial coordinate. '[As terrestrial latitude] *klíma*, however, does not refer to a coordinate as such [...] but to a specific "band" of the earth where the same phenomena (e.g. length of longest daylight) are found. Hence in early Hellenistic times arose the notion of the division of the known world (the *oikouménē*) into 7 standard *climata*. This is reflected in several places in the *Almagest*': Ptolemy, *Almagest*, trans. by Toomer, p. 19.

14 Iranian geographical notions were known to Muslims by the late Umayyad period, yet Iranian sources are almost never mentioned by name in Arabic geographies. One Sasanid geography, the *Šahrestānīhā-ī-Ērānšahr*, is extant and offers a unique glimpse of the type of Iranian materials that early Islamic geographers may have had at their disposal. For instance, it is due to Iranian and Mesopotamian influences that early Islamic geographies display important elements dissimilar to Hellenistic geographies: they introduced 'historical' information and share the tendency to begin descriptions of the world with 'the East'. Hellenistic geographies start with the continent of 'Europe' (rather than 'Asia'), but as early as the tenth century some Muslim geographers preferred to begin with 'Arabia'. Thus,

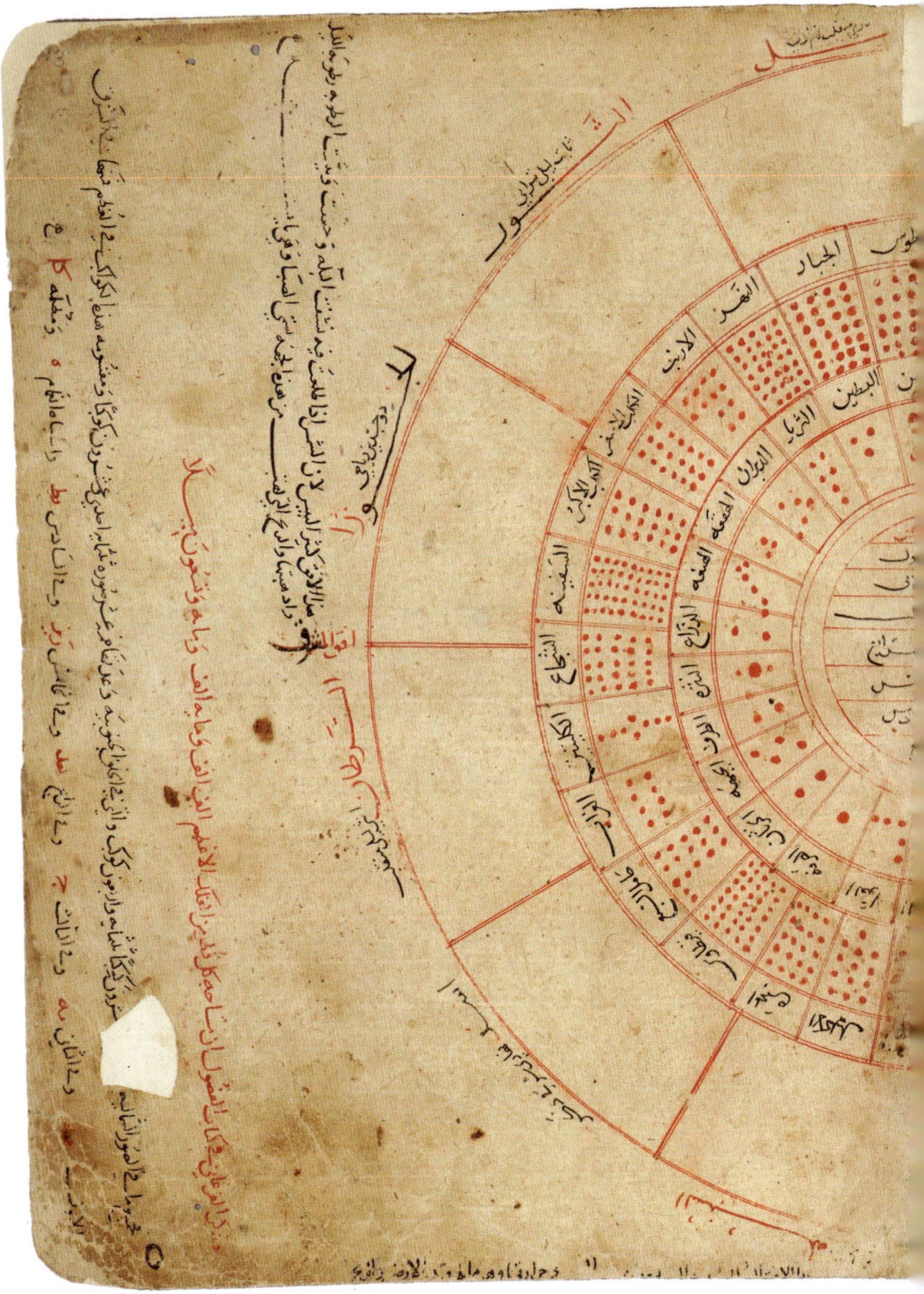

Figure 3.3. 'The Illustration of the Encompassing Sphere and the Manner in Which It Embraces All Existence, and Its Extent', from *Kitāb Gharā'ib al-funūn wa-mulaḥ al-'uyūn* ('Book of Curiosities'), Oxford, Bodl. Lib.,

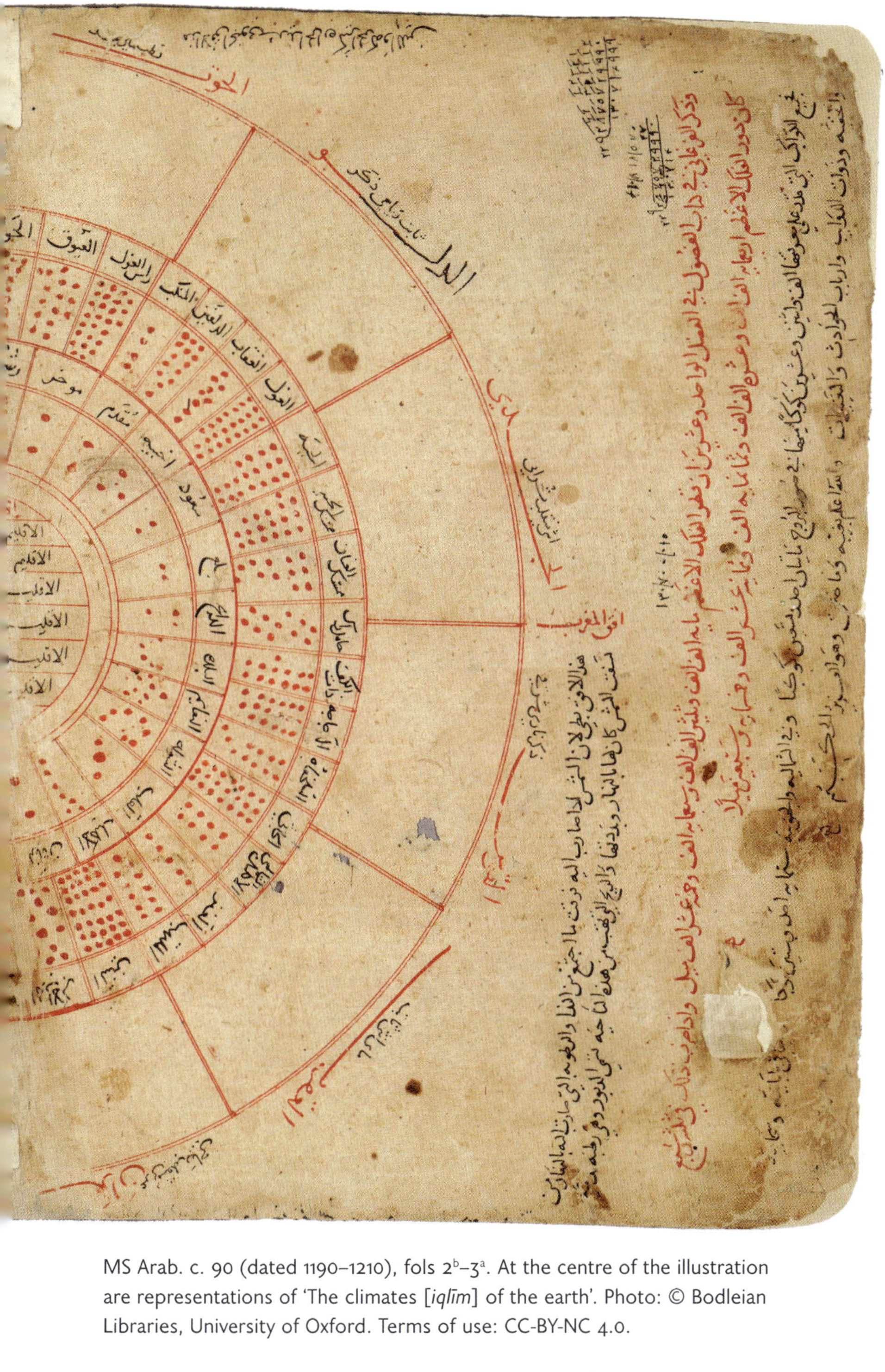

MS Arab. c. 90 (dated 1190–1210), fols 2b–3a. At the centre of the illustration are representations of 'The climates [*iqlīm*] of the earth'. Photo: © Bodleian Libraries, University of Oxford. Terms of use: CC-BY-NC 4.0.

present in early Muslim geographies is the quadripartite division of the world, a division distinct from the Greek divisions into either three continents or seven climates; and a fusion of the Greek and Iranian schemes for dividing the habitable world is found in the work of Ananias of Shirak.

The division of the world into three continents is a notion that the Greeks may have originally borrowed from Near Eastern geographies.[15] Interestingly, it is the shift from ancient to medieval Arabic geography, involving changes in methodology, that deeply affects the content of their topographies. From the late tenth century, geographers held that the method to obtain reliable information required personal observation (*mushāhada*) and eye-witness accounts (*muʿāyana*). A different issue consists in understanding what constituted evidence for them — a difficult question since it could be answered in different ways depending on the geographer's background. As we shall see, this subject linked to the Latin medieval texts examined here. After the tenth century, Islamic accounts underwent a further step in dissociation from ancient geography. While it might appear that the Greek account of seven climates remained in Islamic geographies, the term *iqlīm* signified not 'climate' but 'country', close in meaning to 'region' (*kishwar*).The region system served the purposes of Islamic geographers in that, as Silverstein puts it: 'It could present a specific land or people as being at the centre of the world, unlike the system of climates in which the moderate climate was shared among all peoples who lived along this horizontal slice of the world'.[16]

The Greek climate system necessarily charted the existence of other regions and people in addition to those of the central one, whereas the region system could describe the Muslim world, ignoring the non-Muslim world. While late tenth-century Muslim geographies represented an advance in methodological terms,[17] it seems that there is no Muslim geographer from this period whose work can accurately claim to encompass a worldview, because the focus on the Islamic region marginalised all other regions: 'when Arabic geography became "Medieval" and "Islamic", it ceased to represent a "*World*view"'.[18]

Against this background of ancient and classical transmission of geographical knowledge, we can contextualise the excerpts by Petrus Alfonsi and Adelard of

although early Muslim geographers clearly had alternative models with which to work, many of them chose the Iranian one. See Silverstein, 'Medieval Islamic Worldview', pp. 274–75.

15 Silverstein, 'Medieval Islamic Worldview', p. 282: 'Jewish geography was exposed to both Hellenistic and Mesopotamian ideas that had to be squared with their scripture. Unlike the Muslims, however, Jewish scholars did not come to rationalize these various strands of influence in works dedicated specifically to geography. The fact is that despite their exceptional literary productivity, Jews did not compose geographies until the twelfth century, by which point their immediate sources of inspiration were Islamic'.

16 Silverstein, 'Medieval Islamic Worldview', p. 285.

17 Far from 'armchair' scholars, new geographers promoted personal observation and eye-witness accounts.

18 Silverstein, 'Medieval Islamic Worldview', p. 285; it ceased to represent a '*World*view' in the sense of 'geographical worldview'.

Bath quoted in the first pages, and resume our analysis of what they represent. First of all, they are manifestations of knowledge where Adelard and Petrus reveal an exceptional scholarly level within the context and the audience they are addressing. However, we could provisionally say that they both — and Adelard in particular — represent a transmission of geographical knowledge on the one hand unaware of the newer methodology in the Muslim geographies mentioned above, yet, on the other hand, evidencing knowledge of classical geographical concepts learned from Arab masters.[19] This is predominantly a transmission to a Latin audience connected to a classical worldview.

However, Adelard's curious description of the happy life in the first clime, where in ideal conditions the first person was born,[20] recalls a passage of Ibn Ṭufayl's *Risālat Ḥayy ibn Yaqẓān* which is as beautiful as it is surprising:

ذكر سلفنا الصالح رضى الله عنهم ان جزيرة من جزائر الهند التى تحت خط الاستواء وهى الجزيرة التى يتولد بها الانسان من غير ام ولا اب لانها اعدل بقاع الارض هواء واتمها لشروق النور الاعلى عليها استعدادا وان كان ذلك على خلاف ما يراه جمهور الفلاسفة وكبار الاطباء فانهم يرون ان اعدل ما فى المعمورة الاقليم الرابع فان قالوا ذلك لانه صح عندهم انه ليس على خط الاستواء عمارة بسبب مانع من الموانع الارضية فلقولهم ان الاقليم الرابع اعدل بقاع الارض الباقية وجه وان كانوا انما ارادوا بذلك ان ما على خط الاستواء شديد الحرارة كالذى يصرح به اكثرهم فهم خطاء يقوم البرهان على خلافه.

> (Our forefathers, of blessed memory, tell of a certain equatorial island, lying off the coast of India, where human beings come into being without father or mother. This is possible, they say, because, of all places on earth, that island has the most tempered climate. And because a supernal light streams down on it, it is the most perfectly adapted to accept the human form. This runs counter to the views of most ordinary philosophers and even the greatest natural scientists. They believe the most temperate region of the inhabited world to be the fourth zone, and if they say this because they reason that some inadequacy due to the earth prevents settlement on the equatorial belt, then there is some colour of truth to their claim that the fourth is the most moderate of the remaining regions. But if, as most of them admit, they refer only to the intense heat of the equator, the notion is an error the contrary of which is easily proved.)[21]

19 Despite Adelard's claims about his Arab masters in his *Quaestiones naturales*, his recognisable sources are Greek and Latin. This does not mean that *arabum studia* were in Adelard's rhetorical strategy. His masterly knowledge of scientific texts written in Arabic can be clearly proven through his original works and translation of crucial science texts. See Burnett, *Conversations with his Nephew*, pp. xix and xxvi–xxxi; and see Mantas-España, 'Introducción', pp. 38–39, 41 and n. 97.

20 See the excerpt from Adelard of Bath's *De opere astrolapsus* quoted above at n. 4.

21 Ibn Ṭufayl, *Ḥayy ibn Yaqẓān*, ed. by Léon Gauthier, pp. 20–21; trans. by Goodman, pp. 103–04. Ibn Ṭufayl's reference is to Ibn al-Ḥusayn al-Masʿūdī's *Murūj al-dhahab wa-maʿādin al-jawhar*; for the English translation, see al-Masʿūdī, *Meadows of Gold*, trans. by Sprenger.

Figure 3.4. 'The Waq-Waq tree' from *Kitāb al-Bulhān* ('Book of Wonders'), Oxford, Bodl. Lib. MS Or. 133 (15th c.), fol. 41[b]. Photo: © Bodleian Libraries, University of Oxford. Terms of use: CC-BY-NC 4.0.

The island is said to be located near India, below the equator, in a region with the fairest weather, causing Ibn Ṭufayl to confront the fact that these climatic virtues have been identified by philosophers and physicians as being within the Fourth Region.[22] Although the equator could provide some geographical conditions unsuitable for human habitation, the cause cannot be identified with heat, as this climate is not so hot as to make living there intolerable. Furthermore, Ibn Ṭufayl tells us:

> وقد تبرهن فى علم الهيئة ان بقاع الارض التى على خط الاستواء لا تسامت الشمس رؤس اهلها سوى مرتين فى العام عند حلولها برأس الحمل وعند حلولها برأس الميزان وهى فى سائر العام ستة اشهر جنوبا منهم وستة اشهر شمالا منهم فليس عندهم حر مفرط ولا برد مفرط واحوالهم بسبب ذلك متشابهة.

> (But astronomy proves that in equatorial regions the sun stands directly overhead only twice a year, when it enters the Ram at the vernal equinox and when it enters the Balances at the autumnal equinox. The rest of the year it declines six months to the north and six to the south. These regions, then, enjoy a uniform climate, neither excessively hot nor excessively cold.)[23]

Thus, in Ibn Ṭufayl's account, the Fourth Region shares geographic and climatic conditions with the first clime of the ancients, identified by Adelard as the ideal home for philosophers.

3. Adelard on Climate and the Elements

In Adelard's *Quaestiones naturales*, which is presented as a dialogue, in the context of explanations about the wandering movement of the stars and the moon, the interlocutor (Adelard's nephew) enquires about the Sun, the *aplanos*,[24] and the obliquity of the zodiac:

> Ut quid enim et ipse zodiacus ita distractus sit, queritur. Potuit nempe per medium duci recta linea emisperium. Nam et inde etiam sequeretur

In his translation of Ibn Ṭufayl's *Ḥayy ibn* Yaqẓān, Riad Kocache mentioned 'An island referred to in many stories, said to be situated in the east of the China Sea or above Zanzibar. It was supposed to be inhabited by creatures similar to human beings (females) who cry out "waq waq" and who drop dead if touched. Other ancient sources mention an Indian tree called the Waq Waq which bears fruit that looks like a human head suspended by the hair': Ibn Ṭufayl, *Ḥayy ibn Yaqẓān*, trans. by Kocache, p. 3 n. 2. This marvellous tree is analysed in Malti-Douglas, *Woman's Body, Woman's Word*, pp. 83 and 85–96.

22 In al-Masʿūdī's report of the 'The seven climates', the Fourth Climate includes '*Egypt, Afrikiyah* (Africa provincia), el-Berber, Spain, and the interjacent countries: their sign is the Gemini, and their planet is Mercury': al-Masʿūdī, *Meadows of Gold*, trans. by Sprenger, pp. 199–200.

23 Ibn Ṭufayl, *Ḥayy ibn Yaqẓān*, ed. by Léon Gauthier, pp. 23–24; trans. by Goodman, p. 105.

24 The sphere of the fixed stars. The word *aplanos* is a corruption of the Greek *aplanēs*.

> non parvum humano generi commodum. Obolita etenim foret estatis et hiemis iniocunda distemperantia. Qua excessione caloris ac frigoris eliminata vigeret mortalis res publica mansueta mediocritate exhilarata.
>
> > (Why is the zodiac itself so skewed? - that is the question. Surely it could have been led in a straight line round the middle of the hemisphere? For this, too, would follow a considerable convenience for the human race. For the unpleasant intemperateness of summer and winter would have been destroyed, and if the excess of heat and cold have been eliminated, the human commonwealth would thrive out of joy, having a moderate climate as the norm.)[25]

Adelard's reply is particularly interesting, namely that the consequences would not be joyful at all. Initially he tells us:

> Si enim recta linea, et zodiacus et stelle sue per etherem diducta forent, is semper esset rerum status, quicumque esse potest Sole in equinoctiali degente linea. Ver itaque haberemus perpetuum, sicque numquam estatem vel hiemem. Ablata vero hieme, auferretur et seminum plurima putrefactio. Ea absente, periret eorum vivificatio. Nullum enim semen ad vitam nascitur, nisi prius corrumpatur. At vero estate abrasa. superflue humiditatis rerum periret desiccatio. Ea sublata, nulla earumdem sequeretur maturatio. Abeat igitur hinc in malam rem vernantis incommodi tui prava insinuatio!
>
> > (If both the zodiac and its planets were led through the ether in a straight line, the condition of things would always be whatever condition can occur when the Sun is dwelling on the Equator. We would have a perpetual spring, and thus never summer or winter. Do away with winter, and the putrefaction of seeds, for the most parts, would be removed. When that is missing, their vivification would perish. But wipe out summer and the drying out of the excess moisture in things would perish. Take that away, and no maturing of them would follow. So to hell with the misguided suggestion of your inconvenient spring!)[26]

After this clever and sensible reply, the dialogue between Adelard and his nephew continues. Replying to a new question, Adelard asserts:

> Si enim hec regio grandine turbida, nubibus horrida, feda tenebris, rationem providentiamque sustinere debet, quoniam potest, quanto magis etherea planities omni immunditia purgata, menti ac rationi obediens est?
>
> > (If this world which is churned about with hail, bristling with clouds, and murky with darkness, ought to sustain reason and providence

25 Adelard of Bath, *Quaestiones naturales*, ed. by Burnett, p. 214; trans. by Burnett, p. 215.
26 Adelard of Bath, *Quaestiones naturales*, ed. by Burnett, p. 214; trans. by Burnett, p. 215.

> > — since it can — how much more is the ethereal plane, purged of all uncleanliness, obedient to mind and reason?)[27]

The meaning of the question and what it implies is clear: the world and the skies are not unpredictably ordered and do not operate in a foolish manner; on the contrary, they obey the rules designed by reason and a certain mind. The sense of his words is complemented by a provoking remark which, in fact, is common sense:

> Si in hac ipsa tenebrosa regione nostra ille locus philosophorum fertilior est, qui serenitate quadam a densa fece elimatus purgatior est, qua fronte illam plagam celestem que quanto a nobis remotissima, tanto omni tali sorde intactissima est, animali motu expertem esse dicere sine summa insania quisquam potest? Item, si rerum creatarum animo nichil melius esse potest, isne in tenebris infodiendus est, locusque ei aptissimus ab eo quem nature similitudine vehementer expetit viduandus?
>
> > (If in this shadowy region of ours that place is more productive of philosophers which, in a certain serenity, is the more purified, cleansed of dense pollution, with what impudence can anyone say that celestial realm, which is as much untouched by all such soils as it is remote from us, does not participate in the movement of the soul? That would be the height of folly! Again, if of created things nothing can be better than the mind, should this be buried in darkness, and should the place which is most suitable for it be deprived of that which it strenuously seeks because of the similarity of its nature?)[28]

The nephew has asked Adelard whether the stars are animate or inanimate, and halfway through his explanation Adelard introduces this argument: if an intellectual life can take place in the less polluted human spaces (due to a serene environment), imagine what it would be like in unpolluted celestial regions. Implicitly, then, even in territories with unpleasant living conditions, not every area is so unkind as to deny inhabitants an environment where it would be possible to find serenity and a chance for a productive intellectual life. This kind of judgment was not new in the history of thought. Cicero, in a similar constructive spirit, stressed the need for what resembles the modern concept of climate and seasonal diversity: 'Ita ex quattuor temporum mutationibus omnibus quae terra marique gignuntur initia causaeque ducuntur' (Thus from the changes of the four seasons are derived the origins and causes of all those creatures which come into existence on land and in the sea).[29] On this Adelard and Cicero agree: that weather conditions have a determining influence on our way of life.

---

27 Adelard of Bath, *Quaestiones naturales*, ed. by Burnett, p. 218; trans. by Burnett, p. 219.

28 Adelard of Bath, *Quaestiones naturales*, ed. by Burnett, p. 218; trans. by Burnett, p. 219.

29 Cicero, *De natura deorum*, II, xix, ed. by Mayor, vol. II, p. 19; trans. by Rackham, p. 171.

In the examples above, Adelard and Petrus have drawn a connection between benign weather and good intellectual activity, such that intelligence arises and develops much better in healthy climates. Yet why is this assumed? Is it based on an idea that the mind feels at ease in unpolluted regions whose weather is serene, as is the case with the soul in celestial spaces? The answer to these questions lies in understanding how Adelard sees the influence that living in the first climate, enjoying fresh air and a mild temperature, has upon one's intelligence and good behaviour; and this requires an explanation of Adelard's understanding of the elements. Throughout his *Quaestiones*, Adelard makes use of certain common questions to provide a series of arguments on important topics of physics, for instance: our knowledge of the elements; issues related to motion and matter; and problems linked to essential and accidental qualities. Adelard's approach here deserves our attention. At the beginning of the *Quaestiones*, Adelard's nephew, representing a traditional perspective, asks:

> Licet enim si libet ut pulverem ariduin colligas, subtiliterque cribratum in testeo vel eneo vase reponas, deinde accessione temporum cum herbas inde surgere videas, cui id nisi mirabilis divine voluntatis mirabili effectui imponas?
>
> > (For you can if you like collect dry dust and sieve it finely into an earthenware or bronze container, and then, given the necessary length of time, when you see plants rise up from a pot with dry soil, to what are you to attribute this unless to the wondrous effect of the wondrous divine will?)[30]

Adelard's reply takes a different approach: that using divine will as an explanation is not necessary if our use of reason can provide an explanation:

> Deo non detraho. Quicquid enim est, ab ipso et per ipsum est. Id ipsum tamen confuse et absque discretione non est. Que quantum scientia humana procedit audienda est; in quo vero universaliter deficit, ad Deum res referenda est. Nos itaque quoniam nondum inscitia pallemus, ad rationem redeamus.
>
> > (I am not slighting God's role. For whatever exists is from him and through him. Nevertheless, that dependence <on God> is not <to be taken> in blanket fashion, without distinction. One should attend to this distinction, as far as human knowledge can go; but in the case where human knowledge completely fails, the matter should be referred to God. Thus, since we do not yet grow pale with lack of knowledge, let us return to reason.)[31]

---

30 Adelard of Bath, *Quaestiones naturales*, ed. by Burnett, p. 92; trans. by Burnett, p. 93.

31 Adelard of Bath, *Quaestiones naturales*, ed. by Burnett, pp. 96–98; trans. by Burnett, pp. 97–99.

Adelard explains as follows. The plant in the pot grows because it is fed not from just one element, pure earth, but from a mixture that is of such a nature that each of its parts contains every one of the four elements with their respective qualities. The four elements make up the whole body of the world, even though our senses erroneously conclude that each part of the world is composed of only one of them. Nobody can ever touch earth, water, air, or fire as such. The things of the world are composites of elements, not pure elements. Following Plato's *Timaeus* he explains that we must call them not 'earth' but 'earthy', not 'water' but 'watery', and so on; they get their name from the element that predominates in them.[32] In the soil of the pot, the four elements act as causes: a plant, a herb, or a weed will grow in it. According to Adelard, what grows will also possess all four elements, and will be largely earthy, a little watery, less airy, and least of all fiery. The four elements which are present in the soil and the growing weed are the cause for it to grow, acting from the soil and within the plant. Adelard explains: 'Ut autem id liquido intelligas, in exterioribus elementis suum simile irritantibus et extrahentibus suisque qualitatibus idem exigentibus causam huius processionis pono' (I place the cause of this process in the outer elements stirring and drawing out what is similar to them, and in their own qualities which thrust out the same).[33]

According to Adelard, the explanation lies in the external elements: they stimulate and release in the plant what is similar to them and their qualities. Hence these 'inferior' (internal) elements (Adelard calls them 'inferior' as opposed to 'superior' or 'external') return to their similar elements by dissolution. From the soil in the pot different kinds of plant or seed can grow at the same time since each of them draws from the soil what is fitted to its nature. In general, everything draws what is more appropriate to it; and in the case of plants, the earthy is drawn from the earth, which is why they cannot survive uprooted, given that the earthy component present in the air is too poor to nourish them.[34]

Adelard continues explaining why plants cannot survive uprooted: 'Quam dissolutionem vulgus quidem, quod semper veris rerum vocabulis caret, mortem vocat, cum non mors, set mutatio dicenda sit' (The common people, who always lack the true names for things, call this dissolution death, although it should not be called death, but change).[35] And quoting Plato's *Timaeus*, Adelard concludes, holding a point of view similar to the principle of mass conservation:[36]

32 Plato's *Timaeus* in Calcidius's Latin translation, 49d, ed. by Moreschini, p. 96: 'non est, opinor, ignis sed igneum quiddam, nec aer sed aereum'.

33 Adelard of Bath, *Quaestiones naturales*, ed. by Burnett, p. 98; trans. by Burnett, p. 99.

34 'De eo enim quo magis egebant. minus ibi reperiunt, et de minori indigentia mains' (For there they <plants> find less of what they need more, and more of what they need less): Adelard of Bath, *Quaestiones naturales*, ed. by Burnett, p. 98; trans. by Burnett, p. 99.

35 Adelard of Bath, *Quaestiones naturales*, ed. by Burnett, p. 98; trans. by Burnett, p. 99.

36 This fundamental law establishes that matter is neither created nor destroyed. It is an empirical principle, whose validity rests on experimental observations.

> Et meo certe iudicio in hoc sensili mundo nichil omnino moritur, nec minor est hodie quam cum creatus est. Si qua enim pars ab una coniunctione solvitur, non perit, set ad aliam societatem transit.
>
> > (And, in my judgement, certainly, nothing at all dies in this sensible world, nor is it smaller today than when it was created. For if any part is released from one conjunction, it does not perish but passes over to another association.)[37]

It is later in his commentaries on the elements that Adelard confirms the equivalence between elements and humours. The four humours make up the human body, and these bear the properties of the four elements:[38] 'Ingenium quippe per humiditatem viget; memoria vero per siccitatem [...] Itaque qui humidum habent cerebrum, ingenio quidem pollent' (Intelligence thrives on moisture, memory on dryness [...] Thus, whoever has a moist brain, has a lively intelligence).[39] Maintaining a parallelism between intelligence-moisture and memory-dryness, he states that the brain is moist because its activity in a certain way necessitates this.[40]

In sum, in Adelard's view a balanced climate of proper seasons is linked to balanced environment and individuals, where elements and humours can be reasonably stable, and this equilibrium in turn improves intellectual and social life.[41]

4. Conclusions

I will conclude by addressing the questions introduced early in the discussion above, in order to situate Adelard of Bath in relation to Petrus Alfonsi, and in the long history of thought on the correspondences between climate and humankind.

---

37 Adelard of Bath, *Quaestiones naturales*, ed. by Burnett, p. 98; trans. by Burnett, p. 99. See Plato, *Timaeus*, trans. by Calcidius, 33c.

38 Adelard of Bath, *Quaestiones naturales*, ed. by Burnett, p. 178; trans. by Burnett, p. 179.

39 Adelard of Bath, *Quaestiones naturales*, ed. by Burnett, p. 124; trans. by Burnett, p. 125.

40 Adelard of Bath, *Quaestiones naturales*, ed. and trans. by Burnett, p. 129. Adelard's identification of memory and dryness and what he says about the operations of the soul relies on Aristotelian sources: in Q18, just after the explanation on intelligence and memory, when Adelard is describing the three cells of the brain where imagination, reason and memory are located, he mentions Aristotle and his *Physics*; the reference, however, is not to Aristotle's *Physics* but to Nemesius's *Premnon Physicon* (Ch. XIII, 'De memoria'). His error derives from Nemesius, who in Ch. XIV ('De occulta et manifesta occasione') wrongly quotes 'Aristotle in his *Physics*': Nemesius, *Premnon physicon*, ed. by Burkhard, p. 89.

41 It is not unreasonable to suppose that such conditions could contribute to implementation of a proper *res publica* ruled by philosophers or, in Adelard's historical context, by a philosopher prince. As we read in Adelard's dedication of the *De opera astrolapsus* to the future King Henry II Plantagenet, 'states are blest either if they are handed over for philosophers to rule, or if their rulers adhere to philosophy': Adelard of Bath, *De opere astrolapsus*, ed. by Dickey, pp. 147–48; English translation of this paragraph in Burnett, *Introduction of Arabic Learning*, p. 31.

For any great astrologer from Antiquity to Early Modernity, Petrus Alfonsi and Adelard of Bath among them, the influence of celestial bodies and climate on human beings was unquestionable. For Petrus, both were determinative of human lives, and he had to defend the study of astrology not only as a relevant field but also as an acceptable Christian endeavour. Long before the twelfth century, astrology had been rejected for different reasons. It had often been considered heretical, contrary to Christian doctrine or opposed to the idea of free will - a criticism expressed by Augustine himself:

> Ideoque illos planos quos mathematicos vocant [...] christiana et vera pietas consequenter repellit et damnat. [...] quam totam illi salubritatem interficere conantur cum dicunt, 'de caelo tibi est inevitabilis causa peccandi' et 'Venus hoc fecit aut Saturnus aut Mars'.
>
> > (On the same ground those impostors called astrologers [...] a true Christian piety consistently rejects and condemns this art. [...] Astrologers try to destroy this entire saving doctrine [free will and the will to avoid sin] when they say: 'The reason for your sinning is determined by the heaven', and 'Venus or Saturn or Mars was responsible for this act'.)[42]

Against this kind of accusation, Petrus Alfonsi justified astrology as part of the divine plan. For Petrus, when God created the earth, He granted power to the celestial creatures over things on earth. In his *Epistola ad peripateticos* ('Letter to the Peripatetics (of France)'), Petrus replies to this kind of accusation by appealing to the rigor and predictability of this art:

> Ceterum opinatur pars alia quod nullum prouectum ista ars conferat; qui nimirum inter ceteros et imbecilles et inualidi probantur. Alii autem artem istam contra fidei christiane regulam arbitrantur incedere. Quod quam friuolum sit et ineptum naturalia plenius edocent argumenta. Si enim ars est, uera est. Quod si uera est, non est contraria ueritati. Vnde nee fidei contraire concluditur.
>
> > (It is the opinion of others that this art confers no benefit; these people, more than anyone else, prove themselves to be feeble imbeciles. And then there are others who claim that this art is against the rule of the Christian faith. But natural arguments plainly show how inept and frivolous is their claim. For if it is an art, it is true. If it is true, it is not contrary to the truth. Hence it is concluded that it does not go against the faith.)[43]

---

42 Augustine, *Confessions*, III, 4, ed. by O'Donnell, I, p. 34; trans. by Henry Chadwick, p. 91.

43 Petrus Alfonsi, *Epistola ad peripateticos*, ed. by Tolan, p. 168; trans. by Tolan, *Letter to Peripatetics of France*, p. 176.

As John Tolan has stressed, for Alfonsi, stellar influence was part of a divine plan understandable to human beings through the study of astrology.[44]

Was this kind of celestial determination with clear effects over human affairs assumed by Adelard too? At the beginning of this discussion, we saw Petrus Alfonsi and Adelard of Bath seeming to share similar points of view on the ideal climate for fulfilling human life, with Adelard's remarks in *De opere astrolapsus* about the home of philosophers in the first clime. Yet some of his statements in *Quaestiones naturales* indicate that the two thinkers did not share a position. It seems that Adelard held a more flexible view about ideal climates, in which the pleasantness of spring is not conducive to generation of vegetation, and even territories with unpleasant living conditions can be home to productive intellective life. Is Petrus Alfonsi's characterisation of Aren the same as that of the haven for philosophers in the first clime as described by Adelard in *De opere astrolapsus*? They seem similar, yet there is an important difference. While Petrus's conception of Aren as a living Paradise is deduced from astronomical principles which exclude this status to other living areas extending away from it (whose climates gradually deteriorate into colder environments), according to Adelard's *Quaestiones naturales* the home of philosophers does not exclude reasonable possibilities in different regions and climates, perhaps due to the fact that Adelard's arguments are based not only on astronomy but also on natural philosophy — displaying knowledge derived from a wider perspective on natural sciences.

Petrus's view on the ideal climate is clearly more in tune with the kind of representation that we have found in ancient metaphorical literatures like the story written by Ibn Ṭufayl (albeit in a different location and with different connotations). Adelard's understanding on the topic integrates a more flexible view about ideal climates: the first clime is a desideratum, but nevertheless Adelard sees real possibilities of equilibrium in other climates, even if they are distant to the first. A permanent spring, says Adelard, will ruin the cycle of nature: 'So to hell with the misguided suggestion of your inconvenient spring!'.[45]

Some centuries later, among Early Modern thinkers' reflections on human relations with nature and the possibilities of our adaptation to hard climatic conditions, the adaptive view point was clearly perceived by the opponents to

44 Tolan, 'Reading God's Will in the Stars', p. 20.

45 See n. 26. As Petrus explains to Moses, 'Ibi aromatizas species pulchra coloris et melliflui nascuntur saporis. Corpora quoque hominum non macilenta ibi sunt nimis aut pinguia, sed mediocris succi discretione decora. Temporum quoque temperies hominum corpora sibi consona reddit et pectora, quia ineffabili pollent sapientia et materiali iusticia': Petrus Alfonsi, *Dialogus contra iudaeos*, ed. by Mieth, p. 21. Although Aren (sometimes Arim or Arin) is the Indian city of Udidjayn, the city is treated here as an ideal rather than real city, which Muslim geographers located at the centre of the world. For a discussion on Aren, see Miquel, *La géographie humaine*, pp. 486–87.

deterministic positions. In the opening section of his 'Environment in Utopia',[46] William G. Palmer reminded us that, among other relevant questions, Early Modern thinkers confronted these: is the success of our existence determined by geography and climate? And can mankind fulfil its purpose on earth by bringing order to nature and mastering it? We are now aware that Western thought was involved in questions like these before the Early Modern period. Yet Renaissance thinkers arrived at this question: if climate does exercise a primordial influence, can we nonetheless resist the deterministic influence by understanding the forces it exerts upon us? Their answers were not uniform. The tendency of most intellectuals was toward environmental determinism, that climate does exert a decisive effect on human affairs. Contrasting with this, however, was a powerful idea among some of them, which Palmer sums up eloquently: 'Yet also emerging in Renaissance environmental thought, most visibly in the Utopian visions of Thomas More, Francis Bacon, and Gerrard Winstanley, was a perception of society and history by which mankind was capable of combating the restrictions imposed by climate and geography'.[47] This is precisely what More explains in the Discourse on *Utopia* (Book II) when discussing the limitations of its soil:

> Et quum neque solo sint usquequaque fertili, nec admodum salubri caelo, adversus aerem ita sese temperantia victus muniunt, terrae sic medentur industria, ut nusquam gentium sit frugis pecorisque proventus uberior, aut hominum vivaciora corpora paucioribusque morbis obnoxia.
>
>> (Their soil is not very fertile, nor their climate of the best, but they protect themselves against the weather by temperate living, and improve their soil by industry, so that nowhere do grain and cattle flourish more plentifully, nowhere are people's bodies more vigorous or less susceptible to disease.)[48]

Utopians remain a peaceful and gentle people, even in conditions of unfortunate climate, where they simply adapt to the circumstances they have to deal with. And Adelard, whose place in this history is more than as a transmitter and interpreter of Hellenistic and Muslim learning, and beyond his work as a scientist unlocking the properties of the elements, parts company with Petrus Alfonsi and turns out to be a brother in spirit of the Renaissance utopian thinkers. For Adelard, adaptation and intelligence can be the best way of living in challenging areas: such a place, 'grandine turbida, nubibus horrida, feda tenebris, rationem providentiamque sustinere debet, quoniam potest' — 'churned about with hail, bristling with clouds, and murky with darkness, ought to sustain reason and providence — since it can'.[49]

46 Palmer, 'Environment in Utopia', p. 163.

47 Palmer, 'Environment in Utopia', p. 163.

48 Thomas More, *Utopia*, ed. by Logan and Adams, p. 178; trans. by Logan and Adams, p. 179.

49 Adelard of Bath, *Quaestiones naturales*, ed. by Burnett, p. 218; trans. by Burnett, p. 219.

Bibliography

Manuscripts

Cambridge, St John's College, MS D.11
Cambridge, St John's College, MS E.4
Oxford, Bodleian Library, MS Arab. 356
Oxford, Bodleian Library, MS Laud Misc. 356
Oxford, Bodleian Library, MS Or. 133
Paris, Bibliothèque nationale de France, MS Suppl. lat. 1218

Primary Sources

Augustine, *Confessions*, trans. by Henry Chadwick (Oxford: Oxford University Press, 1991)

——, *Confessions*, ed. and intro. by James J. O'Donnell, 3 vols (Oxford: Oxford University Press, 1992)

Adelard of Bath, *De opere astrolapsus*, ed. by Bruce G. Dickey, in 'Adelard of Bath: An Examination Based on Heretofore Unexamined Manuscripts' (unpublished doctoral dissertation, University of Toronto, 1983), pp. 112–229

——, *Questions on Natural Science*, in *Conversations with his Nephew*, ed. and trans. by Charles Burnett with collaboration by Italo Ronca, Pedro Mantas-España, and Baudouin Van den Abeele (Cambridge: Cambridge University Press, 1998), pp. 81–236

Ananias of Shirak, *Ašxarhacʿoycʿ*, trans. by Robert H. Hewsen, *The Geography of Ananias of Širak: (Ašxarhacʿoycʿ), The Long and the Short Recensions* (Wiesbaden: Reichert, 1992)

Cicero, *De natura deorum*, in *On the Nature of the Gods. Academics*, trans. by Harris Rackham, Loeb Classical Library, 268 (Cambridge, MA: Harvard University Press, 1933)

——, *De natura deorum*, ed. by Joseph B. Mayor and John H. Swainson (New York: Cambridge University Press, 2010)

Ibn Ṭufayl, *Ḥayy ibn Yaqẓān*, ed. by Léon Gauthier, trans. by Lenn Evan Goodman (Chicago: The University of Chicago Press, 2009)

——, *Ḥayy ibn Yaqẓān*, trans. by Riad Kocache (London: The Octagon Press, 1982)

al-Masʿūdī, Ibn al-Ḥusayn, *Murūj al-dhahab wa-maʿādin al-jawhar. The Meadows of Gold and Mines of Gems*, trans. by Aloys Sprenger (London: Oriental Translation Fund of Great Britain and Ireland, 1841)

More, Thomas, *Utopia*, ed. by George M. Logan and Robert M. Adams, Cambridge Texts in the History of Political Thought (Cambridge: Cambridge University Press, 1995)

Nemesius, *Premnon physicon a N. Alfano archipiescopo Salerni in Latinum translatus*, ed. by Karl Immanuel Burkhard (Leipzig: Teubner, 1917)

Petrus Alfonsi, *Epistola ad peripateticos*, ed. by John Tolan, in John Tolan, *Petrus Alfonsi and his Medieval Readers* (Gainesville: University Press of Florida, 1993), pp. 164–72

——, *Letter to the Peripatetics of France*, trans. by John Tolan, in John Tolan, *Petrus Alfonsi and his Medieval Readers* (Gainesville: University Press of Florida, 1993), pp. 172–80

——, *Dialogus contra iudaeos*, ed. by Klaus-Peter Mieth and Spanish trans. by Esperanza Ducay, in *Diálogo contra los judíos*, ed. by María J. Lacarra, Larumbe, 9 (Huesca: Instituto de Estudios Altoaragoneses, 1996)

——, *Dialogus contra iudaeos*, trans. by Irven M. Resnick, *Dialogue against the Jews*, The Fathers of the Church. Medieval Continuation, 8 (Washington, DC: The Catholic University of America Press, 2006)

Plato, *Timaeus*, Latin trans. and commentary by Calcidius, ed. by Claudio Moreschini, *Commentario al Timeo di Platone* (Milan: Bompiani, 2003)

Ptolemy, *Almagest*, trans. and annotated by G. J. Toomer (Princeton, NJ: Princeton University Press, 1998)

Secondary Studies

Burnett, Charles, *The Introduction of Arabic Learning into England*, The Panizzi Lectures, 1996 (London: The British Library, 1997)

——, 'Petrus Alfonsi and Adelard of Bath', in *Petrus Alfonsi and his Dialogus: Background, Context, Reception*, ed. by Carmen Cardelle de Hartmann and Philipp Roelli (Florence: SISMEL Edizioni del Galluzzo, 2014), pp. 77–91

Glacken, Clarence J., *Traces form the Rhodian Shore* (Berkeley: University of California Press, 1967)

Malti-Douglas, Fedwa, *Woman's Body, Woman's Word: Gender and Discourse in Arabo-Islamic Writing* (Princeton: Princeton University Press, 1991)

Mantas-España, Pedro, 'Introducción', in Adelard of Bath, *Cuestiones naturales*, Spanish trans. by José L. Cantón Alonso, intro. and notes by Pedro Mantas-España (Pamplona: EUNSA, 2019), pp. 9–61

Miquel, André, *La géographie humaine du monde musulman jusqu'au milieu du 11e siècle. Géographie arabe et représentation du monde: La terre et l'étranger* (Paris: Mouton, 1975)

Palmer, William G., 'Environment in Utopia: History, Climate, and Time in Renaissance Environment Thought', *Environmental Review*, 8/2 (1984), 162–78

Rajan, Ravi S., 'Clarence Glacken: Pioneer Environmental Historian', *Environment and History*, 25 (2019), 245–67

Silverstein, Adam J., 'The Medieval Islamic Worldview: Arabic Geography in Its Historical Context', in *Geography and Ethnography: Perceptions of the World in Pre-Modern Societies*, ed. by Kurt A. Raaflaub (Malden: Wiley-Blackwell, 2010), pp. 273–90

Tolan, John, 'Reading God's Will in the Stars: Petrus Alfonsi and Raymond of Marseille Defend the New Arabic Astrology', *Revista Española de Filosofía Medieval*, 7 (2000), 13–30

DAG NIKOLAUS HASSE

Avicenna's *On Floods* (*De diluviis*) in Latin Translation

Analysis and Critical Edition with an English Translation of the Arabic

Chapter II.6, the last chapter of Avicenna's *Meteorology* of his great summa *The Cure* (*al-Shifāʾ*) bears the title 'On the Great Events which Happen in the World' (*Fī al-ḥawādith al-kibār allatī taḥduthu fī al-ʿālam*), but is often simply referred to as Avicenna's *On Floods*. It not only treats water floods, but also fire storms, earthy deluges, and destructive winds, as well as the possibility of a deluge that extinguishes a complete genus of living beings, and the possibility of the generation of living beings without procreation after such an extinction. In the Latin West, this chapter had a special transmission and reception history. It was translated independently of the rest of Avicenna's *Meteorology* and travelled in Latin manuscripts under the title *Capitulum in diluviis dictis in Thimeo Platonis* ('Chapter on Floods Mentioned in the Timaeus of Plato'). The Latin title does not contain any reference to Avicenna, but six of the twelve known manuscripts exhibit an additional *titulus* on the first page or a colophon which mentions the author: *capitulum Avicenne* or *tractatus Avicenne*. There does not yet exist a critical edition of the Latin version, which is offered in this article, together with an analysis of the doctrinal content of Avicenna's *On Floods*.

When Albert the Great between 1251 and 1254 wrote his *De causis proprietatum elementorum*, he referred to a major 'controversy' (*altercatio magna*) between Avicenna and Averroes 'in their books on such floods' (*in suis libellis de diluviis istis*) on the question of whether living beings are generated spontaneously after catastrophic floods.[1] At least since the time of Albert, Avicenna's *On Floods*

---

* I am very grateful for advice from Jon Bornholdt, Katrin Fischer, Stefan Georges, Andreas Lammer, Jon McGinnis, Colin Murtha, Eva Sahr and Jens Ole Schmitt.

1 Albert the Great, *De causis proprietatum elementorum*, I, 2, 13, p. 86, ll. 32–36. There is no treatise on floods by Averroes known to us. The reference must be to Averroes's *Long*

Dag Nikolaus Hasse (Würzburg, Germany) is a historian of philosophy focusing on Arabic and Latin philosophy and science and the transmission of knowledge from Arabic to Latin circles.

Mastering Nature in the Medieval Arabic and Latin Worlds: Studies in Heritage and Transfer of Arabic Science in Honour of Charles Burnett, ed. by Ann Giletti and Dag Nikolaus Hasse, CAT 4 (Turnhout: Brepols, 2023), pp. 107–141 BREPOLS PUBLISHERS 10.1484/.CAT-EB.5.134028

was an important reference point for Latin discussions of the extinction of life through floods and the spontaneous generation of animals and human beings: by Pietro d'Abano in the early fourteenth century, for instance, and later in the Renaissance by Antonio Trombetta, Pietro Pomponazzi, Tiberio Russiliano, and Pedro Fonseca.[2] Avicenna's views on spontaneous generation remained a major topic of academic writings until the late sixteenth century.[3] Avicenna was the main authority for the position that the spontaneous generation of animals and human beings is possible, as the result of an interplay between mixtures of elemental qualities and the giver of forms, which is the lowest of the celestial intelligences. Averroes's *Long Commentary on the Metaphysics*, in turn, was often quoted for the opposite position, which denies the possibility of spontaneous generation of human beings. For Averroes, even the animals that seem to arise from matter caused by a specific constellation of celestial bodies are in fact not natural beings but monstruous, unnatural beings.[4] The scholastic discussion of spontaneous generation is only one area where Avicenna's *On Floods* was influential; and its reception within the meteorological tradition proper still needs to be explored.

As Charles Burnett has shown, the Arabic-Latin translation of meteorological writings followed a programme which was to provide the Latin West with the full range of scientific disciplines, apparently after the model of al-Fārābī's *Enumeration of the Sciences*.[5] Gerard of Cremona, the most prolific Toledan translator, translated the first three books of Aristotle's *Meteorology* from Arabic into Latin in the second half of the twelfth century, as part of his attempt to create an Arabic-Latin version of works of Aristotle, which can be regarded 'as a continuation of the Alfarabian Peripatetic tradition'.[6] Gerard did not translate the fourth and final book, because, as his students (*socii*) wrote shortly after his death, 'he surely found that it (i.e. *the fourth book*) had already been translated',[7] namely, from Greek into Latin by Henricus Aristippus in

---

Commentary on the Metaphysics, II (Alpha elatton), comm. 15, and his *Long Commentary on the Physics*, VII, comm. 46, as the scribes of two manuscripts of Avicenna's *De diluviis* note (see the manuscript descriptions below). This passage in Albert probably was the source for later ascriptions of a treatise *De diluviis* to Averroes, as in the influential Renaissance chronicle by Foresti da Bergamo (Hasse, *Success and Suppression*, p. 32).

2 Hasse, 'Arabic Philosophy and Averroism', pp. 125–29; Martin, *Renaissance Meteorology*, pp. 71–72. See Pietro d'Abano, *Conciliator controversiarum*, fols 44^{v}–45^{r}; Trombetta, *Opus in Metaphysica*, qu. V, fol. 58rb; Pietro Pomponazzi, Lecture on the *Physics* in Bologna in 1518, in Nardi, *Studi su Pietro Pomponazzi*, pp. 315–19; Russiliano, *Apologeticus adversus cucullatos*, p. 173; Fonseca, *Commentariorum In Metaphysicorum Aristotelis*, VII, 7, 1, p. 246.

3 See Suarez, *Disputationes Metaphysicae*, Disp. XV, Section II, p. 508.

4 On Avicenna's and Averroes's views on spontaneous generation, see: Freudenthal, '(Al-) Chemical Foundations', pp. 47–73; Freudenthal, 'Medieval Astrologization', pp. 47–73; Hasse, 'Spontaneous Generation', pp. 150–75; Hasse, *Urzeugung und Weltbild*; Bertolacci, 'Averroes against Avicenna', pp. 37–54.

5 Burnett, 'Coherence', pp. 249–88.

6 Burnett, 'Arabo-Latin Aristotle', p. 101.

7 Burnett, 'Coherence', p. 260.

Sicily earlier in the century. Alfred of Shareshill, the Toledan translator of the generation after Gerard of Cremona,[8] was apparently aware of this: he put Gerard's and Aristippus's translations together and added his own Arabic-Latin translation of two chapters on minerals at the end of this meteorological corpus, probably because mineralogy is the sixth part of natural science in al-Fārābī's *Enumeration*, following right upon meteorology. The chapters on minerals Alfred took from the first book of Avicenna's *Meteorology* (Chs I.1 and I.5).[9] This combined Aristotelian-Avicennian corpus of meteorological writings translated by Henricus Aristippus, Gerard of Cremona and Alfred of Shareshill became very successful in the Latin West. It was much read in the arts faculties of later medieval universities as part of the Aristotelian curriculum and had a wide distribution in manuscripts. As a result, Avicenna's two chapters *De mineralibus*, which are extant in more than 130 manuscripts, is by far the most widely known Avicennian text in the Latin Middle Ages.[10]

Avicenna's *On Floods* fits into this story of meteorological translations. Its translator probably was aware of the fact that this latter part of Avicenna's *Meteorology* had not yet been translated.[11] This time, the translator was not Alfred of Shareshill, as has been shown by a quantitative analysis of the translator's style,[12] but a younger contemporary who also started his career as a translator in Toledo: Michael Scot.[13] Michael Scot's interest in extraordinary meteorological phenomena is attested also in his *Liber introductorius*, which treats thunder, lightning, abysses in the ocean, crevices, earthquakes, volcanos, and storms. He relates that some of these topics had been raised by Emperor Frederick II Hohenstaufen,[14] and that the Emperor experimented with the artificial incubation of hens' eggs, i.e., with some form of spontaneous gen-

8 Burnett, 'Shareshill [Sareshel], Alfred of', p. 992.

9 Burnett, *The Introduction*, pp. 71–72; Mandosio, 'Follower or Opponent of Aristotle?', pp. 469–76 (replacing earlier research published in Mandosio and Di Martino, 'La "Météorologie" d'Avicenne'). For the structure of Avicenna's *Meteorology*, or more precisely Avicenna's *On Minerals and Lofty Impressions (al-Maʿādin wa-l-āthār al-ʿulwiyya)*, see the very useful table in Mandosio, 'Follower or Opponent of Aristotle?', pp. 464–65.

10 Kischlat, *Studien zur Verbreitung*, p. 53. Avicenna's *De mineralibus* is often, but confusingly, called by the title of its first chapter, *De congelatione et conglutinatione lapidum* (see Mandosio, 'Follower or Opponent of Aristotle?', p. 470 n. 54).

11 For the context of the other Avicenna translations from Arabic into Latin, see the lists in: Burnett, 'Arabic Philosophical Works', pp. 814–22; Bertolacci, 'Reception of Avicenna', pp. 246–47.

12 Hasse and Büttner, 'Notes', pp. 346–47. On the identity of the translator of *De diluviis*, see also Di Donato, 'Les trois traductions', pp. 335–43. D'Alverny surmised in 1952 that Alfred of Shareshill was the translator: d'Alverny, 'Notes', p. 355.

13 On Michael Scot's life, works and translations, see the fundamental article by Charles Burnett, 'Michael Scot', pp. 101–26; as well as Ackermann, *Sternstunden*, pp. 13–61; Hasse, *Latin Averroes Translations*; Voskoboynikov's introduction to Michael Scot, *Liber particularis*, pp. 5–35.

14 Michael Scot, *Liber particularis*, pp. 187–211 and pp. 219–23. On Michael Scot's views on various meteorological and alchemical matters, see Thorndike, *Michael Scot*, pp. 60–71 and 110–15.

eration.[15] It is probable that Michael Scot's selection of texts for Arabic-Latin translation was influenced by Jewish scholars in Toledo, and by Samuel ibn Tibbon in particular, who visited Toledo between 1204 and 1210. Samuel ibn Tibbon not only shared many interests in meteorology, cosmology and astronomy with Michael Scot. He also quoted and translated into Hebrew a very similar range of Arabic authors: Alpetragius, Averroes, and Avicenna, and employed with much approval Avicenna's *On Floods* for a philosophical explanation of the Book of Genesis.[16] Perhaps it was Ibn Tibbon who made Michael Scot acquainted with this Avicennian text.

The aim of this article is to increase our understanding of Avicenna's *On Floods* by offering an analysis of its content, an English translation of the Arabic, and an edition of the Latin translation, which turns out to be considerably shorter than the Arabic text. Abbreviating the text is a translation technique often practised by Michael Scot. In this case the technique is particularly drastic, as we will see. In what follows, I shall discuss first the contents of the chapter and then the character of the Latin translation.

1. The Content of Avicenna's Chapter

The content of Avicenna's Ch. II.6 on floods loosely corresponds to Ch. I.14 of Aristotle's *Meteorology*.[17] Here Aristotle explains that the distribution of mainland and sea on the earth changes over large periods of time, as a result of excessive rains and deluges, which recur periodically. There is empirical evidence for this, Aristotle explains, in dry areas that bear vestiges of earlier times when they were covered by sea. This process does not affect the entire earth at once: some parts dry up, while others are flooded. We do not have historical records of this, argues Aristotle, because these processes are very slow compared to how long human beings and even entire peoples live. People die out in certain areas that have become inhabitable, but this is a process too long for memory (351a19–353a27).

Avicenna discusses these Aristotelian topics not only in *On Floods*, but also in Ch. I.6 of his *Meteorology*: rivers and, as a consequence, entire seas may dry up and thus disappear, and entire areas of the mainland may be drowned by water, but there are no written records of this.[18] A very similar doctrine is embedded in Avicenna's treatment of the saltiness of the sea in Ch. I.2 of his *Actions and Passions*, an earlier part of *al-Shifāʾ*: 'Nothing in the nature of the

15 Haskins, *Studies in the History of Mediaeval Science*, p. 289.

16 See Freudenthal, 'Medieval Hebrew Reception', pp. 269–311, on Samuel ibn Tibbon and on the later Hebrew reception of Avicenna's *On Floods*. On Samuel ibn Tibbon's works and translations, see Robinson, 'Samuel Ibn Tibbon'.

17 Jean-Marc Mandosio has pointed this out: Mandosio, 'Follower or Opponent of Aristotle?', pp. 465 and 467.

18 Lettinck, *Aristotle's 'Meteorology'*, pp. 144–45.

sea necessitates that it be specific to one location over another and in fact the sea is displaced over periods of the time that lifetimes do not comprehend'.[19] Hence, when Avicenna returns to the topic of floods in the last chapter of his *Meteorology*, he treats the topic of 'great events which happen in the world' from a more specific angle: in addition to watery excesses, he discusses excessive 'floods' or 'storms' (*ṭūfānāt*) originating from the predominance of fire, earth, and air, and he spends the greatest part of the chapter on the specific problem of catastrophic floods and their consequences for the rebirth of life on the earth.[20]

The line of argument of *On Floods* is difficult to understand at times, and unfortunately there does not seem to be a parallel passage in the eight major *summae* of Avicenna, except for a short passage in *The Easterners* (*al-Mashriqiyyūn*), where the beginning of the chapter is repeated almost verbatim from *al-Shifāʾ*: 'great events which happen in the world, which are the dominance of one of the four elements over an inhabited quarter either of the whole of it or of a part of it'. Then follow sentences about the watery, earthy, fiery and airy floods, and Avicenna concludes: 'All this is possible; there is no proof of its impossibility. Rather, these things suffice for the possibility of it'.[21] Then begins the section on the soul in *The Easterners*. We are left therefore with the very text of *On Floods* in *al-Shifāʾ* itself.

Avicenna starts with a definition of *ṭūfān* (flood): flood is the dominance of one of the four elements over an inhabited quarter of the earth. The cause for this dominance is conjunctions of the planets (*ijtimāʿāt min al-kawākib*) in a certain configuration (*hayʾa*), together with the assistance of earthy causes and elemental predispositions. Avicenna then discusses four kinds of flood: watery, fiery, earthy, and airy floods (§§ 2–3). There is historical evidence of the existence of such floods, Avicenna says (§ 4), but their existence can also be proved: on the one hand by proving that extreme increases and decreases of such phenomena are possible (§§ 4–5), on the other hand by recourse to the celestial influence on the sea (§§ 6–7). While Avicenna adds a caveat that the theory of influential astronomical situations may not be convincing (§ 8), he is sure that the first alternative is true: relative increases and decreases of such phenomena exist, and the possibility that increase and decrease reach extreme degrees cannot be denied.

Avicenna's theory of celestial influence (§§ 6–7) belongs to a longer history of linking extreme natural phenomena to rare celestial events. Such theories were current already in Late Antiquity, three of which were particularly influential. The first is Plato's Perfect Year, as expounded in *Timaeus* 39d,

---

19 Avicenna, *al-Afʿāl*, p. 208, ed. by Madkour and Qassem.

20 The possibility that the entire human race may be extinguished is also discussed in Avicenna's *On Animals*; see: Avicenna, *al-Ḥayawān*, p. 386; Latin version *De animalibus*, XV, 1, in Avicenna, *Perhypatetici philosophi*, fol. 59[va].

21 Avicenna, *al-Mashriqiyyūn*, pp. 131–32.

which is the greatest of all cyclical periods and is completed when all planets are in conjunction in one place of the heaven. The second is Aristotle's idea in *Meteorology* I, 14 (352a) that, just as there is a yearly winter, there also happens a 'great winter' at much longer intervals, that goes along with excessive meteorological phenomena. Olympiodorus, the sixth-century commentator on Aristotle's *Meteorology*, spells out in more detail that the 'great winter' arises when all planets are in conjunction at the winter solstice, and likewise for the 'great summer' at the summer solstice.[22] The third influential theory is the astronomical theory of precession, that is, the return of the sphere of the fixed stars to the same position, which takes 36,000 years according to Hipparchus's theory as reported in Ptolemy's *Almagest* VII, 2 (rather than the *c.* 25,800 years in the modern calculation). The main difference between, on the one hand, the Platonic and Aristotelian theories and, on the other hand, the Ptolemaic theory of precession, is that the first astronomical phenomenon concerns the planets, while the second concerns the sphere of the fixed stars. In addition to these theories, Avicenna could also have drawn on Indian, Persian and Islamic astrological traditions of great cycles and rare planetary conjunctions.[23]

Avicenna does not mention a numerical value for the interval in *On Floods*, and his passage about celestial influence remains rather vague. At one point he speaks about 'conjunctions of the planets' (*ijtimāʿāt min al-kawākib*); at another point he speaks of apogee, perigee and places 'close to the equator', which apparently is a reference to the nodes, i.e. the intersections of an orbit with the celestial equator (§ 6). There seem to be at least two possible interpretations of the passage. One is that Avicenna is speaking about the Ptolemaic precessional movement of the sphere of the fixed stars and, together with it, the movement of the tropical and equinoctial points. This would be close to how Avicenna's rough contemporaries, the Ikhwān al-Ṣafāʾ (Brethren of Purity), present the topic.[24] The other possible interpretation is that Avicenna offers a conjunctionist theory, making great floods dependent on the conjunctions of all planets close to the summer and winter solstices, or the equinoctial points. This would be a theory similar to that in Olympiodorus's commentary on *Meteorology*, which is extant in Arabic translation. The phrase 'conjunctions of the planets' makes it more likely that Avicenna is thinking of the second, conjunctionist alternative.[25]

22 Olympiodorus, *In Aristotelis Meteora*, comm. 1, 14, 351a19 et seq., pp. 111–12. See de Callataÿ, *Annus Platonicus*, pp. 1–4 (on Plato), 38–39 (on Aristotle), and 118–19 (on Olympiodorus, with an English translation of the passage).

23 For instance, by drawing on Abū Maʿshar. On such doctrines, see the comments by editors Yamamoto and Burnett in Abū Maʿshar al-Balkhī, *On Historical Astrology*, I, pp. 580–98.

24 On the Ikhwān al-Ṣafāʾ and their views on the topic, see: de Callataÿ, 'World Cycles', pp. 179–93; Ikhwān al-Ṣafāʾ, *Sciences of the Soul*, Epistle 36, pp. 191–96 (Arabic) and 231–33 (English).

25 Celestial influence is a central piece of Avicenna's natural philosophy. Cf. for instance

The astronomical doctrine in § 7 is less difficult to understand. Here Avicenna refers back to what he had said in *On Generation and Corruption* about the orbit of the planets: if the orbit of the planets were not inclined, i.e. inclined against the celestial equator, there would be no variation of the celestial influence, and the ruling quality of the planet would be ruling on a single location on the earth, with the effect of extinction of all life in that location.[26] In *On Floods*, this astronomical situation is curtly expressed with the phrase 'correspondence of the ecliptic with the equator'. Hence, in §§ 6–7, Avicenna describes two different astronomical phenomena, rare planetary conjunctions and the convergence of ecliptic and equator, both of which he finds possible.

In the next section (§ 8), Avicenna reaffirms that meteorological excesses are possible; that there is empirical evidence of the displacement of the seas from north to south, and that such displacement may bring habitation to an end.

The rest of the chapter, §§ 9–23, is devoted to the topic of spontaneous generation of animals and plants, or in other words the generation of life without reproduction after a catastrophic flood. Avicenna not only holds this possible, but presents a full-fledged theory of spontaneous generation.[27] Spontaneous generation comes about because of a rare formation of the stars and a rare disposition of the elements, which arises on account of an elemental first, second or third mixture (§§ 10–11).

One may object, says Avicenna, that this is impossible without uterus and sperm, but he replies that the passive power of the uterus can be replaced by a well-measured meeting of some portions of earth and water, and that the active power of the sperm can be replaced by the deliverance of such power from the Giver of Powers (*wāhib al-qūwā*), when a suitably disposed mixture occurs under the influence of the stars. This is in agreement with what Avicenna will later say in the *Metaphysics* part of *al-Shifāʾ* about the lowest of the celestial intelligences, material disposition, and the influence of the stars on that disposition.[28] Some people may say that the uterus contributes something in addition to the elemental mixture, but the true Peripatetic position, argues Avicenna, is that the cause for the acquisition of the form from the higher principles is a specific mixture (§ 15).

Avicenna's conclusion therefore is that any composite being may be generated spontaneously. This process results from several conditions: a certain elemental mixture; dispositions arising from it; the absence of a contrary

---

Avicenna, *al-Kawn wa-l-fasād* ('On Generation and Corruption'), Ch. 14, p. 192: 'The circular celestial motions causing the higher bodies' powers to be closer and farther away are the primary causes for generation and corruption, and their returns are undoubtedly the cause for the recurrent cycles of generation and corruption' (unpublished translation by Jon McGinnis).

26 Avicenna, *al-Kawn wa-l-fasād*, Ch. 14, p. 193.

27 For literature on Avicenna's theory of spontaneous generation, see n. 4 above.

28 Avicenna, *Metaphysics*, IX, 5, ed. by Marmura, p. 335.

disposition; and the emanation of forms. If this were not the case, it would be possible that whole species, such as human beings or trees, would cease to exist entirely without any return, and this possibility would have already occurred in an infinite universe (which is a version of what today is called the 'principle of plenitude') (§§ 16–19).

Avicenna closes the last chapter of his *Meteorology* with a final argument in favour of the possibility of spontaneous generation of human beings. He says that among the human arts (*ṣināʿāt*, i.e. crafts and skills) some are such that they are needed for the survival of human beings. All arts come to be at some point in time, and their origin is dependent on individuals that have a special property, namely, to be able to exist without these arts, which they invent. Without these individuals, all other individuals would die. Hence, these inventors of the arts cannot be generated by procreation, but only by spontaneous generation (§§ 20–23).

2. The Latin Translation

As noted above, strong stylistic evidence points to Michael Scot as the translator of *De diluviis*. In particular, we encounter three expressions highly characteristic of his translations: *inopinabile* (which appears twice), *semper fuit*, and *non est rectum dicere*; as well as phrases less current in his writings but still characteristic, such as *quoniam quemadmodum* or *et maxime quia*.[29] This finding is corroborated by a salient feature of the translation technique employed for *De diluviis*, which can be seen by comparing the English translation of the Arabic with the Latin text: the translator abbreviates considerably, not by summarizing or paraphrasing, but by omitting half-sentences, full sentences or even groups of sentences. For the rest, he translates almost word for word. This technique is very typical of Michael Scot, who uses it when translating Averroes's long commentaries on *De caelo* and *Metaphysics*, and Averroes's compendium of the *Parva naturalia*.[30] In the case of *De diluviis*, the technique is applied in a particularly drastic fashion: the Latin translation is only half as long as the Arabic original.

What are Michael Scot's motives for abbreviating? This can be best explained by surveying his abbreviations section-wise. Avicenna's first five paragraphs on the different kinds of flood and their proof of existence are transported without much abbreviation into Latin, with the exception of a reference to Arabic philologists (§ 1). Paragraph 6, however, on rare celestial events that cause the displacement of seas, is much shortened. Among other things, the

29 Hasse and Büttner, 'Notes', pp. 346–47. This can be confirmed by searching for these phrases in the quotations contained in Hasse and others, *Arabic and Latin Glossary*.

30 Hasse, *Latin Averroes Translations*, pp. 32–38. For a comparison of Michael Scot's abbreviating technique with that of other translators, see Hasse, 'Abbreviation', pp. 159–72.

mentioning of the astronomical positions 'apogee or perigee or something else in proximity of the equator' is omitted, with the effect that the theory becomes even more unspecific in the Latin version.

In the following paragraphs on spontaneous generation, Michael Scot returns to his usual procedure of omitting material that he finds repetitive, without much interference with the content. A radical version of this strategy can be observed in paragraph 12, where he cleverly shortens Avicenna's reply to the objection that a uterus and sperm are needed for generation. Avicenna's somewhat convoluted sentences are reduced to the bare bones of the argument. In paragraph 13, however, the omission concerns an important ingredient of Avicenna's theory of spontaneous generation. The Latin version does not transport Avicenna's theory that, after a first mixture, it takes a second or third or more mixtures under the influence of the stars until a suitably disposed mixture occurs, which is the condition for the spontaneous generation of a certain being. The Latin readers could grasp this theory only from a half-sentence in paragraph 16, which mentions 'a second composition in a different proportion'.[31]

In paragraphs 16–19, which present Avicenna's conclusion on spontaneous generation, Michael Scot retains most of the essential information, but leaves out an important step in the argument, namely that the possibility of complete extinction would have already occurred in an infinite universe (§ 19). The final section of *De diluviis*, on the spontaneous generation of the inventors of the arts, is translated faithfully into Latin, with the exception of the last paragraph, which is omitted in its entirety (§ 23). This could be due to an accident of transmission, either in Arabic or from the Arabic into Latin, but it seems more likely that Michael Scot found the paragraph superfluous in content, since it does not add anything to the argument: the paragraph offers additional information on the special property of self-sufficiency owned by the inventors of arts.

The most glaring interference of the Latin translator concerns not omission but addition: Michael Scot on seven occasions inserts phrases such as 'as some say', or 'according to some whom you know', or 'according to the astronomers' that are not in the Arabic text, nor in the apparatus of the Cairo edition of the Arabic. Six of the seven additions occur in the later part of *De diluviis* (§§ 9–22) and serve to distance Avicenna from the theory of spontaneous generation. An example is in paragraph 16. Avicenna concludes that, when all conditions are given, 'it is fitting' that generation happens without reproduction; this Michael Scot translates as: 'it seemed correct to some' (*rectum igitur videbatur quibusdam*). Another example is

31 Averroes, who tendentiously describes Avicenna as holding that human beings can be generated from earth, does not mention Avicenna's theory of mixture either; for Averroes's simplifying presentation of Avicenna's theory, see Bertolacci, 'Averroes against Avicenna', esp. pp. 43–44.

the very last sentence in Latin. Michael Scot provides the translation 'And these (i.e., *inventors of the arts*) are believed by some people to be generated without reproduction' (*Et illi putabantur a quibusdam fieri sine gignitione*), where Avicenna had written: 'Since this is the case, it is necessary that <these persons> came to be not by birth'. Michael Scot's purpose is obvious. He turns Avicenna from a protagonist of a theory of spontaneous generation into a person who presents a theory of others. Apparently, Michael Scot was not fully convinced of the possibility of spontaneous generation and of its explanation by Avicenna. The seventh addition appears in paragraph 2 of the treatise: 'And some say (*Et dixerunt quidam*) that the cause of floods is a constellation'. In the Arabic original, Avicenna had spoken of himself: 'We say: The cause for the occurrence of floods is the conjunctions of the planets'. Again, this serves to distance Avicenna from the content of the text, this time from the theory of celestial influence on floods.

The Latin title contains another significant alteration. It reads 'Chapter on the floods mentioned in Plato's Timaeus' (*Capitulum in diluviis dictis in Thimeo Platonis*), where the Arabic had been 'Chapter on the great events which happen in the world'. As is obvious from the apparatus criticus below, the Latin heading is preserved in all manuscripts. Hence, in all likelihood, the translator himself transformed the original title into the new one, which links Avicenna's discussion to Plato's *Timaeus*. It is not surprising that the Latin translator saw this connection in content, given that Calcidius's Latin translation of the *Timaeus* was one of the most popular philosophical treatises of the early Middle Ages,[32] that Michael Scot cites *Plato in Timeo* also in his commentary on Sacrobosco's *Sphera*,[33] and that twelfth-century authors such as William of Conches took their cue from the *Timaeus*'s mention of recurring catastrophic floods (22–23b) when developing their own natural explanation of such catastrophes.[34] The Platonic title chosen by Michael Scot did not misdirect scholastic authors from citing *De diluviis* as a treatise of Avicenna, as did Albert the Great in his *De homine*,[35] *De causis proprietatum elementorum*,[36] and *De IV coaequaevis*,[37] and Thomas Aquinas in his *Scriptum super Sententiis*.[38]

32 Leinkauf and Steel, *Platons 'Timaios'*.

33 Thorndike, *The 'Sphere' of Sacrobosco*, p. 248: 'Nam sicut dici Plato in Timeo'.

34 William of Conches, *Philosophia*, 3 § 39, p. 86.

35 Albert the Great, *De homine*, p. 131: 'Avicenna [...] Item, idem in capitulo de diluviis quae numerantur in Timaeo Platonis'.

36 See n. 1 above.

37 Albert the Great, *De IV coaequaevis*, p. 745b: 'ut dicit etiam Avicenna in libro de diluviis'.

38 Thomas Aquinas, *Scriptum super Sententiis*, III, D. 3, Q. 2, Art. 1 c: 'dicit Avicenna in capitulo de diluviis'.

3. Editorial Note

The Latin translation of *On Floods* is preserved in twelve manuscripts, as far as we know today. Eleven of these are described in Marie-Thérèse d'Alverny's *Codices* volume of the *Avicenna latinus* series, which appeared in 1994; one manuscript was found later in the Universitäts- und Landesbibliothek Tirol of Innsbruck. *De diluviis* first appeared in print in 1949, when Manuel Alonso Alonso published a mildly edited transcription of Seville, MS Biblioteca Capitular y Colombina, 5–6-14.

The present critical edition is based on a collation of all twelve manuscripts. The variant readings of two manuscripts are not recorded in the apparatus, for different reasons. Manuscript P (Palermo, MS Biblioteca Comunale, Qq. G. 31), of the later fifteenth century, written by a Franciscan friar in Brittany, offers not Michael Scot's translation itself, but a thorough stylistic revision of it. Manuscript C (Chicago, MS Newberry Library, 23) contains numerous errors, some of which are shared with manuscript E, while others show that the scribe did not recognize words. Because of the sheer number of faulty readings, this manuscript was excluded from the apparatus.

Marie-Thérèse d'Alverny mentions dates and places of origin for the manuscripts, for the most part without giving reasons. According to this information, which has to be treated with caution, six manuscripts date from the thirteenth century, one from the turn to the fourteenth century, and five from the fourteenth century. The only manuscript of the fifteenth century is the above-mentioned Palermo manuscript with the stylistic revision of the text. As to the places of origin, five manuscripts probably come from Northern France, four from Germany or Austria, one from Italy, and two from unidentified areas. Hence, the manuscript tradition is entirely late medieval and largely central-European. This did not preclude important Renaissance authors from studying and using the text, as was mentioned above. *De diluviis* travelled in the context of other philosophical treatises, mainly of the Graeco-Latin and Arabic-Latin tradition, often together with other twelfth- or early thirteenth-century translations, but, remarkably, not with Plato's *Timaeus*. One manuscript (Krakow, MS Biblioteka Jagiellońska, 1718) stands apart from the group; it also transmits theological works by Thomas Aquinas.

As to the relationship among the manuscripts, two groups share significant variants that bind them together: manuscripts CENaVa and manuscripts KVb, while manuscripts GMNüS are grouped together in a loose way only. This judgement is based on the following evidence, which can only be limited for a short text:

§ 4: potest credi per hanc rationem] per hanc rationem potest credi KVb

§ 4: aut prope medium] *om.* CEKNaVb

§ 7: similiter] si *add.* GIKMNüSVb

§ 7: totam] tota G *om.* KVb

§ 8: sufficere in hoc] in hoc sufficere CEKNaVb sufficere I
§ 8: habitabilis] animalis GMNüS habitabiles I habitatio Na
§ 10: in multis annis] *om.* KVb
§ 11: fuerit possibilis congregatio earum] *om.* CENaVaVb
§ 13: si] non GKMNüSVb sicut E tamen I
§ 16: et facientem aliquam complexionem et componantur secundo in aliam proportionem] *om.* CENaVa
§ 18: generationis] maxime *add.* CENa
§ 19: secundum astronomos non venirent] non venirent secundum astronomos CENa

The two manuscripts KVb are joined together also by a long additional *titulus*, which mentions parallel passages in Averroes and Albert, as is obvious from the manuscript descriptions below. Often, but not always, the GMNüS group offers a cleaner text with fewer copying mistakes than the rest of the manuscripts. For establishing the critical text, however, it was more important to find the correct reading by comparing the literal translation to the Arabic. The present edition therefore does not have a base manuscript. In five cases, I suggest a reading against all manuscripts, often with the help of the Arabic; these are: *nivea*[39] instead of *una* or *viva* (§ 3); *septentrionalis* instead of *habitabilis* (§ 8); *animalia et plante* instead of *et plante* (§ 9); *homines* instead of *omnes* (§ 20); and *sine eis* instead of *eis* (§ 21).

The apparatus does not record orthographic variants, such as *multotiens / multociens, sed / set, imp- / inp-, arena / harena, augmen- / aucmen-, gignitio / gingnitio*, except in names. In such cases of orthographic variety, the reading adopted for the edition follows the more common reading in the manuscripts. The punctuation and the paragraph division of the main text is editorial.

Note that the final paragraph (§23) is not transmitted in Latin.

39 I owe this reading to a suggestion by Stefan Georges.

Manuscripts

C = Chicago, The Newberry Library, 23, fols 181^{r}–182^{r}

First half of the 14th c.

Or. Germany or Austria.

Prov. Monastery of Melk, formerly no. 398 (529, I 47), listed in the 1483 catalogue of Melk as 'F 1'. Note in a seventeenth-century hand on fol. 1^{r}: 'Monasterii Mellicensis L. 35'. London library of E. P. Goldschmidt, from whom acquired in 1938 by The Newberry Library.

Parchment, 182 fols, two columns.

Philosophy: much Aristotle, Averroes, Thomas Aquinas, Avicenna, Costa ben Luca.

Lit. *Aristoteles latinus*, I, no. 60; *Avicenna latinus*, pp. 293–97. Paul Saenger, *A Catalogue of the Pre-1500 Western Manuscript Books at the Newberry Library* (Chicago: University of Chicago Press, 1989), pp. 39–41.

Colophon: *Explicit capitulum de diluviis. Amen.*

E = Erfurt, Universitätsbibliothek, CA. 4° 15, fol. 49^{r-v}

14th c.

Or. probably Germany.

Prov. Amplonius Rating (1363–1435), signature '49 phil. natur.' in Amplonius's 1412 catalogue; Erfurt, Collegium Porta Coeli in 1412.

Parchment, fols 3^{r}–91^{r}, 91^{v} vacat, one column, several German hands, many marginal and interlinear glosses, also on fol. 49^{r-v}, bound together with other manuscripts to a composite volume of 196 fols.

Medicine and Philosophy: Constantinus Africanus, Aristotle, Alkindi, Thomas Aquinas, Albert the Great, several anonymous texts.

Lit. Wilhelm Schum, *Beschreibendes Verzeichnis der Amplonianischen Handschriften-Sammlung zu Erfurt* (Berlin: Weidmannsche Buchhandlung, 1887), pp. 295–98; *Aristoteles latinus*, I, no. 890; *Avicenna latinus*, pp. 202–05.

Titulus on top of page: *Tractatus de diluviis Avicenne.*

Colophon: *Explicit tractatus Avicenne de diluviis. Deo gratias.*

G = Graz, Universitätsbibliothek, 482, fols 241^{v}–242^{r}

End of the 13th c.

Or. probably Northern France.

Prov. Benediktinerstift St Lambrecht, Austria.

Parchment, 242 fols, two columns.

Philosophy: Asclepius, Apuleius, Maimonides, Avicenna, Alexander of Aphrodisias, Alfarabi, Algazel, much Averroes, Costa ben Luca, Alfred of Shareshill, Dominicus Gundisalvi.

Lit. Anton Kern, *Die Handschriften der Universitätsbibliothek Graz: Band 1* (Leipzig: Harrassowitz, 1942), pp. 281–86; *Aristoteles latinus*, I, no. 57; *Avicenna latinus*, pp. 173–80.

Titulus at bottom of page: *Incipit Avicenne capitulum de diluviis enumeratis in Thimeo Platonis.*

I = Innsbruck, Universitäts- und Landesbibliothek Tirol, 302, fols 94v–95v

13th c. (first part of the MS).

Or. probably Northern France (first part of the MS).

Prov. Donated to Kloster Stams, Austria, 14th c. Since 1808 University Library Innsbruck, Austria.

Parchment, two parts, fols 1–95 and fols 96–142, one column, probably only one hand.

Philosophy, mainly natural philosophy: Aristotle, Ps.-Aristotle, Averroes.

Lit. Walter Neuhauser and Lav Subaric, *Katalog der Handschriften der Universitätsbibliothek Innsbruck, Teil 4: Cod. 301–400* (Vienna: Österreichische Akademie der Wissenschaften, 2005), pp. 34–38; not in *Aristoteles latinus*; not in *Avicenna latinus*.

K = Krakow, Biblioteka Jagiellońska, 1718, fols 225v–226r

End of the 13th c.

Or. probably Northern France.

Prov. On the cover a note 'magistri Petri de Zwanow', master of arts and bachelor in theology at Cracow in the fifteenth century.

Parchment, I+227+I fols, two columns, several hands; the text of 226v–227r is added later (14th c.).

Theology: Thomas Aquinas.

Lit. *Avicenna latinus*, pp. 253–54.

Titulus at bottom of page: *C(apitulum) Avi(cenne) de diluviis enumeratis in Thimeo Platonis. Aver(roes) autem uidetur esse contra hoc secundo me(taphisi)ce et 8 phi(si)chorum et de hoc Albertus primo de proprietatibus elementorum tractatu secundo, capitulo tredecimo.*

M = Munich, Bayerische Staatsbibliothek, Clm 8001, fol. 26^{r-v}

End of the 13th c.

Or. perhaps German.

Prov. Cistercian monastery of Kaisheim.

Parchment, 270 fols, two columns, several hands.

Philosophy and Theology: Averroes, Giles of Rome, Dominicus Gundisalvi, Thomas Aquinas, Albert the Great, Alkindi, Alfarabi, Alexander of Aphrodisias, Isaac Israeli.

Lit. *Aristoteles latinus*, I, no. 1035; *Avicenna latinus*, pp. 209–13; David Juste, *Les manuscrits astrologiques latins conservés à la Bayerische Staatsbibliothek de Munich* (Paris: CNRS Éditions, 2011), p. 117.

Na = Naples, Biblioteca Nazionale, XI AA 49 (2), fol. 1^{r}

Second half of the 13th c.

Or. Italian.

Prov. Augustinian monastery San Giovanni a Carbonara in Naples.

Parchment, 100 fols, one column, one hand.

Philosophy: much Avicenna, Thabit ibn Qurra.

Lit. *Avicenna latinus*, pp. 75–76.

Colophon: *Explicit tractatus de diluviis Avicenne. Deo gratias.*

Nü = Nuremberg, Stadtbibliothek, Cent. V 21, fols 181^{v}

14th c.

Or. Unknown.

Prov. Nuremberg, Dominican monastery.

Parchment, I+231 fols, two columns, several hands, composed of two formerly independent manuscripts.

Philosophy: Radulphus Brito, much Avicenna, Costa ben Luca, Aristotle.

Lit. *Aristoteles latinus*, I, no. 1089; *Avicenna latinus*, pp. 213–15; Ingeborg Neske, *Die Handschriften der Stadtbibliothek Nürnberg. Bd.4, Die lateinischen mittelalterlichen Handschriften: Varia 13.-15. und 16.-18. Jh.* (Wiesbaden: Harrassowitz, 1997), pp. 56–60.

P = Palermo, Biblioteca Comunale, Qq. G. 31, fols 199^{v}–201^{r}

Second half of the 15th c.

Or. Written by the Franciscan friar Mauritius Gauffridi, Brittany.

Prov. Franciscan monastery of Cuburien, near Morlaix, Brittany, founded 1458.

Paper, III+202+IV fols, two columns.

Philosophy: Aristotle.

Lit. *Aristoteles latinus*, II, no. 1497 (full description); *Avicenna latinus*, p. 320.

The MS does not offer Michael Scot's translation, but a thorough stylistic revision of it, as d'Alverny noted (*Avicenna latinus: Codices*, p. 320: 'Textus abundanter interpolatus').

Titulus: *Incipit liber de natura diluvii.*

Incipit: *Capitulum de diluviis dictis in thimeo platonis. Dicitur autem diluvium unius victoria elementi super quartam habitabilem aut super unam aliquam regionem ...*

Explicit: ... *hominibus sed in paucis in quibus virtus et fecunditas renovatur.*

Colophon: *Explicit de natura diluviorum.*

S = Seville, Biblioteca Capitular y Colombina, 5-6-14, fols 92^{v}–93^{r}

End of the 13th c.

Or. Unknown.

Prov. Fernando Colón, donor of the MS to the Biblioteca Capitular of Seville.

Parchment, I+164 fols, one column, one hand.

Philosophy: Alexander of Aphrodisias, Alfarabi, Aristotle, Avicenna, Costa ben Luca, Alfred of Shareshill, Avicebron, several anonymous texts.

Lit. *Aristoteles latinus*, II, no. 1181; *Avicenna latinus*, pp. 222–25.

Titulus: *Incipit quoddam Capitulum Avicenne de diluviis.*

Va = Vatican City, Biblioteca Apostolica Vaticana, Vat. lat. 725, fols 36^{r}–37^{r}

13th–14th c.

Or. German.

Prov. Unknown.

Parchment, 69 fols, one column, several German hands.

Philosophy: Albert the Great, Thomas Aquinas, Aristotle, Costa ben Luca, Avicenna, Dominicus Gundisalvi.

Lit. *Aristoteles latinus*, II, no. 1825; *Avicenna latinus* p. 89.

Titulus: *Incipit de diluviis.*

Colophon: *Explicit liber de diluviis generalibus.*

Vb = Vatican City, Biblioteca Apostolica Vaticana, Vat. lat. 4426, fol. $1^{r\text{-}v}$

First half of the 14th c.

Or. Northern France or Flanders.

Prov. Unknown.

Parchment, 70 fols, two columns, one hand of Northern France.

Philosophy: Avicenna, Alfarabi, Alexander of Aphrodisias, Alkindi, Giles of Rome, Thomas Aquinas, Dietrich of Freiberg, Costa ben Luca, Ps.-Augustine, Proclus, Henry of Ghent.

Lit. *Avicenna latinus*, pp. 96–99.

Titulus (rubric): *C(apitulum) Avic(enne) de diluviis enumeratis in Thy(me) o Platonis. Av(er)roys autem uidetur esse contra hoc secundo me(taphysi)ce et octavo ph(ysice) et de hoc Albertus primo de proprietatibus elementorum tractatu secundo, capitulo tredecimo.*

Colophon: *Explicit liber de diluviis Avicenne.*

Editions

A = Latin edition: Manuel Alonso Alonso, 'Homenaje a Avicena en su milenario. Las traducciones de Juan González de Burgos y Salomon', *Al-Andalus*, 14 (1949), pp. 291–319, at 306–08.

Arab = Arabic edition: Avicenna, *al-Shifāʾ: al-Ṭabīʿiyyāt: al-Maʿādin wa-l-āthār al-ʿulwiyya*, ed. by ʿAbd al-Ḥalīm Muntaṣir, Saʿīd Zāyid and ʿAbdallāh Ismāʿīl (Cairo: al-Hayʾa al-ʿāmma li-shuʾūn al-maṭābiʿ al-amīriyya, 1965), pp. 75–79.

Abbreviations

add.	add(s)
coni.	conjecture by the editor
corr.	correct(s)
del.	delete(s)
marg.	margin
inv.	invert(s)
lac.	lacuna
om.	omit(s)
sup. lin.	above the line
< >	added by the editor
( )	suggested reading
{ }	difficult to read

Capitulum in diluviis dictis in Thimeo Platonis.

(1) Et est diluvium victoria unius elementorum super quartam habitabilem aut super unam partem. Et quando est ex aqua dicitur proprie diluvium in ydiomatibus.

(2) Et dixerunt quidam quod causa diluvii est constellatio que facit unum elementum vincere cum causis accidentibus et preparationibus materialibus. Aquosum ergo accidit ex mutationibus marium subito per maximas causas ventosas aut per multas pluvias propter magnam alterationem aeris in aquam.

(3) Igneum autem accidit ex incensione ventorum fortium et istud est fortius. Et terrestre accidit ex multis arenis cadentibus de uno loco in alium aut propter qualitatem terrestrem frigidam congelatam sicut dictum est de una terra. Aereum autem fit ex motibus ventorum fortium.

(4) Et hoc potest credi per hoc quod narratur de diluvio aque. Et potest credi per hanc rationem quia res que suscipiunt

---

Tit. in[1]] de *EMNaVa* dictis in] sumptum de *M* Thimeo] thymeo *EIVa* thym(e)o *M*
1 unius] unle *Vb* elementorum] elementi *I* quartam] totam *I* partem *add. Va* **2** habitabilem] habitabilium *EKMNaVa* habitabile *corr. ex* inhabitabile *Nü* partem] eius *add. I* quando est] quandoque *AGNüSVb* est] *om. I* ex] enim in *E* ab *Na* aqua] est *add. I* **3** dicitur] bene ab *E* proprie] bene *Na* proprie ... in] *om. E* diluvium] *om. INa* in] ab *NaVa* ydiomatibus] proprie diluvium *add. E* omathibus *I* diluvium proprie *Na* **4** est] esset quod *E* esset *Na* quedam *add. KNaVaVb* **5** elementum] altera *add. E* alia *add. Na* vincere] devincere *E* accidentibus] agentibus *EKMNaVaVb* **6** preparationibus] proportionibus *A* ergo] igitur *GKSVb* sibi *del. Va* **7** ventosas] ventorum *Na* per[2]] propter *EI* **8** pluvias] aut *add. I* propter] per *Vb* magnam ... aeris] aeris alterationem magnam *ENa* aquam] aqua *AEGMNaNüSVb* **9** Igneum] ignire *I* incensione] intensione *AEG* intentione *KNa* intencione *MVb* ventorum fortium] *inv. AS* **10** et] ac *A* ex *Na* istud] istis *ANü om. E* id *M* isto *Na* ist{ud} *S* illud *IVaVb* est] eis *add. E* ei *add. Na* Et] *om. ENa* terrestre] autem *add. ENa* et hoc *add. I* **11** de] ex *ANüS* uno] imo *M* alium] aliud *Na* terrestrem] ita *add. Nü* **12** frigidam] et *add. ANa* congelatam] gelatam *A* congelativam *GKMVb* est] *om. I* de] in *Na* una] *coni.* viva *AGKMNüSVb* una *EINaVa* terra] thema *I* Aereum] aerema *I* **13** fit] sit *Na* **14** per] propter *AS* ex *E* de *Na* **14/15** per ... credi] *om. I* **14** diluvio] diluviis *Na* **15** Et] hoc *add. Va* potest ... rationem] per hanc rationem potest credi *KVb* per] propter *A* ex *E* hanc] hac *E* rationem] ratione *E* quia] quod *E* res] rese *S*

Chapter on the Great Events which Happen in the World

by AVICENNA

(Translated from the Arabic)

(1) //75 Among the things which we should discuss in this place is the matter of floods. We say: Flood (*ṭūfān*) is the dominance of one of the four elements over an inhabited quarter, either the whole or a part of it, or one element becoming dominant in this way, according to what the scholars of the <Arabic> language (*ahl al-lugha*) regard as the <proper> usage for the term. Best known among the common people concerning the matters of floods is what derives from water, as if this term were <only> applied to this sense.

(2) We say: The cause for the occurrence of floods is the conjunctions of the planets (*ijtimāʿāt min al-kawākib*) according to one of the configurations which cause the domination of one of the elements in the habitable world, together with the assistance of earthy causes and elemental dispositions. The watery <floods> occur out of sudden displacements of the seas[1] over a large area because of gale-like conditions necessitating this, or because of causes which necessitate a severe rising of flood waters and of continuous rains, or because of an excessive alteration occurring of air into water.

(3) The fiery <floods> happen from inflammations through violently blowing winds, and these have a very great diffusion. The earthy <floods> happen because of an excessive deluge (*sayalān*), which comes about through sands in inhabited open countries, or because of a quality which makes a deluge of cold solid earthy <matter>; this is something that we have talked about. The airy <floods> result from very vehement and destructive motions of the air.

(4) What is convincing about the existence and occurrence of these <floods> is the many uninterruptedly transmitted reports (*al-akhbār al-mutawātira*) relating a flood of water. What is convincing with respect to proving this (*ithbāt*), is that

1 Reading *biḥār* (seas) instead of *bukhār* (vapours).

magis et minus, licet esse in eis est medium aut prope medium inter duo extrema, tamen non exit a possibilitate ut sit in extremis ipsis.

(5) Et sicut accidit multotiens quod sunt anni sine pluvia in aliqua terra, ita potest evenire pluvia subito et potest alterari in aquam subito. Et ita de aliis diluviis.

(6) Et si verum est hoc quod dicunt quod mare mutatur per mutationem celestium in tantum quod omnes aque cooperiunt habitationem, discooperietur unus polus aut duo.

(7) Et similiter hoc quod dicunt de mutatione declinationis, si est verum in tantum quod zodiacus superponeretur equinoctiali, totam cooperiet habitationem.

---

16 magis] maius *I* {...}ag *M* et] uel *AS* esse] *om. E* et *Na* est *Vb* in] inter *ENa* eis] ea *ENa* est] sit *E om. MNa* aut] vel *Va* aut prope] propter *I* aut ... medium[2]] *om. EKNaVb* **17** tamen] cum *I* causam *Va* non] quandoque *E om. I* idem *Na* exit] erit *A om. M* a possibilitate] impossibile *A* apossibilitate *G* ap(otest?)ate *I* apo(ssibili)tate *K* uel aptatam *Na* a pos(sibilitate) *S* apost(erita)te *add. in marg. S* apassibilitate *Va* a p(ossibili)tate *Vb* sit] sic *Na* in] non *Na* cum *Va* extremis] extremum *I* **18** ipsis] ipsius *E* **19** sicut] *om. Na* accidit multotiens] *inv. Na* multotiens] mul{...} *Vb* sunt anni] c{in}mus sunt *Na* pluvia] *om. I* **19/20** in ... pluvia] *om. Na* **20** terra] et *add. E* ita] *om. KVb* evenire] aliqua *add. I* pluvia] in *add. K* cetera *add. Vb* et potest] incipit *Na* alterari] aer *add. E* **20/21** in aquam] *om. AS* in aqua *EGKMNüVb* **21** aquam] ita *add. KVb* subito] *om. Na* ita] est *add. I* aliis diluviis] *inv. Va* **22** hoc] *om. ENa* mare] *corr. ex* in ae *Va* mutatur] permutatatur *(sic)* per terram vel *E* permutatur *Na* per] propter *AENa* **23** celestium] stellarum *ENa* **24** habitationem] habitabilem *I* habita *Na* Iū adh‘ *Va* discooperietur] discooperitur *E* discooperientium *Na* **25** Et] etiam *AM* similiter] in *add. A* si *add. GIKMNüSVb* de mutatione] *om. M* declinationis] decliva *E* detectionis *I* **26** si] hoc *E* sed *IKVb* est verum] *inv. Va* verum] *om. Na* in ... quod] tunc si *E* quod] et ita *add. I* superponeretur] supponeretur *EIVa* supponetur *Na* equinoctiali] et *add. Na* **27** totam] tota *G om. KVb* cooperiet] cooperiret *EVa* cooperiunt *I* habitationem] habitabilem *I*

the things which are receptive of increase and decrease, scarcity and multitude – even if most of the existence in their <case> is an existence in the middle between the two extremes of excess and deficiency, and what comes close to <the middle> – are such that the two extremes //76 are not outside the limit of possibility.

(5) Just as it often happens that years come about for great areas of the inhabited earth in which there is no rain at all, and this is on the side of decrease, likewise it is sometimes possible that rains are excessive in one instant and that the air undergoes alteration into a watery nature suddenly, since that which belongs to these middle <states> differs by increase and decrease, and likewise with the other floods.

(6) If it is true what we surmise about the dependency of the seas upon the direction (*jiha*) of the celestial sphere (*falak*), then it is necessary that <the sea> is displaced by its transfer (*intiqāl*) so that, at some time, it spreads over all of these regions beyond which inhabitation cannot extend; and this is when the position (*mawḍūʿ*) that moves the greater part of the sea by its transfer occurs belonging to the celestial sphere, such as apogee (*awj*) or perigee (*ḥaḍīḍ*) or something else in proximity (*fī qurb*) of the equator. Thus, the water deluges the place which should be inhabited, while one or both poles (*quṭbān*) are uncovered, and land, which is opposed to the sea, is displaced to it (*i.e., to the place of the poles*) and there inhabitation is prevented (*because of cold*). Thus, the earth is divided into land and sea <in a way that> neither of them permits inhabitation by air-breathing animals.

(7) Likewise, if the state of the inclination (*mayl*) and what we surmise about its change and its abating is confirmed truth,[2] so that it is true that there is a correspondence (*inṭibāq*), or something similar to correspondence, of the ecliptic (*falak al-burūj*) with the circle of the equator (*dāʾira muʿaddil al-nahār*), then indeed all this belongs to what necessitates the destruction of inhabitation.

2 Avicenna, *al-Kawn wa-l-fasād* (On Generation and Corruption), Chapter 14.

(8) Et si non est verum illud, illa ratio quam diximus potest sufficere in hoc. Et nos bene scimus quod pars septentrionalis fuit cooperta aquis ita quod montes facti fuerunt et modo maria sunt meridionalia, ergo maria sunt mutabilia. Et mutatio eorum non est determinata, sed possibile est ut ita fiat ut abscidat habitationem. Forte igitur sunt renovationes in annis multe quarum memoria non potuit retineri.

(9) Et non est inopinabile secundum quosdam quos tu scis ut corrumpantur <animalia> et plante et multa genera eorum et post generentur per generationem, non gignitionem, quoniam multa animalia fiunt per generationem et gignitionem et similiter plante. Et ex capillis fiunt serpentes, ex ficubus scorpiones et mus de terra et rane de pluvia. Et omnia huiusmodi gignuntur.

(10) Et quando hec gignitio abscinditur in multis annis, possibile est ut veniat secundum quosdam aliqua constellatio et aliqua preparatio elementorum que faciat ea generari.

(11) Immo dicimus quod quelibet species que fit per complexionem elementorum in quantitatibus scitis, dum elementa

---

28 est] *om. AGKMS* illud] tunc *add. E* istud *GKMNaNü* id *I* illa] ista *NaVa* illa ... quam] {...} *Vb* ratio] tamen *add. E* quam diximus] *om. M* potest] *om. KVb* **29** sufficere] quam diximus *add. M* sufficere ... hoc] in hoc sufficere *EKNaVb* sufficere *I* Et] quod *Na* scimus] propter hoc *add. E* pars] terra *E om. I* hec *Na* vel *add. Va* septentrionalis] *coni.* animalis *AGMNüS* habitabilis *EKVaVb* habitabiles *I* habitatio *Na* **30** fuit cooperta] sunt cooperte *I inv. Va* facti] *om. Na* modo] in *G* **31** meridionalia] distincta *add. Va* mutabilia] mutabila *S* **32** fiat] diluvium *add. E* ut[2]] et *ENa* abscidat] abscidit *E* abscidatur *Na* abscindat *INü* **33** habitationem] habitatio *Na* igitur] ergo *Va* renovationes] revocationes *Na* multe] multis *ENaVa om. I* lis *add. sup. lin. M* **34** quarum] quorum *EINa* potuit] poterat *Va* **35** inopinabile] impossibile *INa* secundum] apud *EMNaVaVb* quos] sed *E* quod *Na* ut *Va* scis] cui scripsi tractatum *add. E* **36** ut] quod *ENaVa* corrumpantur] corrumpatur *M* animalia] *coni. om. AEGIKMNaNüSVaVb* et[1]] ut *E om. INaVa* plante] planate *Na* eorum] earum *EM* **37** post] postea *E* post ea *Na om. Va* generentur] regenerantur *E* generantur *Na* generantur postea *Va* generationem] et *add. E* non] per *add. EVa* non gignitionem] cognitionem *Na* **37/38** quoniam ... gignitionem] *om. Na* **38** animalia] alia *AI* fiunt] *om. E* sunt *I* gignitionem] fiunt simul *add. E* generationem *I* et similiter] *inv. E* **39** plante] planete *Na* fiunt] sunt *I* serpentes] et *add. INaVa* ficubus] *lac. E* ficabus *I* **40** mus] mures *ENaVa* gignuntur] ginguntur *Na* **41** gignitio] generatio *I* abscinditur] absciditur *EMNaVa* in] *om. Na* in ... annis] *om. KVb* **42** est] secundum quosdam *add. Va* veniat] eveniat *Na* secundum quosdam] *om. Va* aliqua] *om. A* **43** aliqua] *om. ENa* elementorum ... generari] *om. Va* faciat] fuerit *Na* ea] animalia *add. E* generari] regenerari *E* **44** Immo] ut *add. E* imo *Na* nos *add. Va* que] *om. EVa* fit] sit *AES* per] ex *Va* complexionem] commixtionem *I* complexione *Va* **45** scitis] suis *ENa* sitis *KVb* dum elementa] duella *M*

(8) If this is not possible either, then <at least> what we have said about the excesses and what we have certified about the possibility of displacement of the seas from one region of a pole to another pole is not beyond possibility. We know by very strong conjecture that the region of the north was covered with water until the mountains came to be; and now the seas are southern, so the seas were displaced. It is not necessary that their displacement is limited, but rather many ways are possible concerning it, some of which verge on bringing habitation to an end (*inqiṭāʿ*). It seems that there are in the world upheavals (*qiyāmāt*), which appear in some years of which histories are not retained.

(9) It is not objectionable that animals, plants or //77 genera of them are destroyed and afterwards come about through generation (*tawallud*) without reproduction (*tawālud*). This is because there is no demonstration at all for the impossibility of the existence and origination of things after their extinction by way of generation without reproduction. Many animals come about by <both> generation and reproduction, and likewise plants: serpents may consolidate from hair, scorpions from chaff and basil, mice are generated from mud, frogs from rain. To all these things there also is reproduction.

(10) It is not necessary, when this generation comes to an end and so is not observed for many years, that it not have some existence rarely, when some rare configuration of the celestial sphere occurs, which is not repeated for some time, and in virtue of a disposition among the elements which chances to happen only at long intervals.

(11) Instead, we say: Whatever is generated from the elements through some mixture leads to the existence of a species due to the occurrence of that mixture by

fuerint et sua divisio secundum illas quantitates et congregatio eorum fuerit possibilis, congregatio earum erit possibilis. Et si prima complexio non sufficit, hec generantur ex secunda et tertia. Quoniam quemadmodum generatur animal ex complexione humorum post complexionem elementorum, non est inopinabile quod fiat complexio secunda sine semine et sine spermate.

(12) Et si quis dixerit quod est impossibile nisi in loco determinato et per virtutem determinatam in matrice et spermate, tunc sermo de istis erit sicut sermo de primo. Omnia enim ista generantur ex complexione elementorum et matrix nichil facit nisi retinere. Et radix est complexio.

(13) Et si est possibile ut una pars terre alteretur cum una parte aque in quantitate determinata, tunc non indiget aliquo retinente et dator formarum dabit uirtutes agentes.

(14) Sed matrix faciet ad meliorationem. Tamen sine ea non erit impossibile hoc accidere ex motibus et aliis causis, nisi quis

---

46 fuerint] fuerunt *ANa* fiunt *M* et[1]] in *EVa* sua] suo *ENaVa* divisio] dm̄o *ENaVa* illas] istas *NaVa* congregatio] cognitio *Na* **47** eorum] earum *A* ist{a}rum *Va* fuerit] erit *K* fuerit ... earum] *om. ENaVaVb* earum] eorum *GINüS* erit] fuerit *G om. Vb* si] *om. I* **48** non] *om. Na* sufficit] ut *add. I* hec] tunc *E* et *Na* generantur] generatur *EM* et] vel *E* **49** generatur] *om. I* **50** complexionem] consuetudinem *E* commixtionem *I* non] nec *EMNaVb* {nñc} *K* **51** inopinabile] impossibile *INa* fiat complexio] *inv. Na* secunda] *om. ENa* sine[2]] *om. I* **52** spermate] et secunda *add. I* proprietate *Na* **53** quis] aliquis *I* est] *om. Va* impossibile] inopinabile *EM* sit *add. Va* determinato] *om. I* **54** per virtutem] parvitatem *Na* **55** de[1] ... erit] erit de istis *Na* erit] est *A* sermo[2]] modo *Na* **55/56** Omnia ... ex] *lac. Na* **56** nichil] in se *add. E* ut *Na* **57** nisi] *om. E* in se *Na* retinere] tenere *KVb* complexio] pulmo *I* elementorum *add. Na* **58** si] non *AGKMNüSVb* sicut *E* tamen *I* alteretur] alteratur *I* ab alia *add. Na* una[2]] altera *AES* **59** in ... determinata] determinata in quantitate *Na* aliquo] *om. I* aliqua *M* **60** et] tunc *add. Va* dator] *corr. ex* corda *Va* **61** faciet] facit *EI* non] *om. ENa* hoc *Va* **62** erit] *om. Va* impossibile] possibile *EINaVa* hoc] *om. Va* et] ex *add. I* quis] aliquis *E*

reason of the combination of the elements according to known proportions. Thus, as long as the elements exist, and <as long as> their being divided into these proportions and their combination is possible, the mixture which arises from them is possible. If the first mixture is not sufficient and instead the generation is only through a second and third mixture, then just as animals are generated from a mixture of humours (*akhlāṭ*) after a mixture of elements, it is not objectionable that a second combination will come to be and a second mixture after the occurrence of the first mixture, without seed or semen.

(12) If someone thinks that this is impossible save in a well-defined place and well-defined power, such as the uterus and the sperm, then <such a> doctrine (*kalām*), once conceded, is based on the mixture which occurs to the uterus, so that what is generated in it is generated, and on that <mixture> which occurs on account of the sperm, so that what is generated from it is generated. Indeed, this doctrine is just like the original doctrine. For all these are in fact generated from a mixture which ultimately goes back to the elements, because its beginning is from the elements, and then it undergoes alteration. The uterus, for example, does nothing but retain, combine and discharge. As to the original <doctrine>, there is the mixing, and the mixing comes from the combination. Just as this combination can happen from combining powers in the uterus and the like, so it is not absurd that it comes about through other causes and by chance.

(13) For it is not impossible that some portion of earth //78 appears together with some portion of water, meeting it with a known measure (*wazn*), and nothing prevents the occurrence of this measure nor is there an obstacle; then there is no need for a receptacle <like the womb>. As for the active powers, the Giver of Powers (*wāhib al-quwā*) delivers them when something <suitably> disposed occurs. After the first mixture, second and third mixtures necessary for the perfection of the species are produced, and the heavenly ordering assists <the mixtures> sufficiently.

(14) Admittedly, if there was, for example, a uterus, this will be easier and more suitable, but if there was not, it will not be impossible conceptually that this occur from motions and other causes. If the uterus contributes something other than the

dicat quod matrix est largiens causam.

(15) Sed hec non est sententia Peripatheticorum, sed forma et virtutes substantiales veniunt a principiis intransmutabilibus.

(16) Si igitur est possibile ut elementa congregentur secundum aliquam proportionem et facientem aliquam complexionem et componantur secundo in aliam proportionem et non obviaverint contrario corrumpenti, tunc dator formarum dabit formas ex principiis eternis. Rectum igitur videbatur quibusdam ut omne compositum posset fieri ex elementis sine ginitione.

(17) Et si hoc non esset, tunc esset possibile secundum astronomos ut cessarent species. Non enim est necesse ut ex quolibet homine fiat homo necessario, sed hoc est ut in pluribus.

(18) Et maxime quia coitus, qui est principium generationis, est voluntarius et casus seminis in terram est naturalis, non necessarium, in maiori autem parte voluntarium. Et res que non

---

63 dicat] dixerit *E* **64** Sed] et *I* hec] hoc *E* homo *M* Peripatheticorum] peripateticorum *ENa* perhypathe(ti)corum *G* perypaticorum *I* pery(patheti)corum *M* perypa(thetic)orum *Nü* perhipathecorum *S* perypa(theti)corum *Va* peripa(theti)corum *KVb* Hic expresse ponit Auic(enna) datorem formarum similiter in phisica sua capitulo x et 9 metaphisice cui consentit Algazel in sua phisica item Rabimoy(ses) parte ii capitulo xiii. Item Auic(enna) in tractatu de anima capitulo penultimo Au(erroes?) uero uadit contra hoc super 7 et v metaphisice *add. K* forma] forme *GIS* **65** virtutes substantiales] *inv. Na* substantiales] s{ubstantia}les *Vb om. GI* principiis] {elt} *add. Na* intransmutabilibus] transmutabilibus *ANaS* **66** est] *om. Na* possibile] impossibile *I* ut] quod *ENa* congregentur] congregent *E* congregantur *INa* **67** aliquam[1]] aliam *KVb* **67/68** et ... proportionem] *om. ENaVa* **67** facientem] faciant *A* faciat etiam *I* fati{en}tem *S* aliquam[2]] a materia *KM* aliam *Vb* **68** componantur] componuntur *M* secundo] illis *I* proportionem] compositionem proportionalem *I* **69** obviaverint] obviavit *AGS* obviaverunt *EM* obviat *I* obviaverit *Nü* obviant *Va* contrario] contraria *I* contrarie *Va* corrumpenti] cor(ruptio)ni *I* **70** igitur] *om. Na* igitur videbatur] *iter. K* videbatur] videbitur *AES* videretur *Na* **71** posset] possit *AE* **73** si] *om. Na* tunc] *om. Na* tunc esset] *om. M* esset possibile] *inv. Na* secundum] proprios *Na* **74** astronomos] astronomios *Na* ut[1]] et *Na* cessarent] restatur *I* cessarent species] *inv. Va* species] *om. K* est] esset *Na* est necesse] *inv. EVa* est *add. I* **75** homine] *om. M* fiat] fiet *Va* homo] *om. I* necessario] *om. Va* **77** maxime] *om. ENa* qui] quod *E* generationis] maxime *add. ENa* **78** voluntarius] voluntarium *E* in] ad *ENa* terram] terra *I* non] nam *EKMVa* **79** necessarium] necessarius *AINa* in] *om. Na* maiori] maiore *A* autem] *om. I* parte] est *add. et del. K* est *add. Va* voluntarium] {i}s *add. E* voluntarius *Na* Et res] enim *I* que] quod *I om. Na*

mixture through which <the mixtures> are disposed for the form, then the uterus will be a contributing cause for the forms.

(15) <But> this is not the way <taken> by the people of the truth among the Peripatetics. Rather, all substantial forms and powers are acquired from the principles – which permanently exist and are not subject to change – <only> when the disposition occurs, the disposition being the mixture.

(16) When it is possible that [1] the basic elements (*al-arkān*) are combined according to some ratio among their parts, which determines whatever mixture there is, and <that> they are combined into a second composition according to whatever ratio there is, and [2] the dispositions occur because of this, and [3] a corrupting contrary need not constantly resist, and [4] there is the emanation (*fayḍ*) giving the forms from the eternal principles, then it is fitting that any composite you wish may be generated from the elements not by way of reproduction.

(17) If this were not the case, it would be possible that an extinction (*inqiṭāʿ*) of the species could occur. That is because it is not necessary that from each human being a human being come to be necessarily, nor from each one of mankind (*nās*), and likewise from every tree; rather, this is possible in most cases (*jāʾiz aktharī*), and it is not impossible to suppose a time in which it chances that things subject to generation pass away without a successor being generated from any of them, because not one of them exists from whom someone else is generated necessarily.

(18) For sexual intercourse, which is the beginning of procreation, //79 is voluntary, not necessary, while the seeds' falling on <fertile> grounds is <something> natural belonging to the category of what is for the most part, not to the category of the necessary or the voluntary. Nothing of the two <i.e., the natural

est necessaria possibile est ut eveniat aliquando suum contrarium.

(19) Et si constellationes secundum astronomos non venirent que facerent individua illarum specierum, esset possibile quod ille species abscinderentur sine reversione.

(20) Et tu cum consideraveris artes, invenies quod omnes sunt invente ex cogitatione aut ex revelatione divina, et suum principium non est nisi cogitatio individui. Et id cuius principium est particulare est novum. Ergo omnis ars est nova, et significat hoc quod in quolibet tempore augmentatur, et hoc quod sunt nove significat quod homines sunt crescentes post abscisionem,

(21) quoniam plures earum sunt tales quod individuum hominis, quod proprie non habet revelationem a deo quam omnes non habent, non potest esse <sine> eis. Ergo homo qui invenit eas non indigebit eis per aliquam proprietatem quam ipse habuit quam nos non habeamus.

(22) Et non est rectum dicere quod illa proprietas semper fuit in primis hominibus inventa et post abscindebatur. Sed illa

---

80 est[1]] *om. Na* necessaria] necessarium *I* possibile] tamen *add. I* est[2]] *om. Va* ut] aliquando *add. G* est aliquando *add. Na* eveniat] veniat *ENa* aliquando] *om. EGMNaVa add. in marg. S* contrarium] si nullus homo generatur *add. E* **82** secundum] s(ecund)us *Na* secundum ... venirent] non venirent secundum astronomos *ENa* venirent] invenirent *I* **83** facerent] fatiunt *Na* individua] in differentia *I* illarum] istarum *Va* esset] esse *A* et *E* **84** ille] iste *Va* abscinderentur] absciderentur *ENaNüVa* abscidentur *KVb* et *add. Va* reversione] resilione *I* **85** tu] *om. INa* tu cum] *inv. E* cum] si *I om. K* consideraveris] considaveris *S* **86** sunt] sint *E* fuerint *I* invente] pauente *Na* cogitatione] cognitione *I* aut] vel *AS* ex[2]] *om. IVa* et] *om. Na* **87** id] *om. ENa* **87/88** principium est] *inv. ENa* **88** est[1]] *om. M* particulare] et novum *add. AEGKMNaNüSVb* quod novum *add. I* est novum] *om. ENa* Ergo] et *add. E* genus *I* Ergo omnis] *inv. KVb* ars] individuorum *add. E* **89** hoc[1]] id *EKMVaVb* illud *Na* quod] ars nova *add. E* quolibet] qualibet *AEI* omni *Va* tempore] specie *AE* tempori *G* parte *I om. Na* augmentatur] augmentantur *AS* argumentatur *K* hoc[2]] *om. I* **90** sunt[1]] sint *ENa* significat] significant *K* homines] *coni.* omnes *AEGKMNaNüSVaVb* omnes artes *I* sunt[2]] sint *Na* post] pot(est) per *Na* **92** quoniam] quam *KVb* quem *Na* plures] per{un}es *Na* earum] eorum *AEGIKMNüS om. Va* quod] quoniam *AEGIKMNüSVb* in *add. Na* individuum] in desiderium *I* **93** non] *om. ENa* habet] habent *I* quam] quod *AS* quoniam *Va* **94** non[2]] *om. I* esse] eis *I* sine] *coni. om. AEGIKMNaNüSVaVb* eis] eadem *E* esse *I om. MNa* euum *Va* Ergo] genus *I* sic *Va* **95** invenit] inveniret *A* eas] res *I* indigebit] invenit *E* indigebat *INaVa* proprietatem] prosperitatem *Na* quam] quoniam *Va* **96** ipse] *om. Na* ipse habuit] *inv. E* non] *om. E* vere *Na* habeamus] habemus *IKNaVb* **97/98** semper fuit] pers{ona}m *Na* **98** fuit] fuerit *INVa* post] postea *Na* abscindebatur] abscidebatur *EKSVa* illa] ista *Va*

and the voluntary> needs to happen necessarily, and for all that does not happen necessarily, it is possible that its contrary happens in rare cases.

(19) If there were no motions and returning relations among the celestial spheres which necessitate that individuals of these species begin, preventing that some species are extincted without any return, then an extinction without any return would be possible. <But> then this possibility would have already occurred in what is infinite by the power of God.

(20) If you consider the arts, you will find them to be created by the reflection of the soul or by God's inspiration and <you will find> that their beginning can only be the reflection of an individual or the inspiration of an individual. For the universal is something imagined which has no existence. Whereas that whose beginning is something particular, is something that comes to be, and so it comes to be after not having existed at all. Each art is something that comes to be, and that they increase at all times indicates that they come to be, and their coming to be indicates that the people are creative after extinction.

(21) <That is> because many of <the arts> are such that the subsistence of an individual human being would not be possible without them, except in the case of a person distinguished by a property of divine inspiration and of divine help different from what belongs to us. Hence, the person who has created <the arts> must have no need of <the arts> because of some special property which belongs to persons who are unlike us.

(22) It would be wrong to say: This special property always belonged to the first people and then became extinct. Rather, this special property belongs to people only individually, and the first person and the first people in this chain

proprietas non est nisi in hominibus paucis notis. Ista igitur proprietas erat in quolibet homine primo in hac continuatione veniente ad nos. Et illi putabantur a quibusdam fieri sine gignitione.

99 hominibus] omnibus *Na* hominibus paucis] *inv. Va* notis] *om. Va* Ista] i(ll)a *K* illa *Vb* igitur] ergo *I* autem *Va* **100** erat] erit *I* primo] *om. A* continuatione] declinatione *Na* **101** putabantur] putabant *I* a quibusdam] fore sine *I* aliquo modo *Va* a ... fieri] fieri a quibusdam *Na* quibusdam] quodam *E* gignitione] Explicit *add. AKM* Explicit tractatus Avicenne de diluviis Deo gratias *add. E* Explicit tractatus de diluviis Avicenne Deo gratias *add. Na* Explicit liber de diluviis generalibus *add. Va* Explicit liber de diluviis Avicenne *add. Vb*

which leads to us were specifically equipped with it. Since this is the case, it is necessary that <these persons> came to be not by birth.

(23) This special property is a self-sufficiency (*istighnā'*) either because of the natural disposition, as <in> animals, resulting in an impulse of the will to produce the art for a reason external to this self-sufficient person, or because of the great superiority (*istiẓhār*) of the self-sufficient person, or <because of> a celestial inspiration which immediately reaches someone who is devoid of it, who <serves as> a storage place until the time of someone else's demand through deliberation and thinking.

Bibliography

Primary Sources

Albertus Magnus, *De IV coaequaevis*, ed. by Auguste Borgnet, Alberti Magni Opera Omnia (Paris: Vivès, 1895), 34

——, *De causis proprietatum elementorum*, ed. by Paul Hossfeld, Alberti Magni Opera Omnia (Münster: Aschendorff, 1980), 5/2

——, *De homine*, ed. by Henry Anzulewicz and Joachim R. Söder, Alberti Magni Opera Omnia (Münster: Aschendorff, 2008), 27/2

Abū Maʿshar al-Balkhī (Albumasar), *On Historical Astrology: The Book of Religions and Dynasties (On the Great Conjunctions)*, ed. by Keiji Yamamoto and Charles Burnett, 2 vols (Leiden: Brill, 2000)

Aristoteles Latinus, Codices, see under Lacombe in Secondary Studies

Avicenna, *Avicenne perhypatetici philosophi [...] opera* (Venice: Octavianus Scotus – Bonetus Locatellus, 1508; repr. Frankfurt am Main: Minerva, 1961)

——, *al-Mashriqiyyūn*, ed. by Ahmet Özcan (Istanbul: Marmara Üniversitesi Sosyal Bilimler Enstitüsü, 1993)

——, *al-Shifāʾ: al-Ṭabīʿiyyāt: al-Maʿādin wa-l-āthār al-ʿulwiyya*, edited by ʿAbd al-Ḥalīm Muntaṣir, Saʿīd Zāyid and ʿAbdallāh Ismāʿīl (Cairo: al-Hayʾa al-ʿāmma li-shuʾūn al-maṭābiʿ al-amīriyya, 1965)

——, *al-Shifāʾ: al-Ṭabīʿiyyāt: al-Kawn wa-l-fasād*, ed. by Ibrahim Madkour and Mahmoud Qassem (Cairo: Dār al-kitāb al-ʿarabī li-l-ṭibāʿa wa-l-nashr, 1969)

——, *al-Shifāʾ: al-Ṭabīʿiyyāt: al-Afʿāl wa-l-infiʿālāt*, ed. by Ibrahim Madkour and Mahmoud Qassem (Cairo: Dār al-kitāb al-ʿarabī li-l-ṭibāʿa wa-l-nashr, 1969)

——, *al-Shifāʾ: al-Ṭabīʿiyyāt: al-Ḥayawān*, edited by ʿAbd al-Ḥalīm Muntaṣir, Saʿīd Zāyid and ʿAbdallāh Ismāʿīl (Cairo: al-Hayʾa al-misriyya al-ʿāmma li-l-taʾlīf wa-l-nashr, 1970)

——, *The Metaphysics of The Healing: A Parallel English-Arabic Text*, ed. by Michael E. Marmura (Provo: Brigham Young University Press, 2005)

Avicenna Latinus, Codices, see under d'Alverny in Secondary Studies

Fonseca, Pedro da, *Commentariorum In Metaphysicorum Aristotelis Stagiritae Libros Tomi Quatuor* (Cologne: Zetzner, 1615–1629; repr. Hildesheim: Olms, 1964)

Ikhwān al-Ṣafāʾ, *Epistles of the Brethren of Purity. Sciences of the Soul and Intellect. Part I, An Arabic Critical Edition and English Translation of Epistles 32–36*, ed. by Paul Ernest Walker, Ismail K. Poonawala, David Simonowitz, and Godefroid de Callataÿ (Oxford: Oxford University Press in association with the Institute of Ismaili Studies, 2015)

Michael Scot, *Liber particularis. Liber physonomie*, ed. by Oleg Voskoboynikov (Florence: Sismel, 2019)

Olympiodorus, *In Aristotelis Meteora commentaria*, ed. by Wilhelm Stüve (Berlin: Reimer, 1900)

Pietro d'Abano, *Conciliator controversiarum quae inter philosophos et medicos versantur* (Venice: Iuntas, 1565; repr. Padua: Antenore, 1985)

Russiliano, Tiberio, *Apologeticus adversus cucullatos*, ed. by Paola Zambelli (Milan: Il Polifilo, 1994)

Suarez, Francisco, *Disputationes metaphysicae* (Paris: Vivès, 1866; repr. Hildesheim: Olms, 1998)

Thomas Aquinas, *Scriptum super Sententiis*, ed. by Pierre Mandonnet and Maria Fabianus Moos (Paris: Lethielleux, 1929–1947)

Trombetta, Antonio, *Opus in Metaphysica Aristotelis Padue in thomistas discussum* (Venice, 1504)

William of Conches, *Philosophia*, ed. by Gregor Maurach (Pretoria: University of South Africa, 1980)

Secondary Studies

Ackermann, Silke, *Sternstunden am Kaiserhof: Michael Scotus und sein 'Buch von den Bildern und Zeichen des Himmels'* (Frankfurt am Main: Lang, 2009)

Alonso, Manuel Alonso, 'Homenaje a Avicena en su milenario. Las traducciones de Juan González de Burgos y Salomon', *Al-Andalus*, 14 (1949), 291–319

d'Alverny, Marie-Thérèse, 'Notes sur les traductions médiévales des oeuvres philosophiques d'Avicenne', *Archives d'histoire doctrinale et littéraire du Moyen Âge*, 19 (1952), 337–58, repr. in Marie-Thérèse d'Alverny, *Avicenne en occident* (Paris: Vrin, 1993), Article IV

d'Alverny, Marie-Thérèse, *Avicenna Latinus: Codices* (Louvain-la-Neuve: Peeters, 1994)

Bertolacci, Amos, 'Averroes against Avicenna on Human Spontaneous Generation: The Starting-Point of an Enduring Debate', in *Renaissance Averroism and its Aftermath: Arabic Philosophy in Early Modern Europe*, ed. by Anna Akasoy and Guido Giglioni (Dordrecht: Springer, 2013), pp. 37–54

——, 'The Reception of Avicenna in Latin Medieval Culture', in *Interpreting Avicenna: Critical Essays*, ed. by Peter Adamson (Cambridge: Cambridge University Press, 2013), pp. 242–69

Burnett, Charles, *The Introduction of Arabic Learning into England*, The Panizzi Lectures, 1996 (London: British Library, 1997)

——, 'Shareshill [Sareshel], Alfred of', in *The Oxford Dictionary of National Biography*, ed. by Henry C. G. Matthew and Brian Harrison (Oxford: Oxford University Press, 2004), <https://www.oxforddnb.com/view/10.1093/ref:odnb/9780198614128.001.0001/odnb-9780198614128-e-345> [accessed 12 May 2023]

——, 'Michael Scot and the Transmission of Scientific Culture from Toledo to Bologna via the Court of Frederick II Hohenstaufen', *Micrologus*, 2 (1994), 101–26, repr. in Charles Burnett, *Arabic into Latin in the Middle Ages: The Translators and their Intellectual and Social Context* (Farnham: Ashgate, 2009), Article VIII

——, 'The Coherence of the Arabic-Latin Translation Program in Toledo in the Twelfth Century', *Science in Context*, 14 (2001), 249–88, repr. in Charles Burnett, *Arabic into Latin in the Middle Ages: The Translators and their Intellectual and Social Context* (Farnham: Ashgate, 2009), Article VII

——, 'Arabic Philosophical Works Translated into Latin', in *The Cambridge History of Medieval Philosophy*, ed. by Robert Pasnau (Cambridge: Cambridge University Press, 2010), pp. 814–22

——, 'The Arabo-Latin Aristotle', in *The Letter before the Spirit: The Importance of Text Editions for the Study of the Reception of Aristotle*, ed. by Aafke van Oppenraay (Leiden: Brill, 2012), pp. 95–107

de Callataÿ, Godefroid, *Annus Platonicus: A Study of World Cycles in Greek, Latin and Arabic Sources* (Louvain-la-Neuve: Université Catholique de Louvain, Institut Orientaliste, 1996)

——, 'World Cycles and Geological Changes according to the Brethren of Purity', in *In the Age of al-Fārābī: Arabic Philosophy in the Fourth/Tenth Century*, ed. by Peter Adamson (London: The Warburg Institute, 2008), pp. 179–93

di Donato, Silvia, 'Les trois traductions latines de la Météorologie d'Avicenne: Notes pour l'histoire du texte', *Documenti e studi sulla tradizione filosofica medievale*, 28 (2017), 331–48

Freudenthal, Gad, '(Al-)Chemical Foundations for Cosmological Ideas: Ibn Sīnā on the Geology of an Eternal World', in *Physics, Cosmology and Astronomy, 1300–1700: Tension and Accommodation*, ed. by Sabetai Unguru (Dordrecht: Kluwer, 1991), pp. 47–73; repr. in Gad Freudenthal, *Science in the Medieval Hebrew and Arabic Traditions* (Aldershot: Ashgate, 2005), Article XII

——, 'The Medieval Astrologization of Aristotle's Biology: Averroes on the Role of the Celestial Bodies in the Generation of Animate Beings', *Arabic Sciences and Philosophy*, 12 (2002), 111–37; repr. in Gad Freudenthal, *Science in the Medieval Hebrew and Arabic Traditions* (Aldershot: Ashgate, 2005), Article XV

——, 'The Medieval Hebrew Reception of Avicenna's Account of the Formation and Perseverance of Dry Land: Between Bold Naturalism and Fideist Literalism', in *The Arabic, Hebrew and Latin Reception of Avicenna's Physics and Cosmology*, ed. by Dag Nikolaus Hasse and Amos Bertolacci (Berlin: De Gruyter, 2018), pp. 269–311

Haskins, Charles Homer, *Studies in the History of Mediaeval Science* (Cambridge, MA: Harvard University Press, 1924)

Hasse, Dag Nikolaus, *Urzeugung und Weltbild: Aristoteles – Ibn Ruschd – Pasteur* (Hildesheim: Georg Olms Verlag, 2006)

——, 'Arabic Philosophy and Averroism', in *Cambridge Companion to Renaissance Philosophy*, ed. by James Hankins (Cambridge: Cambridge University Press, 2007), pp. 113–36

——, 'Spontaneous Generation and the Ontology of Forms in Greek, Arabic, and Medieval Latin Sources', in *Classical Arabic Philosophy: Sources and Reception*, ed. by Peter Adamson, Warburg Institute Colloquia, 11 (London: The Warburg Institute, 2007), pp. 150–75

——, *Latin Averroes Translations of the First Half of the Thirteenth Century* (Hildesheim: Georg Olms Verlag, 2010)

——, 'Abbreviation in Medieval Latin Translations from Arabic', in *Vehicles of Transmission, Translation, and Transformation in Medieval Textual Culture*, ed. by Robert Wisnovsky, Faith Wallis, Jamie C. Fumo, and Carlos Fraenkel (Turnhout: Brepols, 2011), pp. 159–72

——, *Success and Suppression: Arabic Sciences and Philosophy in the Renaissance* (Cambridge, MA: Harvard University Press, 2016)

Hasse, Dag Nikolaus, and Andreas Büttner, 'Notes on Anonymous Twelfth-Century Translations of Philosophical Texts from Arabic into Latin on the Iberian Peninsula', in *The Arabic, Hebrew and Latin Reception of Avicenna's Physics and Cosmology*, ed. by Dag Nikolaus Hasse and Amos Bertolacci (Berlin: De Gruyter, 2018), pp. 314–69

Hasse, Dag Nikolaus, together with Katrin Fischer, Stefanie Gsell, Susanne Hvezda, Barbara Jockers, Reinhard Kiesler, Eva Sahr, Jens Ole Schmitt and Peter Tarras, ed., *Arabic and Latin Glossary* (2009–) <https://algloss.de.dariah.eu/> [accessed 19 April 2023]

Lacombe, George, and others, *Aristoteles Latinus: Codices*, 3 vols (Rome: La Libreria dello Stato, 1945–1961)

Leinkauf, Thomas, and Carlos Steel, eds, *Platons 'Timaios' als Grundtext der Kosmologie in Spätantike, Mittelalter und Renaissance / Plato's 'Timaeus' and the foundations of cosmology in Late Antiquity, the Middle Ages and Renaissance* (Leuven: Leuven University Press, 2005)

Lettinck, Paul, *Aristotle's 'Meteorology' and its Reception in the Arab World* (Leiden: Brill, 1999)

Kischlat, Harald, *Studien zur Verbreitung von Übersetzungen arabischer philosophischer Werke in Westeuropa 1150–1400: Das Zeugnis der Bibliotheken* (Münster: Aschendorff, 2000)

Mandosio, Jean-Marc, 'Follower or Opponent of Aristotle? The Critical Reception of Avicenna's Meteorology in the Latin World and the Legacy of Alfred the Englishman', in *The Arabic, Hebrew and Latin Reception of Avicenna's Physics and Cosmology*, ed. by Dag Nikolaus Hasse and Amos Bertolacci (Berlin: De Gruyter, 2018), pp. 459–534

Mandosio, Jean-Marc, and Carla Di Martino, 'La "Météorologie" d'Avicenne (Kitāb al-Shifā' V) et sa diffusion dans le monde latin', in *Wissen über Grenzen: Arabisches Wissen und lateinisches Mittelalter*, ed. by Andreas Speer and Lydia Wegener, Miscellanea Mediaevalia, 33 (Berlin: De Gruyter, 2006), 406–24

Martin, Craig, *Renaissance Meteorology: Pomponazzi to Descartes* (Baltimore: The Johns Hopkins University Press, 2011)

Nardi, Bruno, *Studi su Pietro Pomponazzi* (Florence: Le Monnier, 1965)

Robinson, James T., 'Samuel Ibn Tibbon', *The Stanford Encyclopedia of Philosophy* (Winter 2019 Edition), ed. by Edward N. Zalta <https://plato.stanford.edu/archives/win2019/entries/tibbon/> [accessed 19 April 2023]

Thorndike, Lynn, *The 'Sphere' of Sacrobosco and Its Commentators* (Chicago: The University of Chicago Press, 1949)

——, *Michael Scot* (Edinburgh: Thomas Nelson, 1965)

ANN M. GILETTI

Latin Scholastics on the Eternity of the World and Eternal Creation on the Part of the Creature

Did They Amount to the Same Thing?

When late medieval Latin scholastics addressed the philosophical question of whether the world was eternal, they drew on ancient Greek, Arabic and early Christian traditions. Two theories based on different principles stood at the core of the discussion. One was Aristotle's theory of the Eternity of the World according to principles of natural philosophy in his *Physics*. The other was the theory of Eternal Creation (*creatio ab aeterno*), based on metaphysics and Neoplatonic principles concerning God's supernatural powers, and drawing on discussions by Augustine and Avicenna. In both medieval Arabic and Christian discussions, the two theories were regularly treated together, as was the problem that they did not fit with the religious belief that the world was newly created by God. Among Christian scholastics, objection to the idea of the world's eternity was grounded on the doctrine of Creation by God in time and *ex nihilo*. However, while both theories' conclusion that the world was eternal was problematic in religious terms, strictly speaking only the Eternity of the World contradicted Christian doctrine, by denying Creation by God. Eternal Creation, by contrast, if argued to consist of eternal production of the world and time by God and out of nothing, did not. This is a distinction to be borne in mind when we examine the implications of how Latin scholastics handled the two theories. What follows is a historical study

* I wish to thank for their generous help Cecilia Trifogli, Dag Nikolaus Hasse, an anonymous reviewer, Nancy Chapple, Richard Taylor, Luis López-Farjeat, Luigi Campi and the librarians at the Bodleian, Oriental Institute and Pembroke (Oxford) Libraries. The material for this article was collected in a project supported by a grant from the European Commission, for which I am very grateful: Marie Skłodowska-Curie Actions – IF – 701523 – BoundSci ('Boundaries of Science: Medieval Condemnations of Philosophy as Heresy').

Ann Giletti (Oxford, UK) is a medieval intellectual historian focusing on Latin reception of Aristotelian philosophy, science-religion conflicts and heresy, and philosophy in inter-faith dialogue.

Mastering Nature in the Medieval Arabic and Latin Worlds: Studies in Heritage and Transfer of Arabic Science in Honour of Charles Burnett, ed. by Ann Giletti and Dag Nikolaus Hasse, CAT 4 (Turnhout: Brepols, 2023), pp. 143–176 BREPOLS PUBLISHERS 10.1484/.CAT-EB.5.134029

about how two controversial philosophical theories about the world's eternity were handled by Latin scholastics. It concerns how these different theories, originating from diverse principles and arguments, came to be treated in the same way, often at the same time, and in terms of their demonstrability as if they were essentially the same.

Given the way Latin scholastics handled the two theories, it is sometimes difficult for us as historians studying their works to isolate whether the writer is addressing one theory or the other, or both simultaneously. Yet although both theories concluded that the world was eternal, they were not the same or logically equivalent. Nor had they always shared a stage in Christian philosophical and theological discourse. While they may have had a shared history in medieval Arabic philosophical thought, in the Latin tradition their histories were distinct. The theory of Eternal Creation had a presence that went back to early theological discourse, most importantly Augustine, to which thinking from Avicenna's *Metaphysics* presented new considerations in the thirteenth century following its translation into Latin. Aristotle's theory entered scholastic discussion after the twelfth-century translation into Latin of his works on natural philosophy, which were joined in the thirteenth century by translations of Averroes's commentaries on them.[1] The two theories had separate origins, histories, systems, principles, and arguments proving them; yet in scholastic treatments of the problem of the world's eternity they were often handled together. This may not be surprising, given that they had in common the conclusion that the world is eternal. What is unexpected is that, in spite of the theories' diverse principles and systems, there was one philosophical case — a set of arguments — that was applied to both theories in the effort to disprove them and uphold the doctrine of Creation in time. That is, the arguments raised against the Eternity of the World to prove its impossibility were much same as those used for the identical purpose against Eternal Creation. This was also the case for the opposite position: scholastics showing instead that the world's eternity was philosophically possible did so by producing rebuttals to this set of arguments, applying the same arguments and counter-arguments to both theories. These arguments addressed what we can say about the world's capacity to be eternal yet, as we shall see, they focused not on natural philosophical principles describing the world, but to a significant extent on metaphysical concepts one would associate with Eternal Creation.

1 For an explanation and history of the late medieval Latin scholastic controversy over the world's eternity, see Dales, *Medieval Discussions of the Eternity of the World*. On the handling of the problem in Arabic philosophical discussion, see Davidson, *Proofs for Eternity*. Regarding Latin translation of Aristotle and his Muslim interpreters, see: D'Alverny, 'Translations and Translators'; Dod, 'Aristoteles Latinus'; a revision of Dod's table of medieval translations in 'Greek Aristotelian Works Translated into Latin'; Burnett, 'Arabic Philosophical Works Translated into Latin'.

While Latin scholastic philosophical discourse often addressed together multiple theories touching on a particular issue, this instance is of particular interest because of its implications regarding deductions that can be made about scholars' positions, and risks they were implicitly taking. That the demonstration against the two theories was the same bears significance with regard to philosophical position. If a scholastic author disproved one theory, he in effect disproved the other. Conversely, if he disarmed the arguments purporting to disprove one theory, he essentially restored the philosophical possibility of both this theory and the other. The two positions had to be the same. Thus, even if a scholastic author presented his view on only one of the theories, his position on the other could be deduced. In the heated environment of this controversy, this meant that a scholar risking his reputation by allowing the possibility of one theory was also justifiably liable regarding the other.

The discussion below shows three links or associations I have noticed in Latin scholastic handling of the theories. The first is that, while the theories were distinct, they were linked together in *quaestiones* and other treatments of the problems by scholastics, and were sometimes conflated. The second is the theories' shared set of arguments in the matter of their demonstrability. The third is a consequence of the second: the implication that a scholastic's position on one theory would necessarily be the same regarding the other. Examination of these associations will involve identification of arguments and evidence of how they were used. The focus in this discussion is mainly on works by scholastics related to the University of Paris and its debates in the late thirteenth century. Most treatments take the form of *quaestiones*, the standard format of scholastic discussion of the time. *Quaestiones* can be described as short essays structured as philosophical demonstrations headed by a question, such as 'whether the world is eternal', and composed predominantly of a set of arguments followed by a set of counter-arguments undermining them. The preliminary set represents the position the author opposes, while the counter-arguments undermine that position and establish the author's (opposing) view. The works we will discuss are philosophical, such as commentaries on Aristotle's *Physics*, and theological, such as commentaries on the *Sentences* of Peter Lombard (the late medieval university exercise for theologians). In the late thirteenth and fourteenth centuries, *Physics* and *Sentences* commentaries took the form of volumes of *quaestiones*. Other works we will see are *quaestiones* which were initially presented orally and publicly at the university, and reworked in written form. The material for this discussion is a first-time collection, the by-product of a project involving a major trawl for late thirteenth- and early fourteenth-century works containing treatments of controversial philosophical theories, including the Eternity of the World and Eternal Creation, mostly connected to activity at the Universities of Paris and Oxford. The examples in this article are illustrative of the findings. The scholastics cited in this study reflect a range of opinions on these theories, from rejection (such as Bonaventure) to acceptance of the theories' philosophical possibility (such as Aquinas, Giles of Rome and Boethius of Dacia). While some modern studies refer both to Aristotle's theory and more loosely to the general problem of the world's eternity (which sometimes includes Eternal Creation) as the 'Eternity of the World', for the sake of clarity in the discussion below I will refer separately to the two theories as the 'Eternity of the World' (Aristotle's theory) and 'Eternal Creation'. References to 'Creation' on its own are to the Catholic doctrine, and references to 'creation' are to the type of production (contrasting with generation, as described below).

In looking at these scholastic arguments, we should bear in mind that the *quaestio* format is derived from university debating techniques, so the counter-arguments are sometimes formulated simply to invalidate the principal (introductory) arguments, not to present a coherent opinion of the author. As the aims of this study are both to show that certain arguments were employed and to highlight the contexts in which they were used (the theories they were applied to), presentation of scholastic use of the arguments and their rebuttals will mostly be limited to indicating their presence, with occasional mention of notable features. Perhaps some arguments will not seem convincing, but it is not the purpose of this historical study to critique their cogency. Limits of space do not permit quotations and analysis of the Latin scholastic arguments cited, but I hope that the multiple citations with respect to each standard argument common to the debate (and the diverse positions represented) will suffice to show how one set of arguments was applied to demonstrating or disproving two different theories.

Before proceeding to this examination, we will look briefly at how the two theories work, simply to highlight how distinct they are in their principles and arguments. Section 1 presents Aristotle's theory of the Eternity of the World and examples of the principles of physics and arguments producing it; and Section 2 presents the theory of Eternal Creation and its central principles and arguments. Section 3 shows how the theories were linked together and sometimes conflated in *quaestiones*. Section 4 turns to the single set of arguments which scholastic writers applied to both theories in demonstrating their impossibility: Section 4-A sets out and explains the core arguments and shows how they were applied to Eternal Creation (specifically Eternal Creation on the part of the creature, as will be explained); while Section 4-B shows how they were applied to the Eternity of the World. Section 5 considers the consequence of this association, that a scholastic's position on one theory would necessarily be the same regarding the other.

1. Aristotle's Theory of the Eternity of the World

The two theories under discussion here were grounded on diverse concepts of the production of things: generation and creation. Generation, Aristotle's account of how things are made, follows the physics principle of motion and involves the production of things out of something already existing.

If we take for an example the making of a bronze statue, the statue's production entails the re-forming of existing bronze (a lump of bronze) into a new shape or form, the statue. In Aristotle's system, generation is a kind of movement or change from one thing to another. Divine creation, in contrast to generation, involves the production of things out of nothing. Production out of nothing is something the scholastics regularly pointed out was not accepted in Aristotle's natural philosophy. It is a supernatural act not subject to the principles of physics, and only one agent can perform it: God. For Latin scholastics adopting Aristotelian physics, both generation and creation were kinds of making in a loose sense of the word, though in a strict sense 'making' (*facere*) referred to natural production (Aristotle's explanation of generation), and only in a metaphorical way to God's act of creating, without the involvement of movement.[2] These two diverse concepts were at the core of the respective theories, both resulting in the world's being eternal.

Let us look first at Aristotle's theory. Generation is one of the central concepts of Aristotle's theory of the Eternity of the World. He argued that the world is eternal in *Physics*, based on principles he established there on generation, time and matter. He showed in *Physics* VIII, 1 that generation must be eternal — that is, that there must be an infinite series of generations. He reasoned that motion must always have a subject — the thing that is/can be moved — and that the subject exists before the motion takes place (*Physics* VIII, 1, 251a8–11). Since every motion or generation of a new thing is from something already existing (we can take as an example the bronze of the statue, borrowed from *Physics* III, 201a28–30), there will always be something existing before any generation. Furthermore, each pre-existing thing itself must have come to be through generation (in the statue example, the lump of bronze must have come from somewhere). Each occasion of generation must have a prior generation behind it, accounting for the existence of the pre-existing subject of each generation. This series of generations is thus infinite; consequently, the world must be eternal (*Physics* VIII, 1, 251a8–251b10 and 251b29–252a5).[3]

2 Prior to the period under discussion here, Peter Lombard addressed the nuances of 'making' and 'creating' in his treatment of Creation in *Sentences*, II, D. 1, Chs 2–3, ed. by PP. Collegii S. Bonaventurae, I, pp. 330–31. In his *Physics* commentary, Aquinas said of God's 'making'/ production of things, 'Et sic "fieri" et "facere" aequivoce dicuntur in hac universali rerum productione': Aquinas, *Commentaria in octo libros Physicorum Aristotelis*, ed. by Fratrum Ordinis Praedicatorum, p. 367, para. 4. See also *Summa contra gentiles*, II, Ch. 37, ed. by Fratrum Praedicatorum, pp. 353–54: 'Et hanc quidem factionem (i.e., *the making of essence or universal being*) non attigerunt primi Naturales, quorum erat communis sententia ex nihilo nihil fieri. Vel, si qui eam attigerunt, non proprie nomen factionis ei competere consideraverunt, cum nomen factionis motum vel mutationem importet'.

3 Aristotle did not accept the idea of an absolute beginning of motion; for an explanation of his thinking on this question, see Davidson, *Proofs for Eternity*, pp. 18–20. For examples of Latin scholastics addressing Aristotle's generation argument, see: Aquinas, *Summa contra*

This is only one of several cases for the world's eternity in Aristotle's *Physics*, which also presents arguments for it according to the principles of time and matter.[4] To take one example, based on the nature of time, we can observe how the physics principle of 'now' results in an eternal world. Aristotle described time as a continuum, with past and future divided by an instant or 'now' (*Physics* IV, 11, 220a5, and VI, 3, 233b32–234a23). 'Now' is not the present or any duration or part of time; nor does Aristotle include the present in describing time (*Physics* IV, 10, 218a5–6). He compared time to a line, with 'now' almost like a point on that line, but with the difference that 'now' is fluid, continuously moving ahead and dividing past from future (*Physics* IV, 10, 218a18–19, and 11, 220a19–21, and VI, 231b6–7). Aristotle presented an argument for the world's eternity based on this principle of 'now'. Since 'now' always has an attendant past and future, even if one posits a beginning or end of time, the 'now' at that theoretical limit must have a past or future beyond it, extending time and making it everlasting (*Physics* VIII, 1, 251b10–28). If time continues forever in the past and future, so does motion or generation. This is because, according to Aristotle, neither time nor motion exists without the other because they define and measure one another (*Physics* IV, 220b15–221a6). We perceive time passing through perceiving changes or motion; while motion is measured by the time in which it occurs. If time and generation/motion are eternal, so are matter (the subject of motion) and the world.

When Latin scholastics spoke of eternity with respect to the world (in the context of both Aristotle's theory and Eternal Creation), they sometimes referred to sempiternity. Sempiternity is a limitless duration in time, an everlastingness (existing with time). It is distinct from atemporality or timelessness (existing outside time), or simultaneity of existence, concepts which are associated with the divine rather than the world.[5] However, most of the texts consulted for this study simply use the word 'eternity' rather than 'sempiternity' when referring to the world's condition, so this is the word used in the discussion below.

The thirteenth-century reaction against Aristotle's theory was fierce. An eternal world clashed with the Christian doctrine of Creation: creation in time and out of nothing. As we shall see, many scholastics argued against the

---

gentiles, II, Ch. 33, ed. by Fratrum Praedicatorum, pp. 346–47, at 346; Boethius of Dacia, *De aeternitate mundi*, ed. by Green-Pedersen, pp. 335–66, at 342 and 361–62, and see pp. 348–50.

4 For a description of the main arguments for the world's eternity, based on the nature of matter, motion and time, and of the handling of these arguments by medieval philosophers in the Muslim and Jewish traditions, see Davidson, *Proofs for Eternity*, pp. 12–30.

5 An example of a scholastic author who did use the term 'sempiternity' is Aquinas, *Commentaria in octo libros Physicorum Aristotelis*, VIII, Lectio 1, ed. by Fratrum Ordinis Praedicatorum, pp. 362 et seq. On the distinction between eternity and sempiternity, see, e.g.: Kukkonen, 'Eternity' (presenting the history of the concept of eternity in ancient and medieval philosophy, and modern discussion of medieval interpretations); Stump, *Aquinas*, pp. 131–58 (focusing on Boethius and Aquinas).

theory. In Paris, the bishop banned the holding of the theory by university members in the famous Condemnations of 1270 and 1277.[6]

To oppose the Eternity of the World philosophically, one could do three things: (1) one could set out rebuttals to Aristotle's arguments, undermining each of his arguments based on time, matter, and generation; (2) one could reply by positing the supernatural creative powers of God; or (3) one could introduce arguments to prove that the world could not possibly be eternal — the approach we will look at in Section 4-B. Scholastic rebuttals to arguments in natural terms (the first approach) sometimes reconsidered the ways one can speak of the principles of physics underlying them; yet often they did not dispute the integrity of Aristotelian principles, but instead introduced God's supernatural powers as transcending them (the second approach).[7] For example, one rebuttal to the generation argument centred on the difference between making in natural terms and creation. One could say that God's kind of production of things, creation, is not restricted by the principles of physics, and so does not require a pre-existing subject. God can create out of nothing, and this production of things can be new. This solution was widely cited, for instance by Bonaventure and Thomas Aquinas.[8] In support of the idea of creation unexplained in natural terms, scholastics often recalled a discussion by Maimonides in *Guide for the Perplexed* about how our knowledge is based strictly on the world in its current state, and cannot address how it may have been as it was coming into being through creation, when God could have produced it new and out of nothing.[9] The approach of introducing God's supernatural power to the problem offered ways to invalidate arguments for the Eternity of the World constructed in natural terms, and to allow for the philosophical possibility of Creation; but, as we shall see, it did not

6 Condemnation of 1270, Art. 5, in *Chartularium universitatis Parisiensis*, ed. by Denifle and Chatelain, I, pp. 486–87, at 487; and Condemnation of 1277, Arts. 88–91, pp. 98 and 205, ed. by Piché, pp. 106, 108 and 142; previously published in *Chartularium*, ed. by Denifle and Chatelain, I, pp. 543–55, at 548–49 and 554–55. On both Condemnations, see: Wippel, 'The Parisian Condemnations of 1270 and 1277'; Wippel, 'The Condemnations of 1270 and 1277 at Paris'. The Condemnation of 1277 is also edited in Hissette, *Enquête sur les 219 articles condamnés à Paris le 7 mars 1277*.

7 For a description of rebuttals against the eternity of matter, motion and time, and how they introduce the supernatural possibilities of the divine, see Davidson, *Proofs for Eternity*, pp. 30–45.

8 Bonaventure, *Commentaria in quatuor libros Sententiarum*, II, D. 1, Part 1, Art. 1, Q. 2, ad 2, ed. by PP. Collegii S. Bonaventurae, pp. 19–24, at 23; Aquinas, *Summa theologiae*, I, Q. 46, Art. 1, ad 5, ed. by Fratrum Ordinis Praedicatorum, 4, p. 479; Aquinas, *Summa contra gentiles*, II, Ch. 37, ed. by Fratrum Praedicatorum, pp. 353–54.

9 Maimonides, *The Guide of the Perplexed*, II, 17, trans. by Pines, I, pp. 294–98. Maimonides, and scholastics adopting this argument, cited a remark by Aristotle in *Topics*, I, 11 (104b12–17), admitting that he had not demonstratively proved the Eternity of the World, but had shown merely its probability: Maimonides, *Guide of the Perplexed*, II, 15, trans. by Pines, pp. 289–93, esp. 292. See on this subject: López-Farjeat, 'Avicenna's Influence on Aquinas' Early Doctrine of Creation', pp. 317–20; Giletti, 'The Journey of an Idea'.

demonstrate the theory's impossibility (the topic of Section 4-B), or prove that Creation actually happened.

2. Eternal Creation (*creatio ab aeterno*)

Although the introduction of God to the problem of the Eternity of the World offers an answer preserving the possibility of Creation, the concept of creation itself presents problems, and precisely because of God's supernatural powers. These powers open another possibility for an eternal world: the theory of Eternal Creation. Eternal Creation was familiar to the Latin world long before the twelfth and thirteenth centuries. Augustine had criticised aspects of it in *City of God* and *Confessions*.[10] It was now considered in conjunction with Aristotle's theory in exploring the ways in which the world could be said to be eternal. Much of Latin scholastic discussion of Eternal Creation focused on Avicenna's concept of an eternal creation involving a world which is eternal, brought into existence by God, and created out of nothing.[11] As we shall see, scholastic authors such as Aquinas drew on important arguments Avicenna made in his *Metaphysics* to support the theory.[12]

Eternal Creation is the theory that God eternally produces the world, and as a result the world is eternal (or, more precisely, sempiternal, as noted above).[13] The reasoning works as follows. That God eternally produces the world is based on the ideas that God is himself eternal, and that he is perfect and unchanging. He does not change because nothing perfect can change without jeopardizing that perfection. Any action of God's must therefore be eternal, so he must eternally produce the world. Consequently, the world is eternal. Latin scholastics sometimes introduced or represented this argument by saying that when a cause is posited its effect is posited, which is how Avicenna explained Eternal Creation.[14] Two metaphors originating with Porphyry were often cited in the context of Eternal Creation: that of

---

10 Augustine, *De civitate Dei*, X, 31; XI, 4–6; and XII, 15–16 (regarding the human race), pp. 308–09, 323–26 and 369–72; Augustine, *Confessiones*, XI, 10, p. 200.

11 Avicenna, *The Metaphysics of the Healing* (*Shifā*), ed. and trans. by Marmura, VI, 2, p. 203 (an eternal cause producing an eternal effect, and giving it complete existence); VIII, 3, p. 272 (the effect/world is brought into being in an absolute sense); and IX, 1, p. 300 (an eternal cause produces an eternal effect).

12 Aquinas was influenced by Avicenna on this question early in his career: see López-Farjeat, 'Avicenna's Influence'.

13 See n. 5 and the text it accompanies.

14 An example of a scholastic author presenting the cause/effect argument is Aquinas ('posita causa ponitur effectus') in *Quaestiones disputatae de veritate*, ed. by Fratrum Praedicatorum, pp. 91–93, at 91. Regarding Avicenna, see n. 11. For an account of the history and variations of the argument concerning change in God, tracing it from Proclus (and Philoponus' report of him) through medieval Muslim authors, see Davidson, *Proofs for Eternity*, pp. 49–51, 56–61 and 76–79 (describing rebuttals).

an eternal sun producing eternal light, and that of a foot forever making a footprint in the dust.[15] Augustine used the foot/footprint metaphor in *City of God*, and was widely cited for this in scholastic texts.[16]

The theory of Eternal Creation received renewed criticism in the thirteenth century. The issue was not about God's eternal creative activity, but about how this implied an eternal world. The conclusion ran counter to the biblical account of the world's having a beginning. Like Aristotle's theory of the Eternity of the World, it was condemned in Paris in 1277.[17]

In scholastic discourse, the theory was confronted from two angles. Separating the aspects of cause and effect, writers addressed the matter from the standpoint of God (what God is capable of, given his powers and eternity), and from the standpoint of the creature/world (what the world, in its limited capacity, can do). That is, even if one argued that God has the power to create the world eternally, this did not necessarily mean that the world could exist eternally as a result, given that the world in itself is restricted to existing and functioning according to certain principles. The two standpoints were reflected in *quaestio* titles, such as 'whether God could create the world eternally' and 'whether the world could be created eternally'. The two aspects of the issue are generally referred to as Eternal Creation on the part of God and Eternal Creation on the part of the creature. The rest of this section looks at arguments objecting to Eternal Creation on the part of God, while Section 4-A will examine arguments challenging Eternal Creation on the part of the creature (arguments which would also be used to demonstrate the impossibility of Aristotle's theory).

Scholastic authors on both sides of the issue generally accepted that God could eternally will the world's existence, but those opposing Eternal Creation on the part of God held that such eternal creative activity did not imply that an eternal world would result. Both sides considered a range of arguments for and against the theory to determine whether it could be disproved. A major argument against Eternal Creation on the part of God, showing that an eternal world was not a necessary result of God's eternal creative power, was that God's action, in the form of willing, could be eternal in the sense that he could will eternally that a non-eternal world come to be. This idea had been presented by Augustine in *City of God*; and Aquinas included it in *Summa contra gentiles* and his commentary on *Physics* (he in fact accepted the possibility of Eternal Creation on the part of God).[18] If, for the sake of explanation, we can borrow

---

15 Regarding the metaphors, see van Veldhuijsen, 'The Question on the Possibility of an Eternally Created World', p. 21; Sorabji, *Time, Creation and the Continuum*, pp. 310–12.

16 Augustine, *De civitate Dei*, X, 31, p. 309.

17 Condemnation of 1277, Arts. 26, 39, 48, 87 and 99, ed. by Piché, pp. 86, 90, 94, 106 and 110; previously published in *Chartularium*, ed. by Denifle and Chatelain, I, pp. 545–46 and 548–49.

18 Augustine, *De civitate Dei*, XII, 17, p. 373, explaining God's fixing in his eternity (cf. Ch. 16) and by predestination the existence of time, human beings, and the 'eternal' life promised to them. Aquinas, *Summa contra gentiles*, II, Ch. 35 (ad 2), ed. by Fratrum Praedicatorum,

concepts of time, we can put it thus: God could preordain that the world come into being at a later moment (Aquinas likens this to a physician's medicine prescription for later consumption). God is always willing this, even once the world exists; but the world exists for a finite time, not eternally. A more scrupulous way to put it is that God's action is beyond time, which comes into existence through creation.[19] The argument did not disprove the theory, but it showed that the conclusion was not necessary, and so made room for the possibility of a non-eternal world.

This is one solution. Yet problems remain which argue for the philosophical necessity of Eternal Creation on the part of God. For instance, if the world did not always exist, one could say that, prior to its existence, something was missing from God (such as the outcome of his action, or his being actually a cause, or something else needed to bring about creation); but God, being perfect, cannot lack anything, so the world must be eternal. Or one could say there was an impediment to the completion of the world's coming to be. If so, there was something (the impediment, whatever it was) which existed before the world, something eternal aside from God. We see versions of these two arguments considered in, for example, Aquinas's *Summa contra gentiles*, Albert the Great's *Physics* commentary, Albert's source Maimonides's *Guide for the Perplexed* and, earlier in the Arabic tradition, Avicenna's *Metaphysics*.[20] One could also ask why, in all of eternity, God would ordain the world's coming into existence when it did: why not before or after? How could God have chosen one moment out of all of eternity for creation?[21] A long tradition stretched back concerning this problem, notably treatments by Aquinas, Albert the Great, al-Ghazālī, Avicenna and Augustine.[22]

---

p. 348; Aquinas, *Commentaria in octo libros Physicorum Aristotelis*, VIII, lectio 2, ed. by Fratrum Ordinis Praedicatorum, pp. 371–72, para. 18; and see nn. 25–26 and the text they accompany. For arguments against the possibility of Eternal Creation on the part of God on the grounds that God's power and will cannot eternally actualise the production of the world (e.g., owing to contradictions which would result, or limitations in the effect), see Matthew of Aquasparta, *Quaestiones disputatae de productione rerum et de providentia*, Q. 9 ('Quaeritur, supposito secundum fidem quod mundus non sit aeternus, sed productus ex tempore, utrum potuit esse ab aeterno vel utrum Deus potuit ipsum ab aeterno producere'), ed. by Gàl, pp. 201–27, at 218–27. The *quaestio* takes a stand against Eternal Creation both on the part of God and on the part of the creature.

19 See n. 23 and the text it accompanies.

20 Aquinas, *Summa contra gentiles*, II, Ch. 32 (4th arg.), ed. by Fratrum Praedicatorum, p. 344; Albert the Great, *Physica*, VIII, Tract 1, Ch. 11, arg. 6, ed. by Hossfeld, p. 572; Maimonides, *Guide of the Perplexed*, II, 14 (6th method), trans. by Pines, p. 288; Avicenna, *Metaphysics*, IX, 1, ed. and trans. by Marmura, pp. 300–04.

21 Aristotle had rejected the world's newly coming into existence on similar (natural) grounds, that there was nothing to distinguish one moment from any other for such an event: *De caelo*, I, 12, 283a11–12; *Physics*, VIII, 1, 252a10–19.

22 Aquinas, *Summa contra gentiles*, II, Ch. 32 (5th arg.), ed. by Fratrum Praedicatorum, p. 345; Albert the Great, *Commentarii in secundum librum Sententiarum*, D. 1, B, Art. 10, ed. by Borgnet, p. 26; al-Ghazālī, *Tahāfut al-falāsifa*, I, Discussion 1, First Proof, first objection, ed.

These arguments supporting Eternal Creation on the part of God had rebuttals which reopened the possibility of a non-eternal world, in line with Christian belief. An example regarding the last argument can be found in Aquinas. He showed how it could be undermined in *Summa contra gentiles* and his *Physics* commentary, on the grounds that we cannot speak of a choice of moments for the outcome of Creation, since moments connote time, and time itself was created with the world.[23]

While the idea that God could eternally create an eternal world was rejected by some thirteenth-century thinkers, such as Matthew of Aquasparta,[24] it came to be regarded as a matter of opinion, as Aquinas and Giles of Rome noted.[25] Aquinas and others held that to say that God could not produce the world eternally would detract from his omnipotence.[26] However, the idea that the world, for its part, could possibly exist eternally as result — Eternal Creation on the part of the creature — underwent sharp debate. This will be the subject of Section 4-A. As Section 4 shows how these arguments were applied to both eternity theories, Section 3 will first explore how the two theories were frequently addressed together in scholastic treatments.

3. Linking and Conflating the Two Theories in Scholastic Writing

From the perspective of Latin scholastics, what precisely was the relationship between Eternal Creation and the Eternity of the World? These authors recognised that the theories were distinct and based on diverse principles. When presenting natural philosophical arguments, they might cite Aristotle's *Physics* or *De caelo* for the source of an argument or principle, or they might identify a set of arguments as being Aristotle's.[27] In presenting arguments

and trans. by Marmura, pp. 13–14; Avicenna, *Metaphysics*, IX, 1, ed. and trans. by Marmura, p. 304; Augustine, *De civitate Dei*, XI, 4–6, pp. 323–26. For this argument's history and analysis of variations of it, tracing it back to Proclus (and Philoponus's report of him), see Davidson, *Proofs for Eternity*, pp. 51–56 and 68–76 (on rebuttals). It should be noted that al-Ghazālī's *Tahāfut al-falāsifa* was unknown to thirteenth-century Latin scholastics; it became available in Latin in 1328, when Averroes's *Tahāfut al-Tahāfut*, which contained and commented on it, was translated from Arabic: see Beatrice Zedler's account in the edition *Averroes' Destructio destructionum philosophiae Algazelis*, pp. 5–6, 21–22 and 24–27.

23 Aquinas, *Summa contra gentiles*, II, Ch. 35 (ad 5), ed. by Fratrum Praedicatorum, p. 349; Aquinas, *Commentaria in octo libros Physicorum Aristotelis*, VIII, lectio 2, ed. by Fratrum Ordinis Praedicatorum, p. 372, para. 19. Augustine had made this argument in *De civitate Dei*, XI, 6, p. 326.

24 See n. 18.

25 Aquinas, *De aeternitate mundi*, ed. by Fratrum Praedicatorum, pp. 85–89, at 85, ll. 17–23; Giles of Rome, *Apologia*, ed. by Wielockx, p. 55.

26 Aquinas, *De aeternitate mundi*, ed. by Fratrum Praedicatorum, p. 86, ll. 71–75; Peter of Auvergne, 'Utrum Deus potuerit facere mundum esse ab eterno', ed. by Dales and Argerami, pp. 144–48, at 144; Guido Terreni, 'Utrum motus sit aeternus', ed. by Giletti, pp. 283–305.

27 Two examples bracketing the time period we are examining are: Albert the Great, *Physics*

about Eternal Creation, they usually excluded arguments from nature and focused on arguments from God or the arguments from the creature we will see in Section 4-A. When Albert the Great presented a certain collection of arguments for the world's eternity in his *Sentences* commentary (completed c. 1246), *Physics* commentary (1251–1252) and *Summa theologiae* (after 1274), he attributed the first arguments to the natural philosophy of Aristotle and Peripatetic followers, and the remainder, regarding Eternal Creation on the part of God, to principles of metaphysics as handled by later Greek and Arab philosophers.[28] He acknowledged Maimonides's *Guide for the Perplexed* as the source of this collection of arguments. Maimonides had specified that the first arguments were drawn from Aristotle and his followers, and the others took God as the starting point.[29] Yet notwithstanding the distinction of principles and arguments underlying the theories, close scrutiny of scholastic handling of the theories in *quaestiones* and other treatments shows them running along the same track: the two were often treated together, or even conflated. We find so many examples of this phenomenon that it is clear that, in the late thirteenth century, they were regarded as closely related in spite of their differences.

For instance, early in his career, Aquinas took up the question of the world's eternity in a *quaestio* in his *Sentences* commentary ('Utrum mundus sit eternus', 1252–1256). Among the many arguments for the world's eternity which he sets out are ones in natural terms (e.g., relating to time and motion/generation), along with several relating instead to Eternal Creation, such as that there cannot be a change in God's will, and that nothing would cause him to create at one time rather than another.[30]

Aquinas covered both eternity issues in a set of chapters, 32–37, in *Summa contra gentiles* II (completed 1265/67), where he divided the arguments by type. He presented three pairs of chapters of arguments and counter-arguments: a pair from the standpoint of God (Chs 32 and 35), regarding Eternal Creation

---

commentary (1251–1252), ed. by Hossfeld, p. 570 ('sunt in universo septem [rationes] collectae a Moyse Aegyptio, Iudaeorum philosopho, quae in diversis locis librorum Aristotelis colliguntur'); and Guido Terreni, 'Utrum motus sit aeternus' (1314–1317), ed. by Giletti, p. 285 ('Hae sunt rationes Aristotelis').

28 Albert the Great, *Sentences*, II, D. 1, B, Art. 10, ed. by Borgnet, pp. 24–30 at 24–26; Albert the Great, *Physics* commentary, VIII, Tract 1, Ch. 11, ed. by Hossfeld, pp. 568–72 at 570–72; Albert the Great, *Summa theologiae*, II, Tract 1, Q. 4, Art. 5, part. 3, ed. by Borgnet, 32, pp. 101–03.

29 Maimonides, *Guide of the Perplexed*, II, Ch. 14, trans. by Pines, pp. 285–89, esp. 287, and Chs 17 and 18, pp. 294–302, esp. 298 and 299. For an example of Albert the Great's citing Maimonides for this argument collection, see n. 27 above.

30 Aquinas, *Scriptum super libros Sententiarum*, II, D. 1, Q. 1, Art. 5, ed. by Mandonnet, II, pp. 27–31. Mark Johnson has argued that Aquinas attributed to Aristotle a doctrine of creation, and shows quotations of works throughout his career to support this understanding. Some of them indicate an interpretation of Aristotle resulting in Eternal Creation; see Johnson, 'Did St. Thomas Attribute a Doctrine of Creation to Aristotle?', esp. pp. 137, 141 and 150. This possibility should be borne in mind when considering the examples of Aquinas presented here and in the discussion below.

on the part of God; a pair from the standpoint of the world (Chs 33 and 36), largely in natural terms; and a pair on the subject of making (Chs 34 and 37), in natural terms.[31]

Aquinas merged the two theories in *Summa theologiae* I, Q. 46, Art. 1 (1266–1268), which asks whether creatures have always existed ('Utrum creaturae semper fuerint'). He considers first the case for the world's eternity, drawing on natural philosophy arguments (involving, e.g., principles of motion, matter, potency, vacuums and time, such as the 'now' argument), as well as arguments to do with Eternal Creation (involving, e.g., a change in God's will, and God's being a sufficient cause). In the response section, Aquinas speaks in quick succession about Eternal Creation and Aristotle's theory: he takes up the theme of God as a cause in Eternal Creation, and then discusses how we should take Aristotle, saying that the Philosopher had not demonstratively proved the Eternity of the World.[32]

Bonaventure, in his *Sentences* commentary (completed *c.* 1253), asked whether the world was produced eternally or in time ('Utrum mundus productus sit ab aeterno, an ex tempore'), and first considered arguments having to do with the natural principles of motion and time. He identified these as arguments of Aristotle obtaining on the part of the world.[33] He then took up arguments on the part of the producing cause, God, such as God's being a sufficient cause, and the problem of his changing from not producing the world to producing it. In his response section, Bonaventure addressed simultaneously Eternal Creation and the Eternity of the World. He refers here to the Eternal Creation metaphors of the co-eternal foot and footprint and co-eternal sun and light. In his next breath he switches to Aristotle's theory, and says that Aristotle never meant that the world never began in any way, but simply that it never began in terms of natural motion. Before this, Bonaventure set out arguments against the Eternity of the World, a demonstration which implicitly also serves to disprove Eternal Creation.[34]

Giles of Rome, in his *Sentences* commentary (a second, *ordinatio* version finished after 1309), asked whether the world was eternal ('An mundus sit aeternus'), and took up common arguments on the Eternity of the World, involving motion, time and prime matter, and citing Aristotle's *Physics*, *De caelo* and *De generatione*.[35] In the same *quaestio* he considered arguments regarding Eternal Creation, such as that God lacked something to be a cause.[36] Similarly, in his commentary on *Physics* (before 1277/78), Giles asked whether motion

31 Aquinas, *Summa contra gentiles*, II, Chs 32–37, ed. by Fratrum Praedicatorum, pp. 344–54.

32 Aquinas, *Summa theologiae*, I, Q. 46, Art. 1, ed. by Fratrum Ordinis Praedicatorum, 4, p. 479. Regarding the non-demonstrability of Aristotle's theory, see n. 9 and the text it accompanies.

33 Bonaventure, *Sentences*, II, D. 1, Part 1, Art. 1, Q. 2, ed. by PP. Collegii S. Bonaventurae, p. 20: 'Hae sunt rationes Philosophi, quae sunt sumtae a parte ipsius mundi'.

34 Bonaventure, *Sentences*, II, D. 1, Part 1, Art. 1, Q. 2, ed. by PP. Collegii S. Bonaventurae, pp. 20–22.

35 Giles of Rome, *Sentences – ordinatio*, II, D. 1, Part 1, Q. 4, Art. 1, pp. 42–54, at 43–45 and 50–52.

36 Giles of Rome, *Sentences – ordinatio*, pp. 45–46 and 53–54.

was eternal ('Utrum hoc ratio simpliciter et absolute concludat motum eternum esse'), a matter regarding Aristotle's theory, but approached here according to God's role and the issue of Eternal Creation, including the problem of why God would create when he did and not before.[37]

The last example is Boethius of Dacia, the Paris Arts Faculty master whom historians often identify as one of the radical Aristotelians of the 1260s and '70s. In his *De aeternitate mundi* (*c.* 1270/72), Boethius compartmentalised approaches to the matter of the world's eternity: according to faith or reason, and within reason according to diverse sciences. His position on the faith-reason distinction is famous: he accepted the philosophical possibility of the world's eternity, but also maintained that as a Christian he accepted (without proof) that the world began. His view has been much discussed in connection with the Double Truth, the position of holding two contradictory philosophical and religious truths.[38] Boethius's separation of diverse rational approaches is also precise. Midway in the treatise, he considers what can be determined about the world's eternity in the distinct fields of the natural philosopher, mathematician and metaphysician. The description of the metaphysician indicates that it is he who handles Eternal Creation.[39] Yet notwithstanding this compartmentalisation, Boethius addresses together arguments in natural terms and arguments relating to Eternal Creation. In the opening remarks, he takes up the matter of the demonstrability of the world's eternity, a theory he ascribes to Aristotle. He then presents sets of arguments representing three possible positions: (1) a set against the world's eternity (that it is not and cannot be eternal); (2) a set for its possibility (that it could possibly be eternal); and (3) a set for its actual eternity (that it is eternal). Leaving aside the first set (for discussion in Section 4-B), the arguments in the second set (for the possibility of the world's eternity) turn largely on Eternal Creation.[40] In the third set (for the world's actual eternity), half are in Aristotelian natural terms (e.g., to do with the nature of motion and time) and half are about Eternal Creation (e.g., that God's effects are immediate, and that there cannot be an impediment to his causing). Following this is an expanded discussion of Eternal Creation on the part of God.[41] Thus, although Boethius distinguishes

37 Giles of Rome, *In libros de physico auditu Aristotelis*, VIII, Lectio III, dubitatio 1, fol. 158r.

38 Boethius of Dacia, *De aeternitate mundi*, ed. by Green-Pedersen, pp. 335–36, 347, 351–53, 356–57 and 365–66. Historians generally conclude that neither Boethius nor any of his contemporaries held the Double Truth, because they held religious truth in an absolute sense and philosophical opinions in a conditional sense, avoiding a contradiction. See, e.g., Bianchi, *Pour une histoire de la 'double vérité'*, pp. 12–13, 17–18, and 43–44; Piché, *La condemnation*, pp. 183–225; John F. Wippel's introduction in his English translation of *Boethius of Dacia: On the Supreme Good, On the Eternity of the World, On Dreams*, pp. 4, 9, 14 and 17; Dales, 'Origin of the Doctrine of the Double Truth'; see also Giletti, 'Double Truth'.

39 Boethius of Dacia, *De aeternitate mundi*, ed. by Green-Pedersen, pp. 347–55.

40 Boethius of Dacia, *De aeternitate mundi*, ed. by Green-Pedersen, pp. 339–40.

41 Boethius of Dacia, *De aeternitate mundi*, ed. by Green-Pedersen, pp. 340–46; rebuttals to the arguments are on pp. 357–64.

between the systems to which natural and Eternal Creation arguments pertain, he addresses together arguments relating to these two domains.

The examples above show how Aristotle's theory and Eternal Creation were frequently linked in scholastic works. While some works linking them may indicate a distinction between them, others seem to conflate them. The next section shows a connection which is more philosophically significant: how one set of arguments was deployed against both theories.

4. Applying the Same Arguments against Both Theories

As mentioned in Section 1, there were three ways to combat Aristotle's theory: (1) one could undermine the arguments Aristotle had made according to time, matter and generation; (2) one could introduce God and his creative powers to posit the possibility of Creation against his theory (the approach shown in Section 1); or (3) one could introduce arguments to prove that the world could not possibly be eternal. The first two routes would succeed in showing that the theory was not demonstrative (not necessarily true), but they would not prove that it was necessarily false. To defeat Aristotle's theory completely, one had to take the third route: demonstrate that the theory was impossible by presenting fresh arguments which built a case against it.

Opposition to Eternal Creation addressed either the role of God or that of the world. In Section 2, we saw arguments opposing Eternal Creation on the part of God. There was a different set of arguments for undermining Eternal Creation on the part of the creature — that is, what is possible for the world itself, with its limited capacity. This was a case aiming to show that, regardless of whether God could produce the world eternally through his eternal action, it was impossible that the world, for its part, could exist eternally.

Most of the arguments for demonstrating the impossibility of Eternal Creation on the part of the creature were the same as those demonstrating the impossibility of the Eternity of the World. Similarly, most of the rebuttals to these arguments proposed by scholars on the other side of the debate were the same for both theories. Section 4-A sets out some of the most common arguments against Eternal Creation on the part of the creature, along with rebuttals to them. Examples of scholastics using these arguments and counter-arguments are cited in each case. Section 4-B examines how scholastics applied these arguments and counter-arguments to the Eternity of the World.

A. The Core Arguments as Applied against Eternal Creation on the Part of the Creature

Included among the arguments below are two which turn on God's nature, rather than the creature's: they are about imposing necessity on God, and the priority of a cause to its effect. They might seem to relate to Eternal Creation on the part of God, but they address God's relationship with the world, and scholastics included them in treatments of Eternal Creation on the part of the creature. The other arguments described below focus strictly on what is possible for a creature: they are about the priority of non-being to being, traversing an infinity, and an infinity of things (e.g., people, stones or souls). Together, these arguments formed a core set used to confront Eternal Creation on the part of the creature. We find the arguments about imposing necessity on God and a cause's priority to its effect grouped with those about the priority of non-being to being and infinities of things in treatments of Eternal Creation on the part of the creature by, for example, Henry of Ghent, Aquinas, Godfrey of Fontaines and John Quidort.[42]

1) Imposing Necessity on God

One argument concerning God and his relation to the world made the case that Eternal Creation would impose necessity on God. It relied on Aristotle's principle in *De interpretatione* 9 that, when something exists, it necessarily exists; that is, it could not *not* exist when it does exist.[43] If God created the world eternally and it existed eternally as a result, he would do so by necessity and not by his will, in that, since it existed eternally, he could never not create it. But as God does have free will and power to do or not do things, this is impossible; so Eternal Creation is impossible. This argument was used against Eternal Creation by, for instance, Richard of Middleton in his *Sentences* commentary (*c.* 1281–1284) and Henry of Ghent, in a quodlibetal *quaestio* of 1286 where he bases his position on remarks on cause/effect relationships in Avicenna's *Metaphysics* VI, 2.[44] In opposition to the position, the argument was addressed by, for example, Aquinas, Godfrey of Fontaines, John Quidort

---

42 Henry of Ghent raises the argument about imposing necessity on God along with the non-being/being argument: Henry of Ghent, *Quodlibet I*, Q. 7–8 ('Utrum creatura potuit esse ab aeterno' and 'Utrum repugnet creaturae fuisse ab aeterno'), ed. by Macken, pp. 27–46, at 29–30 and 33–42. He sets the necessity argument in the context of what is possible in a creature's nature: 'In hac quaestione erat opinio philosophorum quod creatura potuit esse ab aeterno et quod non repugnant eius naturae' (p. 29). Aquinas addresses the argument about God's priority to the world along with those on non-being/being and an infinity of souls: Aquinas, *De aeternitate mundi*, ed. by Fratrum Praedicatorum, pp. 86–89. Godfrey of Fontaines treats the argument about imposing necessity on God, along with those on non-being/being, and infinities of stones and souls: Godfrey of Fontaines, *Quodlibet II*, Q. 3 ('Utrum mundus sive aliqua creatura potuit esse vel existere ab aeterno'), ed. by De Wulf and Pelzer, pp. 68–80, at 68–71 and 76–79. John Quidort handles the necessity argument along with those on non-being/being and an infinity of stones, and discusses the non-being/being and God/creature priority arguments together: John Quidort, *Sentences*, II, D. 1, Q. 4, ed. by Müller, pp. 24–28, at 24–25 and 26–28.

43 *De interpretatione*, 9, 19a23.

44 Richard of Middleton, *Super quattuor libros Sententiarum*, II, D. 1, Art. 3, Q. 4, contra, p. 17; Henry of Ghent, *Quodlibet I*, Q. 7–8, ed. by Macken, pp. 40–42; Avicenna, *Metaphysics*, VI, 2, ed. and trans. by Marmura, pp. 202–03, and see Ch. 3, pp. 214–15.

and Henry of Harclay.[45] Aquinas countered it by saying that God's creating is necessary not in an absolute sense (he has the ability not to create), but rather in a conditional sense; that is, whenever he creates, he can be said to create necessarily only inasmuch as he does create, yet nevertheless he is free not to create.[46]

2) Priority of a Cause to Its Effect

Another argument about God's relation to the world turned on the concept of priority. It took up the relationship between cause and effect, and the principle that a cause is necessarily prior to its effect. It said that, if the world were created by God but were eternal like God, God would not be prior to the world. William de la Mare objected to Eternal Creation on these grounds in his *Sentences* commentary (*c.* 1272–1279) in a *quaestio* asking whether God could make a creature coeternal with him, and in his *Correctorium fratris Thomae* (1278/79) attacking Aquinas (posthumously) in an article rejecting Aquinas's position that God need not be prior to the world in duration.[47] We will examine a rebuttal Aquinas had used against this argument in the description of the next argument.

3) Priority of Non-Being to Being

A different argument about priority addressed what is possible for a creature. The argument considered how, when anything comes into being, it must have non-being (*non esse*) before being (*esse*); that is, before it exists, it must not exist. If the world were eternal, its non-being could not be prior to its being. That a thing not exist before its existence is all the more true for creation, which involves coming into existence out of nothing. This was considered an important objection to Eternal Creation, and was used by, for example,

45 Aquinas, *Summa theologiae*, I, Q. 46, Art. 1, resp., ed. by Fratrum Ordinis Praedicatorum, 4, p. 479; Godfrey of Fontaines, *Quodlibet II*, Q. 3, ed. by De Wulf and Pelzer, p. 68; John Quidort, *Sentences*, II, D. 1, Q. 4, ed. by Müller, pp. 25 and 28; Henry of Harclay, *Ordinary questions XV–XXIXI*, ed. by Henninger, Q. 18, pp. 732–73, at 750; previous edition: Dales, 'Henricus de Harclay *quaestio* "Utrum mundus potuit fuisse ab aeterno"', pp. 223–55.

46 Aquinas, *Summa theologiae*, I, Q. 46, Art. 1, resp., ed. by Fratrum Ordinis Praedicatorum, 4, p. 479. Aquinas refers to his earlier discussion on the difference between necessity that is absolute and necessity by supposition, where he shows that God's willing things apart from himself is not absolutely necessary: Aquinas, *Summa theologiae*, I, Q. 19, Art. 3, resp., ed. by Fratrum Ordinis Praedicatorum, 4, pp. 234–35, at 235.

47 William de la Mare, *Scriptum in secundum librum Sententiarum*, Q. 3 ('Utrum Deus potuerit facere creaturam sibi coaeternam'), contra, ed. by Kraml, pp. 10–11 at 10; idem, *Correctorium fratris Thomae*, Art. 7 (addressing Aquinas's position 'Quod non sequitur si Deus est causa activa mundi quod sit prior mundo duratione'), ed. by Glorieux, pp. 40–45, at 41. See also the in-depth study of the Dominican responses to the *Correctorium* in relation to the Eternity of the World in Hoenen, 'The Literary Reception of Thomas Aquinas' View'.

Bonaventure in his *Sentences* commentary and *Breviloquium*, and *quaestiones* by Henry of Ghent, John Pecham and Matthew of Aquasparta.[48]

Both objections to Eternal Creation according to priority (of a cause and of a creature's non-being) could be rebutted. Aquinas showed this in his *De aeternitate mundi* (*c.* 1270/72).[49] In the relationship between cause and effect, priority need not be related to time. The cause could be prior simply by essence or nature, in the sense that the effect could not exist without it. The effect could depend on the cause for its existence, and yet exist eternally with that eternal cause without perverting the relationship between cause and effect. Aristotle had made the distinction between priority in time and priority in nature or being.[50] The source for this solution was Avicenna's *Metaphysics* (VI, 2; VIII, 3; and IX, 1), which was cited in making the rebuttal by, for example, John Quidort and Aquinas in their *Sentences* commentaries.[51]

One could use the same solution for the problem of non-being having to be prior to being in the production of the world. Non-being could precede being not in time but by nature. Thus the world could be created *ex nihilo* eternally. When Aquinas made this argument in *De aeternitate mundi*, he explained that *ex nihilo* ('out of nothing') does not mean 'after nothing'; and that in Eternal Creation nothingness is prior to the world by nature, not in time.[52] Godfrey of Fontaines, in a quodlibetal *quaestio* of 1286, explored the non-being/being problem in depth and, following Aquinas, came to the same conclusion;[53] so had Giles of Rome, according to a *reportatio* of his early *Sentences* teaching (1269–1271).[54]

The scholastics had this rebuttal from Avicenna's *Metaphysics*.[55] Avicenna described creation as being unlike generation. Generation involves matter and potentiality as prior in time to the newly existing thing; but creation involves the absolute bringing into existence of a thing, not from something pre-existing. Avicenna explained that non-being was prior to being in

48 Bonaventure, *Sentences*, II, D. 1, Part 1, Art. 1, Q. 2, Sed ad oppositum, arg. 6, ed. by PP. Collegii S. Bonaventurae, p. 22; Bonaventure, *Breviloquium*, II, 1, ed. by PP. Collegii S. Bonaventurae, p. 219; Henry of Ghent, *Quodlibet I*, Q. 7–8, ed. by Macken, p. 30; John Pecham, 'Utrum mundus potuit fieri ab aeterno', ed. by Brady, pp. 170 and 175; and see Pecham's discussion of the being/non-being problem in several ways in a *quaestio* on creation *ex nihilo* ('Utrum aliquid factum sit vel fieri potuit de nihilo ordinaliter') on pp. 156–57; Matthew of Aquasparta, *Quaestiones disputatae de productione rerum*, Q. 9, ed. by Gàl, pp. 206–07.

49 Aquinas, *De aeternitate mundi*, ed. by Fratrum Praedicatorum, pp. 86–87 and 88.

50 *Categories*, 12; *Metaphysics*, V, 11; *Physics*, VIII, 7, 260b17–19.

51 Avicenna, *Metaphysics*, VI, 2, VIII, 3, and IX, 1, ed. and trans. by Marmura, pp. 202–03, 272 and 305–07; John Quidort, *Sentences*, II, D. 1, Q. 4, ed. by Müller, p. 27; Aquinas, *Scriptum super libros Sententiarum*, II, D. 1, Q. 1, Art. 5, ed. by Mandonnet, II, p. 29.

52 Aquinas, *De aeternitate mundi*, ed. by Fratrum Praedicatorum, pp. 87–88.

53 Godfrey of Fontaines, *Quodlibet II*, Q. 3, ed. by De Wulf and Pelzer, pp. 69–71.

54 Giles of Rome, *Sentences – reportatio*, II, D. 1, Q. 7, ed. by Luna, pp. 204–07.

55 Avicenna, *Metaphysics*, VI, 1; VIII, 3; and IX, 1, ed. and trans. by Marmura, pp. 194–200, 272–73 and 305–07.

creation, but that this was not a temporal priority. Instead, the world had ontological dependence on God, whose priority was in essence or nature, not time. Thus it could not be demonstrated that the world began in time, and it was philosophically possible for it to be eternal. Avicenna was cited in connection with the non-being/being argument by scholastics on both sides of the issue of Eternal Creation. Henry of Ghent, who opposed Eternal Creation, cited and critiqued Avicenna repeatedly in analysing the argument in the quodlibetal *quaestio* of 1286 mentioned above.[56] John Quidort, who by contrast accepted Eternal Creation in his *Sentences* commentary (1292–1296), relied on Avicenna for his rebuttal.[57]

4) Traversing an Infinity

Other arguments involved concepts of infinity. One was that an infinity cannot be traversed. This is because, according to Aristotle, infinities exist only potentially, not in actuality, so the traversing of an infinity could never be completed.[58] The argument is that, if the world were eternal, there would be an infinity of days to traverse, which is impossible. One could also argue that, with an infinity of days in the past, the past could not be completed, so we could not arrive at today. The argument about traversing an infinity was a commonplace, used, for example, by Bonaventure in his *Sentences* commentary and Matthew of Aquasparta in his *Quaestiones disputatae de productione rerum* (*c.* 1279–1284).[59] It was rebutted, for instance, in Aquinas's *Summa contra gentiles*, where he said that the days (or 'revolutions of the sun') would not exist in actuality (not all at the same time) but rather in succession, each one a finite element which could be completed, allowing the infinite series to be traversed.[60]

56 Henry of Ghent, *Quodlibet I*, Q. 7–8, ed. by Macken, pp. 33 and 35–37.

57 John Quidort, *Sentences*, II, D. 1, Q. 4, ed. by Müller, p. 27.

58 *Physics*, III, 4, 204a3–6; and 6, esp. 206b13–16; *Posterior Analytics*, I; 3, 72b10; and 22, 82b37–39; *Metaphysics*, XI, 10, 1066a35-b1. For an account of the argument about traversing an infinity and variations of it, tracing its history from Philoponus through Muslim philosophers, see: Davidson, *Proofs for Eternity*, pp. 87–89, 117–20 and 127–34 (on rebuttals); Sorabji, *Time, Creation and the Continuum*, pp. 210–24.

59 Bonaventure, *Sentences*, II, D. 1, Part 1, Art. 1, Q. 2, Sed ad oppositum, arg. 3, ed. by PP. Collegii S. Bonaventurae, p. 21; Matthew of Aquasparta, *Quaestiones disputatae de productione rerum*, Q. 9, ed. by Gàl, p. 207. See also John Pecham, 'Utrum mundus potuit fieri ab aeterno', ed. by Brady, p. 172. Regarding whether Bonaventure regarded this argument (and the next infinity argument) to be demonstrative against the world's eternity, see *Baldner*, 'St Bonaventure on the Temporal Beginning of the World'.

60 Aquinas, *Summa contra gentiles*, II, Ch. 38 (3rd arg. and ad 3), ed. by Fratrum Praedicatorum, p. 355 (as will be seen in Section 4-B, this chapter also addresses the Eternity of the World). Grounds for this solution can be found in *Physics*, III, 6, 206a19-b14. Another example is Henry of Harclay, 'Utrum mundus potuit fuisse ab aeterno', ed. by Henninger, pp. 750 and 756–58.

5) Infinity of Things

Another infinity argument worked on the objection against an infinity of things existing at the same time, an actual infinity. According to what were widely accepted Aristotelian terms, an actual infinity is impossible because infinities exist only potentially.[61] Arguments against the world's eternity made propositions such as that, if the world were eternal, the human race would also be eternal (because there is no reason to posit a first or last generation), resulting in an infinity of people; and such as that, if the world were eternal, God could produce on each day a stone or other object, so that there would already exist an actual infinity of these things, which is impossible. The most common version of the argument was that an eternal world and eternal human race would result in an actual infinity of souls of people who had died, which is impossible. While one could counter the first argument by saying the people would exist in succession rather than at the same time (avoiding an actual infinity), this could not be said about stones (theoretically producing an infinite mass) or souls. This argument, too, had a prior history among philosophers writing in the Arabic philosophical tradition, including al-Ghazālī and Maimonides.[62] Among Latin scholastics opposing Eternal Creation, this argument was presented by Richard of Middleton in his *Sentences* commentary, John Pecham in a *quaestio* on Eternal Creation (*c.* 1270), and Matthew of Aquasparta in his *Quaestiones disputatae de productione rerum*.[63] Matthew employed in his treatment a series of arguments about infinities to demonstrate the impossibility of Eternal Creation on the part of the creature, including those of celestial rotations, souls and human beings, and returned to them later to refute possible objections to them, declaring that they demonstrated that the world could not possibly be eternal, and any counter-arguments to them were sophistic.[64]

61 See n. 58.

62 al-Ghazālī, *Maqāsid al-falāsifa*, I, Tract 1, Div. 6, ed. by Muckle, pp. 40–41; al-Ghazālī, *Tahāfut al-falāsifa*, I, Discussion 1, First Proof, first objection, and Discussion 4, ed. and trans. by Marmura, pp. 19–20, 80 and 82–83; Maimonides, *Guide of the Perplexed*, I, 74 (7th arg.), trans. by Pines, pp. 220–22. Latin scholastics often cited al-Ghazālī's *Maqāsid al-falāsifa* for solving the infinite souls problem by accepting the possibility of an actual infinity; al-Ghazālī in fact did not accept the world's eternity, as he showed in his *Tahāfut al-falāsifa*: see n. 22 above regarding this work's unavailability to Latin readers before 1328. For an account of the argument about an infinity of things and variations of it, tracing its history from Philoponus through Muslim philosophers, see Davidson, *Proofs for Eternity*, pp. 86–89, 122–25 and 127–34 (on rebuttals); Sorabji, *Time, Creation and the Continuum*, pp. 225–31.

63 Richard of Middleton, *Super quattuor libros Sententiarum*, II, D. 1, art. 3, Q. 4, p. 17; John Pecham, 'Utrum mundus potuit fieri ab aeterno', ed. by Brady, pp. 171–72; Matthew of Aquasparta, *Quaestiones disputatae de productione rerum*, Q. 9, ed. by Gàl, pp. 207 and 210–12.

64 Matthew of Aquasparta, *Quaestiones disputatae de productione rerum*, Q. 9, ed. by Gàl, p. 210: 'Quamvis autem rationes illae de infinitate animarum, de infinitate revolutionum et de infinitate generationum sufficiant ad improbandam mundi aeternitatem vel [ad probandam] impossibilitatem exsistendi ab aeterno sub ista universitate et forma in

On the other side of the debate, accepting the theoretical possibility of Eternal Creation, Giles of Rome cited and countered the argument in the *reportatio* of his *Sentences*, as did Peter of Auvergne in a quodlibetal *quaestio* ('Utrum Deus potuerit facere mundum esse ab eterno', 1296), and Henry of Harclay in an ordinary *quaestio* ('Utrum mundus potuit fuisse ab aeterno') which he wrote while serving as chancellor of the University of Oxford (1312–1317).[65] The standard rebuttal to the argument was problematic. It played with various theories about the soul, listing several ideas which avoid the impossibility of an infinity of souls, such as the highly controversial theory of the Unicity of the Intellect. With the unicity theory, an infinity of souls would not result because individual souls would be reabsorbed by the single intellect after death of the body. The rebuttal was useful in showing how the argument against the world's eternity did not stand; but at the same time it was distasteful in that most scholastics putting forward the rebuttal were appalled at the hint of accepting such an idea. Peter of Auvergne, for instance, called the Unicity of the Intellect 'heretical' when he mentioned it in describing the solution.[66] When Aquinas discussed and rebutted the infinity of souls argument in accepting the possibility of Eternal Creation in his *De aeternitate mundi*, he said as a rebuttal that God could create an eternal world that did not necessarily have an eternal human race. This answer offers a possibility, not a conclusive solution (it has as much force as the argument that God could eternally will the existence of a non-eternal world). Aquinas conceded that this argument was a tough challenge.[67]

qua mundus est: probant enim demonstrative mundum nec fuisse nec esse potuisse ab aeterno. Demonstrative, inquam, non demonstratione dicente "propter quid", nec a priori seu ostensive, sed demonstratione "quia", a posteriori et ducente ad impossibile, quibus responderi non potest nisi sophistice'.

65 Giles of Rome, *Sentences – reportatio*, ed. by Luna, pp. 204 and 205–07; Peter of Auvergne, 'Utrum Deus potuerit facere mundum esse ab eterno', ed. by Dales and Argerami, pp. 144 and 146–47; Henry of Harclay, 'Utrum mundus potuit fuisse ab aeterno', ed. by Henninger, pp. 734 and 772 (regarding stones), and 750 and 756–58 (regarding souls).

66 Peter of Auvergne, 'Utrum Deus potuerit facere mundum esse ab eterno', ed. by Dales and Argerami, p. 146. A set of four theories about the soul was usually mentioned as possible (but unacceptable) rebuttals: that an actual infinity is possible; that there is a single, unified intellect; that a finite number of souls migrate to new bodies; and that souls die with the body's death. This formulation appeared already in the early 1230s, but as absurd consequences of an eternal world proving its impossibility: Robert Grosseteste, *Hexaëmeron*, I, Ch. 8, 7, ed. by Dales and Gieben, p. 62; see also Dales, 'Robert Grosseteste's Place', pp. 557–58.

67 Aquinas, *De aeternitate mundi*, ed. by Fratrum Praedicatorum, p. 89, ll. 297–99; cf. *Summa contra gentiles*, II, Ch. 38 (ad 6), ed. by Fratrum Praedicatorum, p. 355.

B. Applying the Core Arguments to the Eternity of the World

Most of the arguments described above for demonstrating the impossibility of Eternal Creation on the part of the creature were also used for showing the impossibility of the Eternity of the World. Likewise, in countering this position, most of the rebuttals outlined above were applied to Aristotle's theory. We find this not just in *Physics* commentaries addressing Aristotle's teaching, but also in *Sentences* commentaries and other theological works. Apart from occasionally posing the question as one of eternal movement or time, there is often little to distinguish *Physics* treatments from theological ones. In some examples below, of both philosophical and theological works, the *quaestio* is about Aristotle's theory, but includes arguments relating to Eternal Creation. This is characteristic of treatments of the issue at this time, though writers were conscious of the distinction between the principles/arguments of the theories, as the examples below show.[68] Many of the works cited below treat both theories together in the manner described in Section 3. In these instances, the arguments demonstrating the impossibility of the world's eternity (or the opposing view) apply to both theories. The examples are presented in two groups, according to whether the author opposed or accepted the philosophical possibility of the Eternity of the World.

Starting with opponents to the Eternity of the World, the first example is a *quaestio* by the theologian William of Baglione investigating whether Aristotle's theory is demonstrably false ('Utrum mundum non esse aeternum sit demonstrabile', 1266–1267). He attributes the theory to *Physics* VIII, *De generatione* II, and Ps.-Aristotle *De plantis* I, and explores how it can be demonstrated to be false through two arguments: those about priority of non-being to being and about an infinity of souls.[69] This discussion is expanded in the response section, where William considers types of causes, treating the two arguments at length under the subject of material causes and the world's limited state. Following this, he considers formal, efficient and final causes. Under efficient causes, where he introduces the topic of God as a cause, he presents a variation of the argument that an eternal world would impose necessity on God (the argument that God could not *not* make the world). In concluding, he returns to Aristotle and states that it can be demonstrated that the world is not, and could not possibly be, eternal.[70]

Geoffrey of Aspall, working probably at the Arts Faculty at the University of Oxford, wrote commentaries on Aristotle's books on natural philosophy

68 Examples from works already cited are Albert the Great's *Physics* commentary and Aquinas's *Summa contra gentiles*: see nn. 28 and 31 above and the text they accompany.

69 William of Baglione, 'Utrum mundum non esse aeternum sit demonstrabile, ita quod per rationes necessarias possit istud probari', according to the *quaestio* incipit), ed. by Brady, pp. 367–70.

70 William of Baglione, 'Utrum mundum non esse aeternum sit demonstrabile', ed. by Brady, p. 370.

and metaphysics, including a *Physics* commentary (between 1250 and 1263) with a *quaestio* asking whether time (and hence the world) is eternal. In it, he sets out principal arguments purporting to demonstrate the world's eternity in natural terms, such as the argument about 'now', as well as arguments for Eternal Creation on the part of God working on variations of the idea that God cannot change, and the argument asking how Creation could take place at one particular moment rather than another.[71] For the opposing position, demonstrating that time and the world are not eternal, Geoffrey presents five arguments, including those about traversing an infinity and an actual infinity of souls.[72] He closes by rebutting the principal arguments for the eternity of time/the world, leaving the case against their eternity to stand. As the principal arguments include both natural and Eternal Creation arguments, the case against the world's eternity applies to both theories.

Bonaventure, in the *Sentences* commentary *quaestio* discussed in Section 3 ('Utrum mundus productus sit ab aeterno, an ex tempore'), also addressed both Aristotle's theory and Eternal Creation. He set out six arguments against the world's eternity: one about priority of non-being to being, and five about infinities, such as that an infinity cannot be traversed, and that an eternal world would result in an infinity of souls. This set of arguments opposes simultaneously Aristotle's theory and Eternal Creation.[73]

Peter of Tarentaise (later Pope Innocent V) took a stand against the world's eternity in his *Sentences* commentary (completed by 1257), where his case addressed both theories. He opened by considering the possibility of the world's eternity, first with arguments from the standpoint of the world, followed by a set from that of a cause. The first set is predominantly in natural terms, such as arguments concerning matter and time (e.g., the argument about 'now'). Included here is the argument about priority of the creator to the world in time/nature, which he attributes to Avicenna. The second set is a mix of arguments in natural terms (to do with the eternity of motion and generation/corruption) and arguments for Eternal Creation on the part of God (mostly variations on his eternally causing because he does not change). Following these is a set of arguments for the opposite view, that the world cannot possibly be eternal. While the first two sets of arguments are later rebutted, this set is allowed to stand. It includes the arguments about imposing necessity on God, traversing an infinity (two on not arriving at today or any 'present'/'now') and an infinity of souls.[74]

---

71 Geoffrey of Aspall, *Questions on Aristotle's 'Physics'*, VIII, Q. 5 ('Utrum tempus sit aeternum'), ed. by Donati and Trifogli, I, pp. 652–63, at 654. In the *quaestio*'s opening argument (p. 652), Geoffrey offers a version of the non-being/being argument, in this instance for the side of the world's eternity: that non-being and being of time could not be in the same instant.

72 Geoffrey of Aspall, *Questions on Aristotle's 'Physics'*, VIII, Q. 5, ed. by Donati and Trifogli, I, p. 656.

73 Bonaventure, *Sentences*, II, D. 1, Part 1, Art. 1, Q. 2, ed. by PP. Collegii S. Bonaventurae, pp. 20–22; and see nn. 33–34 above and the text they accompany.

74 Peter of Tarentaise, *In IV libros Sententiarum commentaria*, II, D. 1, Q. 2, art. 3, pp. 10–12, at 11;

There are also examples on the other side of the debate, by scholastics who accepted the philosophical possibility of the Eternity of the World. In his *Physics* commentary, Albert the Great confronted the position against the world's eternity in a chapter considering standard arguments, including those about traversing an infinity, an infinity of souls, and that a cause must precede its effect. This case presumably applies to both the Eternity of the World and Eternal Creation, as the chapter is the counter-position to that of the previous chapter, discussed above in Section 3 as distinguishing between natural and Eternal Creation arguments but grouping them together.[75]

John of Jandun, in his *Physics* commentary of 1315 (during his regency in the Paris Arts Faculty), treated a *quaestio* on whether motion is eternal ('Utrum motus sit eternus'). He showed that in natural terms the Eternity of the World was philosophically possible and an accurate account of the teaching of Aristotle and Averroes. Before proceeding to the case for the eternity of motion (and hence the world), he opened the *quaestio* by considering standard arguments used to disprove the position, among them those about traversing an infinity, an infinity of souls, and the priority of non-being to being.[76]

Theologian Guido Terreni, who held that both Aristotle's theory and Eternal Creation could not be disproved, examined the same arguments against both theories. In a *Physics* commentary posing the question of whether motion is eternal ('Utrum motus sit aeternus', 1314–1317), he set out arguments purporting to demonstrate the impossibility of Aristotle's theory, including ones about infinities of days, stones and souls, and those about imposing necessity on God and the non-being/being problem.[77] He presented much the same arguments in a *quaestio* on Eternal Creation on the part of the creature ('Utrum mundus potuerit creari ab aeterno', 1313–1318): among the principal arguments, he included those about infinities of days and souls, and imposing necessity on God; and later in the *quaestio*, in another section of arguments against Eternal Creation on the part of the creature, he included arguments about infinities of asses and celestial rotations, and the non-being/being problem.[78]

While Aquinas's treatment of the problem of the world's eternity in his *De aeternitate mundi* focused on Eternal Creation on the part of the creature (discussed Section 4-A regarding priority in time or nature), in *Summa theologiae* and *Summa contra gentiles* his exploration of the arguments for and against the world's eternity covered both Eternal Creation and Aristotle's theory (as discussed in Section 3). *Summa theologiae* I, Q. 46, Art. 1 turns on the question of whether Aristotle's theory is demonstrative, and opens with

---

quaestio also ed. by Dales and Argerami, *Medieval Latin Texts*, pp. 61–68, at 65.

75 Albert the Great, *Physics* commentary, VIII, Tract 1, Ch. 12, ed. by Hossfeld, pp. 572–74; and see n. 28 above and the text it accompanies.

76 John of Jandun, 'Utrum motus sit eternus', ed. by Dales and Argerami, pp. 182–83.

77 Guido Terreni, 'Utrum motus sit aeternus', ed. by Giletti, p. 288.

78 Guido Terreni, 'Utrum mundus potuerit creari ab aeterno', ed. by Giletti, pp. 293 and 296–97.

natural philosophical arguments for the world's eternity such as ones to do with the eternity of matter, motion and time; these are joined by arguments for Eternal Creation on the part of God regarding God as a cause. Following this, Article 2 presents and refutes arguments showing that the world's eternity is impossible. Included here are the arguments about the priority of non-being to being, traversing an infinity, and an infinity of souls.[79] As Article 1 features arguments for both Aristotle's theory and Eternal Creation, the case in Article 2 against the demonstrability of the non-eternity of the world applies to both theories.

Similarly, in *Summa contra gentiles* II, Ch. 38, we find both theories targeted. This chapter immediately follows the pairs of chapters identified in Section 3 as being from the standpoint of God, the created world, and making. The arguments from the standpoint of God (Chs 32 and 35) concern Eternal Creation on the part of God; while some of the arguments from the standpoint of the creature (Chs 33 and 36) and all of those from the standpoint of making (Chs 34 and 37) are in natural terms, including ones for the eternity of time, matter and motion.[80] These pairs of chapters show that it cannot be demonstrated that the world is eternal. Chapter 38 shows conversely that it also cannot be demonstrated that the world is *not* eternal — whether one's starting point is the Eternal Creation or natural terms of the preceding chapters. The chapter lists and rebuts a catalogue of arguments Aquinas's contemporaries used to demonstrate the world's non-eternity, including those about the priority of a cause to its effect, priority of non-being to being, traversing an infinity, and several about infinities of days, people and souls.[81]

Giles of Rome, in the *Sentences* commentary (the *ordinatio*) discussed in Section 3 for linking the eternity theories, treated another *quaestio* on the world's eternity, this time asking whether it can be demonstrated that the world is not eternal. In setting out the case that this can be demonstrated, Giles presents (and later refutes) nine arguments regarding infinities, including that an infinity cannot be traversed and that an eternal world would result in an infinity of souls or asses/horses, as well as six arguments from the standpoint of making, including that the eternal existence of the world would impose necessity on God.[82] Given the treatment of both the Eternity of the World and Eternal Creation in the previous *quaestio*, the arguments regarding the demonstrability of the world's eternity in this *quaestio* address both theories.

Boethius of Dacia's *De aeternitate mundi* is the last example. As discussed in Section 3, although Boethius distinguished between the approaches/arguments of the natural philosopher and the metaphysician, he used both

79 Aquinas, *Summa theologiae*, I, Q. 46, Art. 2, 2 and ad 2, 6 and ad 6, and 8 and ad 8, ed. by Fratrum Ordinis Praedicatorum, 4, pp. 481–82.

80 Aquinas, *Summa contra gentiles*, II, Chs 32–37, ed. by Fratrum Praedicatorum, pp. 344–54.

81 Aquinas, *Summa contra gentiles*, II, Ch. 38, ed. by Fratrum Praedicatorum, pp. 355–56.

82 Giles of Rome, *Sentences – ordinatio*, II, D. 1, Part 1, Q. 4, Art. 2, pp. 55–56 and 66–69.

natural and Eternal Creation arguments in presenting the case for world's eternity.[83] On the opposite side, in the set of arguments against the world's eternity, there are variations of the arguments about priority of causes to effects, priority of non-being to being, and several infinities arguments including the problems of an infinity of souls and traversing an infinity.[84] What is interesting about this example is that Boethius, who so carefully distinguishes the ways of thinking of a natural philosopher, mathematician and metaphysician, and between a philosopher and a Christian (or both aspects in an individual thinker), as well as between the actual and possible eternity of the world (giving them separate sets of arguments), does not make a distinction between natural philosophy and Eternal Creation in terms of arguments demonstrating their impossibility.

Taking the examples in this section together, we find that scholastics on both sides of the debate over the Eternity of the World applied to it the arguments and rebuttals employed in the same way for Eternal Creation on the part of the creature. Yet there is an important distinction to be made regarding those arguments/rebuttals involving a role for God. As noted at the start of Section 4-A, in confronting Eternal Creation on the part of the creature, two arguments turning on the role of God were used side-by-side with the arguments about priority of non-being to being, traversing an infinity, and infinities of things. These were the arguments about imposing necessity on God and about a cause's priority to its effect. The grouping of the arguments fit the purpose of confronting Eternal Creation on the part of the creature, in the sense that God's role in these arguments is in relation to the world. However, strictly speaking such arguments do not suit confrontation of the Eternity of the World, which is based on principles to do with nature, not the divine.

Indeed, in the examples cited in this section, scholastic use of the two arguments relating to God occurs where both eternity theories are addressed, or the role of God as a cause has been introduced to a discussion of Aristotle's theory in natural terms. Among the four examples employing or rebutting the argument about imposing necessity on God, the *Sentences quaestiones* by Peter of Tarentaise and Giles of Rome (the *ordinatio*) address both theories together.[85] The two other examples where this argument appears do so only after a discussion in natural terms evolves into one taking the role of God into account. William of Baglione's *quaestio* on whether the Eternity of the World is demonstrable opens by focusing on the natural philosophical origins of Aristotle's theory, responding with the non-being/being and infinity of souls arguments in this context. It is not until the response section, which takes up the subject of causes, that the argument about imposing necessity on God

83 See n. 41 and the text it accompanies.

84 Boethius of Dacia, *De aeternitate mundi*, ed. by Green-Pedersen, pp. 336–38.

85 For the examples of Peter of Tarentaise and Giles of Rome, see nn. 74 and 82, respectively, and the text they accompany.

is put forward.[86] The argument appears following a similar route in Guido Terreni's *Physics quaestio*. The *quaestio* presents arguments for the Eternity of the World as establishing the position of Aristotle and Peripatetics, and then rejects the position as 'false' and 'heretical', introducing the role of the divine to say that God can do as he wishes in causing the world. When Guido later sets out the case for the non-eternity of the world, he includes the argument about imposing necessity on God along with arguments about infinities.[87]

The three examples where we encounter the argument about the priority of a cause to its effect treat the Eternity of the World and Eternal Creation together. Albert the Great's *Physics quaestio* includes this argument as part of the case against the world's eternity according to a previously presented set of arguments grouping natural and Eternal Creation arguments. In setting out variations of the arguments about infinities and the priority of a cause to its effect, Albert describes the infinities arguments as reputed among philosophers of the Arabic tradition to demonstrate the impossibility of the world's eternity, whereas the cause/effect arguments have been regarded as merely probable.[88] Aquinas's *Summa contra gentiles* II, 38 also addresses both eternity theories, with this single chapter considering the case for the non-eternity of the world, following six chapters testing the case for its eternity according to natural philosophical and divine principles.[89] Similarly, Boethius of Dacia's *De aeternitate mundi* includes the cause/effect argument in showing the case for the non-eternity of the world, which is set alongside cases characterized as for its eternity and the possibility of its eternity, in which we again encounter arguments relating to both theories.[90]

As the two arguments presupposing a creator fit with Eternal Creation but not with the natural basis of Eternity of the World, the arguments which most accurately can be said to serve a dual purpose in addressing both theories are those of priority of non-being to being, traversing an infinity, and infinities of things. Strictly speaking, then, this is the core set of arguments which was applied to both theories.

86 For the example of William of Baglione, see nn. 69–70 and the text they accompany.

87 For the example of Guido Terreni, see nn. 77–78 and the text they accompany.

88 For the example of Albert the Great, see n. 75 and the text it accompanies. Regarding the reputed demonstrability or probability values of the infinities and cause/effect arguments, see Albert the Great, *Physics* commentary, VIII, Tract 1, Ch. 12, ed. by Hossfeld, p. 572.

89 For the example of Aquinas's *Summa contra gentiles*, see nn. 80–81 and the text they accompany.

90 For the example of Boethius of Dacia, see n. 84 and the text it accompanies.

5. Does Employing the Same Arguments Mean Having the Same Position for Both Theories?

At this point, we might ask what lay behind the conflating of the two theories, and whether the shared set of arguments about their demonstrability indicates a harmonisation of the theories despite their diverse principles. The conflation of the theories seems to be a product of personal outlook. While to the scholastics the distinction between the principles/systems underpinning the theories was clear, their personal conception of the world was Christian: that the world was created by God out of nothing and had had a new beginning. To address the matter of the world's eternity only in natural philosophical terms would not take the full matter into account as they saw it, which meant including God and Creation in the discussion, as well as the eternity problems introduced by positing an eternal, unchanging cause of the world. Even radical philosophers defending conclusions in purely natural terms — or the right to come to such conclusions — said they believed in Creation.[91]

This is not to say that the theories were generally regarded as compatible with the doctrine of Creation.[92] This is also not to say that they and the principles establishing them were regarded as working in harmony with each other. A hard distinction lay between natural making involving a pre-existing subject and divine production *ex nihilo*. Exploring this area meant pursuing further matters such as: accounting for prime movement (which is responsible for eternal movement of the heavens) along with God's absolute production

91 This was the stance taken by John of Jandun and Boethius of Dacia, both of whom have been studied in modern scholarship in connection with the Double Truth: John of Jandun, 'Utrum motus sit eternus', ed. by Dales and Argerami, pp. 192–93, which shows that motion is eternal according to natural philosophy, but concludes with a defence of creation *ex nihilo* according to faith and without rational demonstration; regarding Boethius of Dacia, see n. 38, above.

92 Thirteenth-century scholastics had diverse views as to whether, or to what degree, Aristotle's thinking and other philosophical opinions on the world's duration/eternity might be made compatible with the doctrine of Creation. In the 1230s, Philip the Chancellor and Alexander of Hales took Aristotle as teaching that movement was commensurate with the duration of time (it was perpetual, not eternal) and thus had an absolute beginning, an interpretation that harmonised Aristotle's *Physics* and Creation. Robert Grosseteste fiercely attacked this view as an incorrect representation of Aristotle owing to poor understanding and corrupted texts: Robert Grosseteste, *Hexaëmeron*, I, Ch. 8, 4, ed. by Dales and Gieben, pp. 58–59 and 61; and see the analysis of Philip the Chancellor, Alexander of Hales and Grosseteste in Dales, 'Early Latin Discussions', pp. 177–84; Dales, 'Robert Grosseteste's Place', pp. 547–52. Following this period, scholastics generally agreed that Aristotle had not posited Creation: see Noone, 'The Originality of St. Thomas's Position'. In *De potentia*, Aquinas set out the generally recognised distinction between what Aristotle had posited (the eternity of motion and therefore the world) and what later thinkers who believed in God posited (eternal production through God's will, or Eternal Creation): Aquinas, *Quaestiones disputatae de potentia*, Q. 3, art. 17, resp., ed. by Bazzi and others, 2, pp. 90–96, at 93–94. See, however, n. 30 above regarding Aquinas on Aristotle and creation.

of the world out of nothing; and whether an efficient cause only brings about its effect through motion or change in something pre-existing (as Aristotle and Averroes taught), or can produce the very being of the effect (as Avicenna held), and thus whether God is an efficient cause as well as a final cause of the world. Such matters were the subject of discussion and diverse opinion, and did not resolve in a consensus blending the theories in terms of their underlying principles.[93]

Thus, while the conflating of the two theories may have been a natural outcome of discussion carried out by Christian thinkers, the use of a core set of arguments against both theories, in a discussion that differentiated between the principles of physics and metaphysics, presents a striking philosophical phenomenon which goes beyond blurring argumentative lines. In particular, it prompts the question of where a scholastic author's conclusion on the demonstrability of one theory put him with regard to the other, and what the implications of a link in these positions might be. That is, if the two theories were treated as the same in terms of philosophical demonstrability, would that make one's position on both the same? In short, yes. If one demonstrated that Eternal Creation on the part of the creature was impossible, this case would be equivalent to demonstrating the impossibility of the Eternity of the World. Likewise, if one showed instead that the case demonstrating the impossibility of Eternal Creation on the part of the creature could be undermined, thus leaving the theory philosophically possible, by implication one held that Aristotle's theory was also possible. In the latter case, the implication that a scholar held both theories were possible is not a logical necessity — given the diverse principles and arguments establishing the theories — but rather a striking consequence of the association of the theories in contemporary discussion. That a writer's position on both theories would likely be the same can be seen among the examples cited above. This was clearly the case with Bonaventure,[94] who held that the two theories could be demonstrated to be

93 While Latin scholastics generally identified the Prime Mover as God (e.g., Aquinas, *Summa theologiae*, I, Q. 2, Art. 3, resp., 1st way), Aristotle's Prime Mover's eternal act was on an eternally existing subject that it did not create. One could posit, as Aquinas did, that God was an efficient cause and, following Avicenna, that his efficient causality was a metaphysical causality transcending natural movement/making and capable of producing being. John of Jandun, by contrast, a close adherent to Aristotle and Averroes, did not see God as an efficient cause, and placed efficient causality of the eternal movement of the heavens not with God but with the intelligence moving the outermost sphere of the universe. See: John of Jandun, 'Utrum motus sit eternus', ed. by Dales and Argerami; and analysis in Maurer, 'John of Jandun and the Divine Causality', which includes an edition of a *quaestio*, 'Utrum aeternis repugnet habere causam efficientem'. See also: Kukkonen, 'Creation and Causation'; and Davidson, *Proofs for Eternity*, pp. 67 and 237–40 et seq. on the history of the complexities involved in identifying the Prime Mover with God.

94 Bonaventure, *Sentences*, II, D. 1, Part 1, Art. 1, Q. 2; and see nn. 33–34 and 73 and the text they accompany.

impossible, and with Aquinas[95] and Guido Terreni,[96] who both held the opposite view. This is also the case with the other examples of linking and conflating of the issues, where the case for the demonstrability/undemonstrability of the world's eternity applies to both theories.

It would thus have been reasonable to suppose that a scholar accepting the philosophical possibility of Eternal Creation on the part of the creature would also have accepted the possibility of the Eternity of the World, even in the absence of evidence of his stand on both issues — and even if he did not believe that either scenario actually happened. In the highly-charged atmosphere at the University of Paris in the late thirteenth century, this could bear significance regarding the kinds of accusations made against scholars, by the authorities and by colleagues. In Christian terms, both theories were unorthodox, owing to their acceptance of an eternal world, and holding them as philosophically possible could cause scandal, while holding them as necessarily true invited accusations of maintaining errors against faith. Yet, bearing in mind the principles establishing the theories, could it be sustained that they both offended against faith? In principle, a scholar questioned about his acceptance of the possibility of Eternal Creation and the theory's conflict with the doctrine of Creation in time and out of nothing could argue that Eternal Creation involved the (continuous) new production of time and the world *ex nihilo*. No such compatibility existed between Aristotle's natural philosophical theory and Christian doctrine. In physics terms, time is eternal and nothing can be made without a pre-existing subject. Indeed, Aquinas, in the introduction to his *De aeternitate mundi*, took great care in arguing that the possibility of Eternal Creation did not risk heresy.[97] By distinguishing between production involving something pre-existing (he cited the pre-existence of passive potency) and absolute production into being, Aquinas could assert that the idea that there was something eternal and not made by God was heretical, whereas the idea that God could eternally and wholly bring about an eternal world was not. Thus, from the standpoint of the faith-reason problem, there is reason to distinguish between the two theories, as the linking of them through their demonstrability has implications beyond philosophy.

95 Aquinas, *Summa contra gentiles*, II, Chs 32–38; Aquinas, *Summa theologiae*, I, Q. 46, arts. 1–2; and see nn. 31–32 and 79–81 and the text they accompany.

96 Guido Terreni, 'Utrum motus sit aeternus' and 'Utrum mundus potuerit creari ab aeterno', ed. by Giletti; and see nn. 77–78 and the text they accompany.

97 Aquinas, *De aeternitate mundi*, ed. by Fratrum Praedicatorum, pp. 85–86. Aquinas argued that his opinion on Eternal Creation did not risk heresy on two grounds: because it posited Creation by God (not the eternity of the world without production by God), and because he held it as a philosophical possibility, not a necessary conclusion.

Bibliography

Primary Sources

Albert the Great, *Commentarii in secundum librum Sententiarum*, ed. by Auguste Borgnet, Alberti Magni Opera Omnia, 27 (Paris: Vivès, 1894)

——, *Summa theologiae*, ed. by Auguste Borgnet, Alberti Magni Opera Omnia, 31–33 (Paris: Vivès, 1894–1895)

——, *Physica*, ed. by Paul Hossfeld, Alberti Magni Opera Omnia, 4/1–2 (Munster: Aschendorff, 1987–1993)

Augustine, *De civitate Dei*, Corpus Christianorum Series Latina, 47–48 (Turnhout: Brepols, 1955)

——, *Confessiones*, Corpus Christianorum Series Latina, 27 (Turnhout: Brepols, 1981)

Averroes, *Averroes' Destructio destructionum philosophiae Algazelis in the Latin Version of Calo Calonymos*, ed. and intro. by Beatrice H. Zedler (Milwaukee: Marquette University Press, 1961)

Avicenna, *The Metaphysics of the Healing*, ed. and trans. by Michael E. Marmura (Provo: Brigham Young University Press, 2005)

Boethius of Dacia, *De aeternitate mundi*, ed. by Niels Jørgen Green-Pedersen, Boethii Daci Opera, 6/2 (Copenhagen: Gad, 1976)

——, *Boethius of Dacia: On the Supreme Good, On the Eternity of the World, On Dreams*, trans. and intro. by John F. Wippel (Toronto: Pontifical Institute of Mediaeval Studies, 1987)

Bonaventure, *Commentaria in quatuor libros Sententiarum*, II, ed. by PP. Collegii S. Bonaventurae, Doctoris Seraphici S. Bonaventurae Opera Omnia, 2 (Quarrachi: Typographia Collegii S. Bonaventurae, 1885)

——, *Breviloquium*, ed. by PP. Collegii S. Bonaventurae, Doctoris seraphici S. Bonaventurae Opera Omnia, 5 (Quarrachi: Typographia Collegii S. Bonaventurae, 1891)

Chartularium universitatis Parisiensis, ed. by Heinrich Denifle and Emile Chatelain, 4 vols (Paris: Delalain, 1889–1897)

Geoffrey of Aspall, *Questions on Aristotle's 'Physics'*, ed. by Silvia Donati and Cecilia Trifogli, trans. by E. Jennifer Ashworth and Cecilia Trifogli, 2 vols (Oxford: Oxford University Press, 2017)

al-Ghazālī, *Algazel's Metaphysics: A Medieval Translation*, ed. by J. T. Muckle (Toronto: St Michael's College, 1933) (= *Maqāsid al-falāsifa*)

——, *The Incoherence of the Philosophers*, ed. and trans. by Michael E. Marmura (Provo: Brigham Young University Press, 2000) (= *Tahāfut al-falāsifa*)

Giles of Rome, *In libros de physico auditu Aristotelis* (Venice: Zilettum, 1502)

——, *In secundum librum Sententiarum quaestiones* (Venice: Zilettum, 1581) (= *Sentences – ordinatio*)

——, *Apologia*, ed. by Robert Wielockx, Aegidii Romani Opera Omnia, 3/1 (Florence: Leo S. Olschki, 1985)

——, *Reportatio lecturae super libros I–IV Sententiarum*, ed. by Concetta Luna, Aegidii Romani Opera Omnia, 3/2 (Florence: SISMEL, 2003) (= *Sentences – reportatio*)

Godfrey of Fontaines, *Quodlibet II*, in *Les quatre premiers Quodlibets de Godefroid de Fontaines*, ed. by Maurice De Wulf and Auguste Pelzer (Louvain: Institut Supérieur de Philosophie de l'Université, 1904)

Guido Terreni, 'Guido Terreni: Two Quaestiones on the Eternity of the World', ed. by Ann Giletti, *Textes et Études du Moyen Age*, 78 (2015), 283–305

Henry of Ghent, *Quodlibet I*, ed. by Raymond Macken, Henrici de Gandavo Opera Omnia, 5 (Leuven: Leuven University Press, 1979)

Henry of Harclay, 'Henricus de Harclay *quaestio* "Utrum mundus potuit fuisse ab aeterno"', ed. by Richard C. Dales, *Archives d'Hisotire Doctrinale et Littéraire du Moyen Age*, 50 (1983), 223–55

——, *Ordinary questions XV–XXIXI*, ed. by Mark G. Henninger (Oxford: Oxford University Press, 2008)

John of Jandun, 'Utrum motus sit eternus', ed. by Richard C. Dales and Omar Argerami, *Medieval Latin Texts on the Eternity of the World* (Leiden: Brill, 1991), pp. 180–96

John Pecham, 'Utrum mundus potuit fieri ab aeterno', *see under* Brady in Secondary Studies

John Quidort, *Jean de Paris (Quidort) O.P. Commentaire sur les Sentences, Reportation*, ed. by Jean-Pierre Müller, 2 vols (Rome: Herder, 1961–1964)

Maimonides, *The Guide of the Perplexed*, trans. by Shlomo Pines, 2 vols (Chicago: University of Chicago Press, 1963)

Matthew of Aquasparta, *Quaestiones disputatae de productione rerum et de providentia*, ed. by Gedeon Gàl (Quaracchi: Typographia Collegii S. Bonaventurae, 1956)

Peter Lombard, *Magistri Petri Lombardi Parisiensis Episcopi Sententiae in IV libris distinctae*, ed. by PP. Collegii S. Bonaventurae, 3rd edn, 2 vols (Grottaferrata: Editiones Colegii S. Bonaventurae ad Claras Aquas, 1971–1981)

Peter of Auvergne, 'Utrum Deus potuerit facere mundum esse ab eterno', ed. by Richard C. Dales and Omar Argerami, *Medieval Latin Texts on the Eternity of the World* (Leiden: Brill, 1991), pp. 144–48

Peter of Tarentaise, *In IV libros Sententiarum commentaria* (Toulouse: Colomerium, 1649–1652; repr. Ridgewood, NJ: Gregg Press, 1964)

——, 'Questio de eternitate mundi', ed. by Richard C. Dales and Omar Argerami, *Medieval Latin Texts on the Eternity of the World* (Leiden: Brill, 1991), pp. 61–68

Richard of Middleton, *Super quattuor libros Sententiarum* (Brescia: De consensu superiorum, 1591)

Robert Grosseteste, *Hexaëmeron*, ed. by Richard C. Dales and Servus Gieben (Oxford: Oxford University Press, 1982)

Thomas Aquinas, *Commentaria in octo libros Physicorum Aristotelis*, ed. by Fratrum Ordinis Praedicatorum, Thomae Aquinatis Opera Omnia, 2 (Rome: Typographia Polyglotta, 1884)

——, *Summa theologiae*, I, ed. by Fratrum Ordinis Praedicatorum, Sancti Thomae Aquinatis Opera Omnia, 4–5 (Rome: Typographia Polyglotta, 1888)

——, *Summa contra gentiles*, II, ed. by Fratrum Praedicatorum, Sancti Thomae Aquinatis Opera Omnia, 13 (Rome: Garroni, 1918)

——, *Scriptum super libros Sententiarum*, ed. by Pierre Mandonnet, 4 vols (Paris: Lethielleux, 1929–1947)
——, *Quaestiones disputatae de potentia*, in *Quaestiones disputatae*, ed. by P. Bazzi, M. Calcaterra, T. S. Centi, E. Odetio, and P. M. Pession, 2 vols (Turin: Marietti, 1953)
——, *Quaestiones disputatae de veritate*, ed. by Fratrum Praedicatorum, Sancti Thomae de Aquino Opera Omnia, 22.1, fasc. 2 (Rome: Editori di San Tommaso, 1970)
——, *De aeternitate mundi*, ed. by Fratrum Praedicatorum, Sancti Thomae de Aquino Opera Omnia, 43 (Rome: Editori di San Tommaso, 1976), pp. 85–89
William of Baglione, 'The Questions of Master William of Baglione, O.F.M., *De aeternitate mundi*', ed. by Ignatius Brady, *Antonianum*, 47 (1972), 362–71
William de la Mare, *Correctorium fratris Thomae*, edited with the response of Richard Knapwell in *Le Correctorium Corruptorii 'Quare'*, ed. by Palémon Glorieux (Kain: Le Saulchoir, 1927)
——, *Scriptum in secundum librum Sententiarum*, ed. by Hans Kraml (Munich: Verlag der Bayerischen Akademie Der Wissenschaften, 1995)

Secondary Studies

Baldner, Stephen, 'St Bonaventure on the Temporal Beginning of the World', *The New Scholasticism*, 63 (1989), 206–28
Bianchi, Luca, *Pour une histoire de la 'double vérité'* (Paris: Librairie Philosophique Vrin, 2008)
Brady, Ignatius, 'John Pecham and the Background of Aquinas's *De aeternitate mundi*', *St Thomas Aquinas: 1274–1974 Commemorative Studies*, ed. by Armand Maurer, 2 vols (Toronto: Pontifical Institute of Mediaeval Studies, 1974), II: 141–78, at 165–78; repr. in John Pecham, *Quaestiones disputatae*, ed. by Girard J. Etzkorn, Hieronymus Spettmann, and Livarius Oliger (Rome: Editiones Collegii S. Bonaventurae ad Claras Aquas, 2002), pp. 578–94
Burnett, Charles, 'Arabic Philosophical Works Translated into Latin', in *The Cambridge History of Medieval Philosophy*, ed. by Robert Pasnau, 2 vols (Cambridge: Cambridge University Press, 2010), II: 814–22
Dales, Richard C., 'The Origin of the Doctrine of the Double Truth', *Viator*, 15 (1984), 169–79
——, 'Robert Grosseteste's Place in Medieval Discussions of the Eternity of the World', *Speculum*, 61 (1986), 544–63
——, 'Early Latin Discussions of the Eternity of the World in the Thirteenth Century', *Traditio*, 43 (1987), 171–97
——, *Medieval Discussions of the Eternity of the World* (Leiden: Brill, 1990)
D'Alverny, Marie-Thérèse, 'Translations and Translators', in *Renaissance and Renewal in the Twelfth Century*, ed. by Robert L. Benson and Giles Constable (Oxford: Clarendon Press, 1982), pp. 421–62; repr. in Marie-Thérèse D'Alverny, *La transmission des textes philosophiques et scientifiques au moyen âge* (Aldershot: Ashgate, 1994), Article II
Davidson, Herbert A., *Proofs for Eternity, Creation and the Existence of God in Medieval Islamic and Jewish Philosophy* (Oxford: Oxford University Press, 1987)

Dod, Bernard G., 'Aristoteles Latinus', in *The Cambridge History of Later Medieval Philosophy*, ed. by Norman Kretzmann, Anthony Kenny, and Jan Pinborg (Cambridge: Cambridge University Press, 1982), pp. 45–79

——, 'Greek Aristotelian Works Translated into Latin', revised in *The Cambridge History of Medieval Philosophy*, ed. by Robert Pasnau, 2 vols (Cambridge: Cambridge University Press, 2010), II: 793–97

Giletti, Ann, 'The Journey of an Idea: Maimonides, Albert the Great, Thomas Aquinas and Ramon Martí on the Undemonstrability of the Eternity of the World', in *Pensar a natureza. Problemas e respostas na Idade Média, séculos XII–XIV*, ed. by José Meirinhos and Manuel Lázaro Pulido (Porto: Faculdade de Letras da Universidade do Porto, 2011), pp. 239–67

——, 'The Double Truth: How Are We to Look at It?', *Recherches de Théologie et Philosophie Médiévales*, 88 (2021), 89–141

Hissette, Roland, *Enquête sur les 219 articles condamnés à Paris le 7 mars 1277* (Louvain: Publications Universitaires, 1977)

Hoenen, M. J. F. M., 'The Literary Reception of Thomas Aquinas' View on the Provability of the Eternity of the World in De La Mare's *Correctorium* (1278–79) and the *Correctoria Corruptorii* (1279-ca 1286)', in *The Eternity of the World in the Thought of Thomas Aquinas and His Contemporaries*, ed. by Jozef B. M. Wissink (Leiden: Brill, 1990), pp. 39–68

Johnson, Mark F., 'Did St. Thomas Attribute a Doctrine of Creation to Aristotle?' *The New Scholasticism*, 63 (1989), 129–55

Kukkonen, Taneli, 'Creation and Causation', in *The Cambridge History of Medieval Philosophy*, ed. by Robert Pasnau (Cambridge: Cambridge University Press, 2010), pp. 232–46

——, 'Eternity', in *The Oxford Handbook of Medieval Philosophy*, ed. by John Marenbon (Oxford: Oxford University Press, 2012), pp. 525–46

López-Farjeat, Luis Xavier, 'Avicenna's Influence on Aquinas' Early Doctrine of Creation in *In II Sent.*, D. 1, Q. 1, A. 2', *Recherches de Théologie et Philosophie médiévales*, 79 (2012), 307–37

Maurer, Armand, 'John of Jandun and the Divine Causality', *Mediaeval Studies*, 17 (1955), 185–207

Noone, Timothy B., 'The Originality of St. Thomas's Position on the Philosophers and Creation', *The Thomist*, 60 (1996), 275–300

Piché, David, *La condamnation parisienne de 1277* (Paris: Vrin, 1999)

Sorabji, Richard, *Time, Creation and the Continuum: Theories in Antiquity and the Early Middle Ages* (London: Duckworth, 1983)

Stump, Eleonore, *Aquinas* (London: Routledge, 2003)

van Veldhuijsen, Peter, 'The Question on the Possibility of an Eternally Created World: Bonaventura and Thomas Aquinas', in *The Eternity of the World in the Thought of Thomas Aquinas and His Contemporaries*, ed. by Jozef B. M. Wissink (Leiden: Brill, 1990), pp. 20–38

Wippel, John F., 'The Condemnations of 1270 and 1277 at Paris', *The Journal of Medieval and Renaissance Studies*, 7 (1977), 169–201

——, 'The Parisian Condemnations of 1270 and 1277', in *A Companion to Philosophy in the Middle Ages*, ed. by Jorge J. E. Gracia and Timothy B. Noone (Madden, MA: Blackwell, 2003), pp. 63–73

DAVID PORRECA

Whitewash for 'Black Magic'

Justifications and Arguments in Favour of Magic in the Latin Picatrix

This article addresses the broad question of how texts explicitly magical in nature justify to their readers (and, in some cases, to actual literate practitioners) the effectiveness and moral or theological acceptability of the rituals and ceremonies they contain. It uses as a basis the Latin version of the *Ghāyat al-Ḥakīm* ('The Goal of the Sage'), known as the *Picatrix*, since in many ways this text is the best example of both ceremonial and astral magic (*nigromancia/siḥr*) as it was known during the Middle Ages and Renaissance due to the combination of its sheer length and the wide range of sources that were drawn upon to compose it. The article examines twelve interrelated and mutually reinforcing themes that manifest in the *Picatrix* in order to demonstrate the pro-magic apologetic nature of the text. These themes are: (1) direct claims to the effectiveness and power of magic; (2) exhortations to secrecy with regard to the knowledge imparted in the text; (3) claims that magic is a legitimate part of human knowledge worthy of being sought; (4) appeals to *auctoritates* who are said to have been successful at magical operations; (5) the appropriation of the positive reputation of other intellectual disciplines which lie at the core of how magic was thought to work; (6) the coherence, rationality, and orderliness of magic; (7) direct references to God's power being at the root of magic; (8) the discipline's complexity and the consequent need for precision in magical operations; (9) the need for the practitioner to be focused fully in will, belief and devoutness; (10) the associated need for purity in the operator; (11) the author's choice of narrative mode (first-person perspective); and finally (12) his explicit expression of good intent. These themes are not all equally important. Nevertheless, each contributes to a web of rhetorical and intellectual support that, especially for a reader inclined to think along Platonic and mystical lines (as was common in Islamic, Jewish, and Christian thought circles at the time), would erect a sturdy, self-reinforcing

David Porreca (Waterloo, Canada) is a medieval intellectual historian specialising in Christian reception of the pagan Classical tradition, particularly magic and the Hermetic tradition.

Mastering Nature in the Medieval Arabic and Latin Worlds: Studies in Heritage and Transfer of Arabic Science in Honour of Charles Burnett, ed. by Ann Giletti and Dag Nikolaus Hasse, CAT 4 (Turnhout: Brepols, 2023), pp. 177–206 BREPOLS PUBLISHERS 10.1484/.CAT-EB.5.134030

platform of arguments not only against potential critics and sceptics, but also against any doubt that might arise spontaneously in the mind of the reader/practitioner resulting from the failure of any of their operations.

The *Picatrix* itself, originally composed in Arabic in the 950s by the Andalusian *ḥadīth* scholar Maslama ibn Qāsim al-Qurṭubī[1] and translated first into Castilian Spanish in the 1250s and from Castilian into Latin shortly thereafter, is a carefully curated compilation drawn from an impressive number of sources — 224 different books, according to the author himself (*Picatrix*, 3.5.4).[2] These sources were gathered from a wide range of traditions, including Arabic, Babylonian, Egyptian, Greek, Indian, Persian and Syrian. Too little of the Castilian version survives to assess the entire Latin text's fidelity to its vernacular source, but those passages that do survive suggest that the two versions were very close.[3] The Latin text we have to work from, however, is not an entirely faithful rendition of the original Arabic. Many passages in the Arabic version are not present in the Latin or appear there in a simplified form, in many instances seemingly perhaps to render the text more acceptable to a Christian audience; the Latin version also contains a number of lengthy interpolations absent from the original Arabic.[4] Considering the liberties known to have been taken in the trilingual translation process, the Latin *Picatrix* can be said to reflect the concerns and expectations of its Spanish or Latin translators at least as much as those of Maslama al-Qurṭubī himself, since it results from a careful triage-and-supplementation process that reflects the perceived expectations of its Christian audience, and only indirectly retains al-Qurṭubī's original material. Although the Latin version appears not to have circulated widely during the later Middle Ages, the fact that eighteen complete manuscripts of the text remain extant — all produced between the late 1450s and the early seventeenth century — suggests a substantial level of interest in the *Picatrix* during the Renaissance. The absence of any printed editions from the same period hints at the controversial nature of the book's content. The latter consideration is likely a root cause behind the need for the extensive scheme of justifications of magical practice examined below.

For the purposes of this discussion, passages from the *Picatrix* that amount to 'pro-magic apologetic' include any statement or subtext that would defend

1 On the identity of the author of the *Ghāyat al-Ḥakīm*, see: Fierro, 'Bāṭinism in al-Andalus'. For a brief biography of Maslama ibn Qāsim al-Qurṭubī, see: Boudet and Coulon, 'Version arabe', pp. 68–73. For a history of the Arabic text, see de Callataÿ and Moureau, 'Again on Maslama Ibn Qāsim al-Qurṭubī', pp. 330–36.

2 Internal references to the Latin *Picatrix* in this article are expressed as trinomials, reflecting the Book, Chapter, and Paragraph numbers as they appear in David Pingree's edition. All Latin quotations are from Pingree's edition. All English translations are from the translation by Dan Attrell and David Porreca of Pingree's Latin edition.

3 For an assessment of the differences among the Arabic, Spanish, and Latin versions, see Pingree, 'Between the *Ghāya* and the *Picatrix* I'; for a more recent re-assessment, see Burnett, 'Magic in the Court'.

4 For a list of these interpolations, see Bakhouche, Fauquier, and Pérez-Jean, *Picatrrix*, pp. 27–31.

or justify both the legitimacy and the effectiveness of magic itself, or the belief in either. Many of these passages come in the form of pre-emptive explanations or rationalizations that address unstated but common critiques of magic, such as that it is evil, that its practice is antithetical to accepted religious beliefs, or that it simply does not work. Whether these passages were motivated by the widespread opprobrium attached to the practice of magic in the Latin West (especially when it involved supernatural entities as intermediaries), or by the perceived potential for scepticism among the book's readership in the face of the practice of magic, the abundance of such passages strewn throughout the text indicates that the translator clearly felt the need to retain such justifications from the original Arabic. Hence, the arguments are originally Arabic and react to a widespread mistrust of magic in the Islamic world, yet they were retained in the Latin (via Castilian) version because their appeal would readily cross over to the latter's Christian audience. The apologetic nature of any given passage is not always stated explicitly, but often must be discerned by reading between the lines to figure out why the translator wrote what he did and in the way he did. Finally, most of the themes outlined below — and the mutually reinforcing links between them — are not exclusive to the *Picatrix*: they can be found in most other magic books in some form (e.g., the *Munich Handbook*),[5] since the need to justify the belief in and practice of magic was omnipresent for those involved in the production and use of such texts.

Theme 1: Direct Claims of the Effectiveness and Power of Magic

This first and most obvious of the themes that serve to defend magic are the repeated and explicitly stated claims of its effectiveness. The frequent promises of 'wondrous effects' or seeing 'wondrous things' resulting from particular rituals that appear throughout the book reflect a significant level of insecurity regarding the question of the effectiveness of magic.[6] Indeed, in several instances, the translator claims that the rituals he is compiling have been 'tested' and 'proven' to work as described,[7] with one case where the writer himself performed some testing 'que nostris temporibus [...] in effectibus figurarum, signorum et planetarum' (regarding the effects of the figures, signs, and planets) (2.12.59). Elsewhere, regarding a prayer containing divine names, he claims: 'vidi experienciam istorum nominum' (I saw those

5 *Munich Handbook*, ed. by Kieckhefer.

6 *Mirabiles effectus*: *Picatrix*, 1.5.36, 2.2.6, 2.5.6, 2.6.4, 2.7.5, 2.10.39–40, 2.10.85, 2.12.58–59, 3.9.9, 3.10.1, and 4.9.1. *Mirabilia*: *Picatrix*, 1.5.15, 1.5.19, 1.5.23, 2.12.59, 3.7.27, 3.10.1, 3.11.54, 4.2.26, and 4.7.44. *Virtus mirabilis*: *Picatrix*, 4.8.7.

7 Examples in the *Picatrix* include the following: *experimento comprobatur* (2.6.1); *per experienciam comprobantur* (4.1.6); *vidi experienciam* (1.5.27); *multis racionibus comprobarunt* (2.11.1); *multos probantes vidi* (3.11.96); *percipiuntur experimento* (4.1.3).

names put to the test) (1.5.27); that he tested a stone whereby 'qui tenuerit eum postea numquam passus est' (whoever carries it never suffers) (2.12.39); and 'Ego autem predictas computaciones prout sunt multociens sum expertus, et inveni veritatem in eis' (I myself have tested these calculations many times and I have found truth in them) (3.11.125). The latter four examples also serve to connect this theme to Theme 11 below regarding the rhetorical choice of using first-hand accounts. In other instances, he reports that reputable authorities have done the testing and proving (e.g., 1.5.23, 2.9.1, 2.12.55, 3.5.3, 3.11.123, 4.7.3, and 4.7.23–24), which provides a link between these direct claims of magic's effectiveness and Theme 3 that involves the text's reliance on the reputation of its ancient sources.

The *Picatrix* provides theoretical justifications for the functioning of divination in particular (2.5.4–5), as well as for magic in general (2.12.59), both serving to link the effectiveness of magic to the coherence of the theories underlying it (Theme 5 below). This, in turn, can help the reader overcome any scepticism that might arise in the absence of such justifications, which is crucial for maintaining an appropriately focused will in the practitioner (Theme 9 below).

Impressive claims regarding the extraordinary power of the magical rituals described in the *Picatrix* also serve to reinforce belief in magic, since along with immense power comes great risk, and the risk involved is reflected in the repeated exhortations to secrecy that permeate the text (as well as the entire genre of magical manuals).[8] One salient example from the Prologue confirms the link between the power of magic and the need for secrecy: 'si hec sciencia esset hominibus discooperta, confunderent universum' (If this knowledge [about the science of magic] were revealed to humankind, it would destroy the universe) (Prologue.3).[9] Elsewhere, the power and associated danger of magic is revealed through the legal sanctions against it, as when we hear 'fuit a lege prohibitum profundum studium artis astronomie propter quod eius sciencia profunda attingit ad scienciam magice artis perscrutandum' (the profound study of the art of astronomy was prohibited by the Law, since a profound knowledge of it overlaps with the study of the magical art) (2.3.17).[10]

The mortal risks associated with the power of magic apply to the effects of some of the physical operations on the practitioner as well, which manifests in the *Picatrix*'s occasional warnings about the deadliness of certain formulas, which are then immediately followed by recipes for their antidotes (3.10.5–6, 4.6.9–12). If the former were not truly powerful and dangerous, the latter would not be necessary.

---

8 On secrecy in magic, see: Theme 2 below; and Porreca, 'How Hidden Was God?', pp. 143–45.

9 An equivalent sentiment is expressed in the original Arabic as well; see *Picatrix*, German trans. from the Arabic by Ritter and Plessner, p. 2.

10 'The Law' here is to be understood as Islamic law, especially since this passage also appears in the original Arabic; see *Picatrix*, German trans. from the Arabic by Ritter and Plessner, p. 81.

Theme 2: Exhortations to Secrecy

The reader of the *Picatrix* is repeatedly enjoined to keep secret the information contained therein, or to reveal it only to trustworthy individuals: 'Et rogamus Deum omnipotentem quod sua pietate et mercede liber iste non perveniat nisi ad sapientum et bonorum virorum manus. Tu autem custos esse debes operis predicti ut nemini reveles indigno' (We beseech God omnipotent that, by His piety and grace, this book may reach the hands of none but those who are wise and good. You must, therefore, be the custodian of everything written herein; never reveal it to any unworthy person) (1.4.33);[11] 'Et debes esse cautus omnino ut nemini predicta reveles et ostendas nisi intelligenti in ipsis et studenti' (Be very cautious not to reveal or show the things discussed here to anyone, except to those who understand and study these matters) (2.3.17). In these instances, we could be in the presence of a sort of psychological play of self-aggrandizement or esteem-boosting upon the reader: if the text came into their hands legitimately, it is because they are to be considered among the 'wise and good', and worthy of receiving the secrets. A key component of salesmanship is to make the prospective buyer feel good about themself and the product on offer. Here, the product is magic (including the book that contains it), and the buyer is the reader, presumably a less advanced practitioner, to whom a great secret is being revealed (2.7.4). Another interpretation of these exhortations to secrecy could be that they represent disclosure tactics for sustaining and maintaining a community of readers by generating a sense of *esprit de corps* among those to whom the secrets are known.

The *Picatrix* even attributes the obscure manner in which the ancient sages wrote about magic to the need for secrecy (1.6.5). It claims that the sages of old hid their knowledge on purpose in difficult language, which the text aims to clarify for its readership (2.6.6). These examples serve to link the theme of secrecy with that of references to ancient authorities, which is subject of Theme 3 below. The *Picatrix*'s emphasis on the need to maintain secrecy is quite typical of texts of its genre, and indeed in cognate disciplines such as alchemy, and even regular texts of natural philosophy, in which the 'secrets of nature' were revealed to the reader.[12] 'Books of Secrets' represent an entire sub-genre of medieval writing.[13] By using the same rhetorical schemes of secret-revealing, the *Picatrix* juxtaposes and assimilates itself with texts that were part of the mainstream of learning in the Latin West from the twelfth century onward, which serves to link this Theme with Theme 4 below, 'Appropriation of Other Scientific Disciplines' Legitimacy'.

---

11 Also note here the earnest invocation of God, a manifestation of the phenomena raised in Theme 7 below.

12 For a general discussion of the intersection between secrecy in scientific and esoteric writing in the Latin Middle Ages, see: Fanger, 'Secrecy II: Middle Ages'; Kieckhefer, *Magic in the Middle Ages*, pp. 140–44.

13 Eamon, *Science and the Secrets of Nature*, pp. 15–37.

Theme 3: Reference to Revered, Ancient, and/or Foreign *Auctoritates*

The tendency to accord great respect and attention to authors from the ancient past was not exclusive to the field of magic. Indeed, reflexive reference (and deference) to such authorities was one of the defining characteristics of scholarly activity at the time of the *Picatrix*'s translation into Latin. The fact that numerous venerable authorities are invoked to buttress the intellectual edifice of magic being erected helped place the text's methodology well within the mainstream of the age. References to authorities in turn helped reinforce the legitimacy of the contents of the book and the practices it describes. Indeed, expressions like 'the ancient sages say' are ubiquitous in the *Picatrix*, and where they are absent, it is the identity of specific ancient scholars that often takes their place.[14] The named authors in question were usually famous for their intellectual work in general, but one particular section explains in specific terms (including the titles of works) why the likes of al-Rāzī, Geber Abnehayen (Jābir ibn Ḥayyān), Pythagoras and Plato are important to the study of magic (2.12.58–59). The text's frequent reliance on generally ancient authorities who were respected for their contributions to human knowledge beyond magic (such as Plato and Aristotle in philosophy and the natural sciences, and Galen in medicine) serves to link this theme with Theme 4 below, which deals with the *Picatrix*'s regular efforts to transpose to magic the generally positive reputations of those other disciplines.

In some cases, the text is explicit about why certain authorities are cited. In section 3.8.3, there appears a page-long prayer invoking Saturn that is to be spoken above a magical image for the purpose of removing the planet's ill effects from the operator and his people. This sort of image-magic involving prayers is one of the more frowned-upon magical practices, so the immediately following paragraph (3.8.4) opens with the following disclaimer that shunts any potential opprpbrium onto to the ancient authority:

> Predicta autem verba Abenrasia in Agricultura Caldea [...] composuit. Quam oracionem nunc sumus hic recitati nisi ad patefaciendum communem concordanciam sapientum antiquorum erga opera planetarum et in cunctis temporibus suorum corporum protectionem planetarum naturis. Et ipsam oracionem sumus in hoc libro recitati ut huius operis nil comprehensibile sapientum deficiat antiquorum. Et hunc nostrum librum tradimus completum prout in principio promisimus tractaturi. Et quia hec oracio est in nostra lege prohibita, hic recitamus eam solum ad discooperienda sapientum antiquorum secreta, quia contra hoc opus antiquitus antedatam legem faciebant. Et ideo nullus predicta a se expellere debet, ut eciam quantum in hoc opere sum locutus et de cetero sum et ero

14 David Pingree has provided an analysis of the sources of the *Picatrix* in 'Some of the Sources'.

bona intencione professor. Cunctos autem hunc librum videntes rogo et audientes ut nemini insensato revelent. Et si fuerit necesse ipsum alicui patefacere, hoc non fiat nisi bonis sapientibus intellectu illustratis et eorum vitam secundum ordinis iura ducentibus. Omnipotenti Deo supplico: hoc opus nostrum a manibus insipientum defendat, et mihi cunctorum hic dictorum parcat, cum bona intencione predicta omnia sum locutus.

> (Abenrasia wrote these words in the *Chaldean Agriculture* [...] We have included this prayer here solely for demonstrating the common agreement between the ancient sages concerning the planetary operations and the constant protection of their bodies by means of planetary natures. We include that prayer in this book such that nothing intelligible in this work of ancient sages be lacking. We pass along our book complete with what we promised when we began to write. Now, since this prayer is forbidden in our law, we included it here merely for uncovering the ancient sages' secrets, since they performed this ritual in ancient times before the law was given. For this reason, no one must reveal these things since, despite everything I have said about this and other rituals, I am and always shall be a teacher with good intentions. Ultimately, I ask that all those who see and hear this book never reveal it to fools. If it were necessary to reveal it to someone, do not do so except to good sages enlightened by intellect and leading their lives according to the laws of good conduct. I pray to God omnipotent: May He defend our work from the hands of fools and forgive me for everything mentioned here, for I have spoken everything above with good intentions.) (3.8.4)

This passage is probably where one can witness the greatest interplay between all of the various themes discussed in this chapter. Indeed, we see not only a reliance on the work of an ancient sage, but also an exhortation to secrecy as insistent as any cited under Theme 2 above, a strong nod toward the internal coherence of the work itself via self-reference (Theme 5 below), emphatic claims of authorial good intentions (Theme 12 below), and a culminating invocation of God himself (Theme 7 below). It is noteworthy that this passage is one that was maintained largely intact from the Arabic to the Latin versions.[15]

In terms of signalling specific links between ancient authorities and the intellectual disciplines (Theme 4 below) to which they contributed, the *Picatrix* relies on the *Chaldean Agriculture*, whose translation from the Chaldean language the text variously attributes to 'Abenrasia' (as in the passage above), 'Abenvasia' (4.2.26), 'Abubaer Abenvaxie' (4.7.1), and 'Zeherith' (4.7.9, 4.7.24–42, and 4.7.60–61) for specific magical rituals and prayers. The text relies on: al-Khwārizmī for references that are to do with

15 *Picatrix*, German trans. from the Arabic by Ritter and Plessner, pp. 244–45.

the properties of specific numbers (1.5.6 and 3.11.125); al-Rāzī for astrological conditions for magic (2.12.55) and for a prayer to Jupiter to avoid a storm at sea (3.7.20); al-Ṭabarī for instructions on how to receive planetary powers (3.7.1); Aristotle as author of a book on magical images (1.5.37), on the figures of the heavens (2.2.6, as 'First Teacher'), as advisor to king Alexander (2.3.17), as an authority on the 'Perfect Nature' in his book entitled *Aztimehec* (= *Kitāb al-Istamāṭīs*, 3.6.1), as an authority on the powers of the planets and their spirits in the *Book of Antimaquis* (3.9.1–10), and from the same book on the topic of stones that can influence spirits (3.10.1–5), on the separation and union of the spiritual and the material (3.11.24), on the relation between sensation and spirit (4.1.10), on the governances of each of the planets (4.4.59), and a reference to Aristotle's *Metaphysics* (4.5.9); Dorotheus (of Sidon) on the properties of the Moon (2.3.10); Empedocles on the definition of 'nature' (3.12.2), on the corruptibility of perceptible things (4.1.3), and on sense perception (4.1.11); Enoch on the properties of the constellation Sagittarius (2.12.48); Galen on medicines and theriacs (2.6.2), on a recipe against curses and enchantments (3.11.53), and on the definition of 'nature' (3.12.2); Geber Abnehayen on magic (including a list of books attributed to him, 2.12.58), on rational and irrational animals (2.12.59), on the wonders associated with the human body (3.11.58–112), and on the working of magical images (4.4.64); Hermes (Trismegistus) on constructing magical images (of a lion: 1.5.32; of a fox: 2.10.46; for each of the zodiacal signs: 2.12.39–51), on summoning spirits to achieve one's 'Perfect Nature' (3.6.1); on 'Perfect Nature' itself (3.6.5), on a magical ring to invoke Mercury (3.7.32), as king, prophet and sage (3.7.35), for a ritual sacrifice pertaining to Saturn (3.7.38), on the qualities and powers of the planets (3.11.52), and on the wonders associated with the human body (3.11.54); Hippocrates on the relationship between the shape of human bodies and where they are born (3.3.32), and on the influence of Jupiter on earthly life (4.4.58); Johannitius (Ḥunayn ibn Isḥāq) on medicine (2.6.2), and as translator of a book by Aristotle (4.4.61); Mercurius on the figures of the planets (2.10.10 and 3.3.2–10), on the colours, cloths, and suffumigations associated with each of the planets (3.3.11) and their powers (3.3.25), and as author of a book called *The Secret of Secrets*, from which an extensive list of philosophical, magical, and astrological aphorisms are extracted (4.4.1–45); Plato on friendship (1.2.2), on the power of words (1.5.35), on forms from the *Timaeus* (1.6.5), on magic (2.12.59), on the definition of 'nature' (3.12.2), on astrology and the bodily humours (4.4.57), on the opposition between bodies and spirits (4.4.65–66), and on the relationship between dryness and sickness (4.5.12); Plinio (who seems different from Pliny the Elder) on the twenty-eight mansions of the Moon (4.9.29–56); Ptolemy on the figures of the zodiacal signs (1.5.5), on celestial forms (2.1.1), a series of astrological and magical aphorisms (4.4.46–56), all three drawn from his *Centiloquium*, and on the image of Venus (2.10.29); Pythagoras on the figures of the heavens (2.12.59), and a series of aphorisms that span between magic and ethical behaviour (4.9.28); Socrates on the 'Perfect Nature' as source of light for sages (3.6.5), and on rules of life for his disciples (4.9.27); Thebit ben Corat on astrological images (1.5.36); Thoos on the links between words, images and spirits (1.5.40); Tintinz the Greek on focusing the mind for ritual practice (3.6.4); Tymtym on celestial figures and degrees (2.2.4); and Zadealis on the knowledge of magic (1.6.5). From this extensive list of authorities, it is clear that the realms of knowledge attributed to these authors in the *Picatrix* offers no more than a haphazard match with what is known of their authentic corpora of works, when their names can even be identified. Nevertheless, it is revelatory in terms of its diversity and of the position these authors held in the mindsets of the author and the translator.

In one case, claims to the antiquity of the sources themselves are supplemented by a chronologically specific history of the origins of magical practice, itself ascribed to an ancient authority. Section 3.6.3 attributes to Aristotle a story about Caraphzebiz as the 'primus qui fuit cum ymaginibus operatus et cui spiritus primo apparuere' who also 'fuit qui primo magicam artem invenit' (the first man to whom the spirits first appeared [...] [who also] was the first to discover the magical art) (3.6.3). The story continues: 'Et ab hoc sapiente Caraphzebiz usque ad alium sapientem nominatum Amenus (qui Amenus fuit secundus in spiritibus et magicis operibus operatus) fluxerunt anni 1260' (Between this wise Caraphzebiz and another sage named Amenus (this Amenus was the second one to work on spirits and magical operations), 1260 years passed) (3.6.3). Since both of these characters must pre-date Aristotle, the origins of magic are projected backwards in time to an extremely remote antiquity. The fact that modern scholars have identified neither of the two names with otherwise known historical figures does not remove from the impression of historicism conveyed by this anecdote. If the venerable theories and practices of magic were worth transmitting through such an illustrious line of authorities, there must be some value to them. This, in turn, would contribute to maintaining a belief in the effectiveness of magic for the *Picatrix*'s intended readership.[16]

The antiquity of authorities is not the only factor that could contribute to their prominence and reputation. The well-recognized phenomenon of attributing talent in magic to neighbouring but 'foreign' cultures manifests widely in the *Picatrix*, with 'very great miracles' being attributed to 'the Copts, the Nabateans, the Egyptians, the Greeks, the Turks, and the Indians',[17] and with 'the sages of India' being cited on numerous other occasions for similar reasons.[18] 'Ancient Greek sages' were known to have 'requested their desires

16 For a discussion of the nature of the *Picatrix*'s readership based on the intent of the rituals contained therein, see the introduction to *Picatrix*, trans. by Attrell and Porreca, pp. 16–26.

17 *Picatrix*, 3.5.2: 'secundum Capteos, Nepteos, Egipcios, Grecos, Turcos et Indos [...] et isto modo faciunt mirabilia magna valde'.

18 *Sapientes Indi*: *Picatrix*, 1.4.1–2, 2.2.2–3, 2.5.1–3, 2.5.6, 2.11.2, 2.12.1, 2.12.52–53, 3.10.7, 3.11.124–25, 3.11.127–29, 3.11.131, 4.2.18, 4.6.1, 4.6.10, 4.6.13, and 4.7.23.

and fulfilled them' thanks to their skill in magic.[19] The text reports that a certain magical ritual is still 'in use among [the Indians] in all matters',[20] and that the Chaldeans (or, Babylonians) of all social classes, both men and women, made ubiquitous use of magic.[21]

The Latin version's Prologue reveals explicitly that the translation was commissioned by the Christian King Alfonso X (r. 1252–1284), 'rex Hispanie tociusque Andalucie' (king of Spain and all of Andalusia) (Prologue.1). At least at first glance, this royal patronage generates an aura of legitimacy around the *Picatrix*, on the grounds that such an erudite monarch would not sponsor a truly heinous work.

The *Picatrix* even features examples of authorities who combine all three of the characteristics described in this theme, namely that they are simultaneously ancient, foreign, and royal figures (albeit pre-Christian). King Behentater is one of these ancient foreign sage-kings (3.10.125), as is Hermes, whose epithet *Trismegistus* (Thrice-Great) is justified by his achievements as 'rex, propheta et sapiens' (a king, a prophet, and a sage) (3.7.35).

The *Picatrix*'s frequent references to ancient and/or foreign authorities is partly a by-product of the nature of the work itself, since it is explicitly described as a compilation. Reliance on such authorities, however, is closely linked to and reinforces several of the other themes of pro-magic apologetic examined below. These links are based on the observation that if the ancient sages were so successful at magic, it must be because they had amassed the requisite knowledge of the orderly universe (Themes 5 and 6) with a sufficiently focused will (Theme 9) and being in a ritually pure and spiritually attuned state (Theme 10). If they could accomplish assorted 'wondrous things' by meeting these conditions, so could the readers of the *Picatrix*, as long as they also met the stringent requirements of precision for success (Theme 8). The ancients' reputed successes at magic, juxtaposed with their record of accomplishment in all the other arts and sciences (Theme 4), serves to reinforce readers' and practitioners' belief in the existence and effectiveness of magic. Moreover, the text describes what it means to be a magician as a way of life by means of exemplary figures.

---

19 *Picatrix*, 1.5.48: 'Antiqui vero sapientes Graecorum [...] petebant quod volebant ipsumque obtinebant'.

20 *Picatrix*, 4.6.13: 'opus est continue in cunctis negociis apud eos'.

21 *Picatrix*, 4.7.43: '[O]mnes Caldei tam magni quam parvi et tam viri quam mulieres [...] Cuncti autem Babilonici nollo deficiente opeerabantur predicta'.

Theme 4: Appropriation of Other Scientific Disciplines' Legitimacy

On one level, this theme is largely self-explanatory, with parallels being drawn repeatedly in the text between the functioning of medicine and magic.[22] A prime example of such parallels reads as follows:

> Et ego dicere intend post ista illud quod habet quilibet planetarum ex metallis, animalibus, arboribus, tincturis, suffumigationibus atque sacrificiis cuiuslibet eorum. Et iuvabis te de quolibet eorum in omnibus operibus tuis quemadmodum phisicus multis medicinis et speciebus operatur et obediencia infirmi in dietis observandis et medicinis capiendis; per hunc modum phisicus attingit intentum.
>
>> (Hereafter, I intend to relate which things are governed by each of the planets, from metals to animals, trees, colors, suffumigations, sacrifices, and so forth. From these you will benefit in all your operations, just as the physician works with many medicines and diagnoses in addition to the cooperation of the patient toward observing diets and taking medicines. This is how a physician reaches his intent.) (1.5.48)[23]

In addition, we hear that one of the prerequisites for being successful at magic is to have mastered the liberal arts, in particular the quadrivium (2.1.3 and 2.8.2–8). In another section, there appears a list of 'decem sciencie huic arti necessarie' (ten sciences necessary for this art [of magic]) (4.5.1), which are:

> agricultura, marinaria et gubernare populum; [...] ars ducendi milites, gubernandi exercitus, lites et prelia faciendi, animalia et aves vociferandi et ipsas decipiendi; [...] grammatica, divisio idiomatum, raciocinare iudicia, facere raciones et iura intelligere et ea que sequuntur ista; [...] arismetica; [...] geometria; [...] astronomia; [...] musica; [...] dialectica; [...] phisica; [...] ars nature.
>
>> (agriculture, sea travel, and governing people; [...] leading soldiers, commanding armies, waging skirmishes and battles, calling animals and birds, and deceiving them; [...] grammar, the division of languages, legal jurisprudence, devising explanations, understanding laws and those who follow them; [...] arithmetic; [...] geometry; [...] astronomy; [...] music; [...] dialectic; [...] medicine; [...] natural philosophy.) (4.5.1–10)

22 For discussions on the relationship between medicine and magic, see: Ockenström, 'Demons, Illness, and Spiritual Aids'; Saif, 'Between Medicine and Magic'; Weill-Parot, *Les 'images astrologiques'*, pp. 389–588; Kieckhefer, *Magic in the Middle Ages*, pp. 56–69.

23 Other analogous examples can be found at *Picatrix*, 2.6.1–2, 2.7.8, and 2.9.39.

Indeed, 'ad istam scienciam nemo pervenire poterit complete nisi philosophus perfectus' (none can grasp that science [of magic] perfectly except the perfect philosopher) (2.1.3). If the entire intellectual background considered necessary for practicing magic is made up of legitimate and uncontroversial pursuits, by extension, magic must also be considered legitimate. The process of assimilating magic to other legitimate pursuits is stated succinctly toward the beginning of the *Picatrix* when magic is counted among the crafts that distinguish humans from animals (1.6.1). Finally, the mastery of all these realms of knowledge would require a keen thirst for knowledge on the part of the prospective magician, which links this theme with Theme 6 below that focuses on the quest for knowledge as an intrinsic good.

On another level, the extensive theorizing about magic that occurs throughout the text represents in and of itself a sturdy-looking intellectual armature not unlike the theorizing that appears in texts devoted to other disciplines, such as medicine or astrology. The presence of all the theorizing on magic places the *Picatrix* on a par with standard, mainstream texts in those other disciplines, thereby enhancing the general aura of legitimacy the text wished to project. Moreover, the combined complexity and coherence of the theorizing serves to connect this theme to Themes 5 and 8 below.

Theme 5: Internal Coherence, Rationality, and Orderliness of Magic

Although at first glance, the *Picatrix* may come across as an impressive yet disorganized compilation from sources too improbably varied to produce a unified system of thought, we nevertheless see multiple strategies being deployed to emphasize that magic has a legitimate place in an internally consistent, intelligible, orderly, and rational conceptual structure of how the universe is constituted and operates. In *Picatrix* 2.7.1, *coadunacio* (unification) is identified as a guiding principle for magic, which represents a good rhetorical rallying point to demonstrate the coherence of the system that explains how and why magic works, that is, by the magician's yoking of the world.

According to the cosmology of the *Picatrix*, the very spiritual fabric of the universe makes the functioning of magic inevitable. Here magic is not 'supernatural'. It is a part of nature resulting from God's creation (4.1.12; and cf. Theme 7 below on references to God). The text thus provides frequent summaries of how and why magic was thought to operate, which function as steady reminders of the reality, rationality, and effectiveness of the practice (e.g., 2.8.1, 3.3.25, 3.5.5, and 4.4.64).

In more practical terms, the fact that there are several cross-references within the *Picatrix* to other portions of the text demonstrates a degree of intentionality in the curation of the contents that goes beyond what one might expect from a mere anthology or compilation.[24] Maslama al-Qurṭubī worked hard to tease together a coherent system from his broad selection of sources, and the translators of the text retained these features purposefully.

Theme 6: Pursuit of Knowledge as a Good

In a telling passage, after an aside on the importance of ritual purity (Theme 10 below), we read: '[D]eviavimus ab intencionibus huius libri quia ista et istis sensus sunt radices rerum in quibus liber iste fundatur, qui est ex sciencia magice' ([W]e have deviated from the intent of this book. This intent — knowledge of magic and the awareness of it — are the very foundations of this book) (1.6.5). Elsewhere, the pursuit of knowledge is described as a good thing — 'sciencia est quid valde nobile et altum' (knowledge is most excellent and exalted) (1.6.1) — through which a human is perfected, 'nec aliquid maius eo poterit invenire quoniam nihil melius quam fortunam completam inquirere poterit et habere' (and one can find no greater thing since there is nothing better than a perfect lot) (2.5.5). The opening sentence of Book 1 addresses the reader directly on this topic: 'Scias, o frater carissime, quod maius donum et nobilius quod Deus hominibus huius mundi dederit est scire quia per scire habetur [...] que sunt cause omnium rerum huius mundi' (Know, dearest brother, that the ability to possess knowledge is a very great and noble gift that God bestowed on humankind since through knowledge one can become acquainted with [...] the causes of everything in this world) (1.1.1), including magic. Later in the same paragraph, we read: 'scire est res summa et nobilis, et quotidie studere debes in Deo [...] quia sciencia, sensus et bonitas ab ipso procedunt' (Understand that knowledge is the utmost noble thing, and strive in God every day [...] since knowledge, sense perception, and goodness all proceed from Him) (1.1.1). Elsewhere, it is stated, 'qui in scienciis non laborat est defectuosus et debilis auctoritatis' (one who does not pursue knowledge is defective and lacking authority) (1.6.1). In sum, the quest for knowledge about magic represents the culmination of a person's duty in fulfilling God's plan: 'Et hoc est maximum donum quod ipse Deus hominibus dedit, ut studeant scire et cognoscere. Nam studere servire Deo est' (Dedication to knowledge and understanding — this is the greatest gift that God gave to humanity. Therefore, to study is to serve God) (1.1.1), which links the pursuit of knowledge with Theme 7 (below) on the importance of references to God in rendering the *Picatrix* more palatable to the text's audience.

24 Examples of explicit cross-references include *Picatrix*, 2.5.1, which refers to material first introduced in Book 1; *Picatrix* 4.6.2 and 4.6.9 refer to the subjects discussed in *Picatrix*, 3.6.

Theme 7: References to God

There are numerous references to God throughout the text, many of which would not have been seen as inappropriate to a casual medieval reader. These help justify the practice of magic by associating it with theologically acceptable notions of God, by pointing out that magic is part of God's creation, and that certain sages saw magical practice as a means of understanding God.

The majority of these references to God are positioned at the beginning and/or end of important sections of the text, i.e., precisely those sections that would catch the attention of a casual reader. This positioning seems intended to appease a potentially critical reader by whitewashing the more inappropriate passages that are strewn throughout the text (e.g., those rituals that invoke spirits and/or are intended to cause harm). Thanks and praise to God are the very first things that appear in the Prologue — a common occurrence in Arabic texts, and maintained in the Latin version. The beginning of Book 1 offers a lengthy description of the nature of God that is generic enough to be non-controversial:

> [P]er eum omnia dissolvuntur, et per ipsum omnia nova et vetera sciuntur. Ipse enim est in veritate primus, et nihil in eo deficit nec aliquo alio indiget cum ipse sui ipsius et aliarum rerum sit causa, nec ab alia recipit qualitates. Ipse vero non est corpus nec ex aliquo corpore compositus, nec est mixtus cum aliquo alia extra se sed totus est in se ipso. Et ideo dici non potest nisi unus. [...] [I]pse solus perfectus est. Nec veritas perfecta absque sua est nec unitas; sola autem eius veritas vel unitas perfecta dici potest. Omnia vero sunt sub eo, et ab ipso veritatem et unitatem, generacionem atque corrupcionem tamquam a sua causa recipiunt. [...] Et scias quod ipse Deus est factor et creator tocius mundi omniumque rerum existencium in ipso et quod iste mundus et omnia in ipso existencia ab ipso altissimo sunt creata.
>
> > (Through God all things are reconciled, and through God all things both old and new are discerned. He Himself is first in truth. Nothing is lacking in Him, nor does He lack anything. He is the cause of Himself and everything else, and He received qualities from no other. God is neither a body, nor composed of any matter, nor mixed with anything outside of Himself. Rather, God is whole in and of Himself. He cannot be described except as *The One*. [...] God alone is perfect. Truth is incomplete without God's unity and vice versa, for only His truth or unity can be considered perfect. All things exist below God and from Him receive their truth, unity, generation, and corruption. [...] Know that God Himself is maker and creator of the entire universe and that everything existing within it was created by Him, the most exalted.) (1.1.1)

This sequence of characterizations of God and explanations of causality act as a *captatio benevolentiae* for the reader, since the first mention of magic in the text itself (beyond the Prologue) is subsequent to it (1.2.1). Book 1 ends by pointing out: 'predicta verba et dicte raciones sunt [...] verba que recipit Adam a Domino Deo' (the words and explanations discussed are [...] the very words that Adam received from the Lord God) (1.7.4). Other significantly positioned (and therefore possibly gratuitous) mentions of God include the following: the closing statements of *Picatrix*, 1.4.33 mentioned above under Theme 2; the beginning of Ch. 1.7, where the reader is reminded that 'ipse Deus qui est factor et creator omnium' (God is the maker and creator of all) (1.7.1); at the beginning of Ch. 3.12, which is entitled *De regulis in hac sciencia necessariis* ('On the Rules Necessary to This Science') (3.12), i.e., magic, we see the reader being enjoined 'ad tuam animam salvandam et in amore Dei quantum poteris assidue cogitabis' (Seriously consider, as much as you can, the salvation of your soul and the love of God) (3.12.1); and at the end of Ch. 4.7, we hear, 'intellectus est ipse Deus a quo totum celum et natura dependet. Qui sit benedictus per infinita secula seculorum. Amen' (mind exists through God Himself upon whom all heaven and nature depend. Let Him be praised through the infinite cycle of the ages. Amen) (4.7.62). In one instance, a sentence that occurs at the end of a lengthy prayer to Jupiter acts as a sort of colophon that defangs the planet worship involved in the prayer itself: 'Amen. Et ad mundam voluntatem et salvam erga dominum Deum nostrum habentes amen' (Amen! To a pure and redeemed will toward our Lord God, amen) (3.7.21). The prayer in question includes multiple statements addressed to Jupiter that could themselves be applied without modification to either the Christian or the Muslim God, and this concluding statement serves to link the prayer with acceptable religious sentiments. Each of these examples gives the casual reader the impression that the *Picatrix* is more legitimate (or less objectionable) than it truly is.

The *Picatrix* itself is described as a godly endeavour. In a self-referential section describing what the Indian sages believe, we read:

> Et in hoc libro sunt magna secreta que veritatem rerum appellant, ex quibus sciunt Deum altum et ipsum cognoscunt factorem omnium et creatorem. Et asserunt quod omnia que fuere hoc opere usi non faciebant nisi ut ad ipsum Deum et ad eius unitatem cognoscendam attingere prevalerent ut eius possent lumine illustrari.
>
> > (In this book are contained the great secrets that [the Indian sages] call the truth of things through which to understand God on high and recognize Him as maker and creator of all. They assert that everyone using this work did so only to come to an understanding of God Himself and His unity, that they be illuminated by His light.) (2.12.53)

Here again, we see the interrelation between the prioritization of knowledge in the *Picatrix* (Theme 6) and the importance of God, all of which serves to reinforce the acceptability of the work as a whole.

Certain passages even allude to sacred scripture, such as the statement 'Et hoc est quod ait Deus quando dixit: exaltemus eum in altum' (This is what God meant when He spoke: 'Let Us exalt man on high') (1.2.5), which resonates with both *Sura* 19. 58 in an Islamic context, but it could also be understood as an allusion to II *Corinthians* 12. 3–4 to a Christian reader. At 4.1.9, we hear that 'legis conditor sic ait: primum a Deo creatum fuit sensus, qui aspiracione divina motum suscepit' ([t]he Founder of the Law [i.e., Mohammad] spoke thus: sense perception was first created by God who exalted motion with a divine aspiration) (4.1.9), a sentiment which would be broadly agreeable to a Christian audience nursed on Platonic philosophy.

Certain symbols and concepts central to the Muslim and Christian faiths also appear in the *Picatrix*. The text claims more than once that the power of magic resides in the power of the word (1.5.35 and 1.6.5), and even that 'verba non habent tale posse nisi hoc esset ex precepto Dei gloriosi et alti' (words do not have such power except at the behest of the glorious and high God) (1.5.39). The passage does not go so far as to use the statement that God is the Word, but elsewhere we hear that magical rituals 'verbis et nominibus nigromanticis non perficiuntur nisi cum Dei virtute, mandato et gracia fuerint adiuncta' (are not accomplished by words and magical means unless they are joined with the power, command, and grace of God) (4.4.61).

A final note regarding references to God: notwithstanding the above, they need not be exclusively apologetic in nature. An additional explanation for their presence can also be that they serve as expressions of pride in the practice of the discipline. Indeed, much like in Ptolemy's introduction to the *Almagest* where he states that studying astronomy is the highest and most divine form of knowledge and brings one closer to God, or Averroes stating that metaphysics amounts to worshipping God, both the author and the translator of the *Picatrix* are eager to associate their subject with divinity to demonstrate its excellence over other disciplines.[25]

Theme 8: Complexity of Magic Generates Need for Precision in Its Operations

The system of thought that encompasses both astrology and magic deflects criticism and buttresses the impression of its internal coherence (Theme 5 above) by attributing any failure to the practitioner rather than to the system as a whole. In *Picatrix*, 4.4.57, the following is attributed to Plato: 'illud propter quod in sciencia astronomie veritas deficit nec omni tempore inventa extitit

25 I am grateful to Dag Nikolaus Hasse for pointing out the parallels with Ptolemy and Averroes.

et nec ubique nonnisi ex diversitate actoris accidit et ex errore' (when the science of astrology seems false or entirely a sham, this arises purely from a mistake or error on the astrologer's behalf) (4.4.57).[26] The complexity of the operations necessary for successfully performing magical rituals is signalled as the main source of error by the practitioner, and thence of any failure of the operation. One expression of the requirement for precision arises in relation to the chanting of divine names during a prayer: 'Et caveas tibi ne erres in ipsis nominibus nec in formis nec in figuris eorum ut error non cadat in eis' (Beware lest you make a mistake in those names, in their forms, or in their figures; they must be transcribed accurately) (1.5.27). The very next sentence discusses the problematic existence of variants between different versions of the same prayer: 'Et nomina que sunt scripta bohorim vidi sapientem scribere nohorim. scilicet cum n. sed ego recollegi ea cum b ut superius dixi' (Among the names written, I have seen a wise man write *bohorim* or *nohorim* with an 'n', but I recall it with a 'b' as I said above) (1.5.27). The requirement of performance without error is restated in Book 3, where we read: 'Qui autem predicta absque errore fuerit operatus ad quod voluerit poterit evenire' (Whoever performs this without error will be able to achieve their desire) (3.5.6), and again in Book 4, where — prior to providing an illustration of a series of thirty magical symbols 'invente sunt in libro Folopedre regine' (found in the book of Queen Cleopatra) — the reader is enjoined thus: 'optime advertas ut in his figuris nullo modo falles; quod si aliquid fefellerit in eis, nil operaretur effectu' (pay very close attention that you make no mistake in any way in these figures, for if anything is lacking in them, nothing will arise from the effect) (4.8.36). A similar sentiment arises within a quotation attributed to Aristotle, who is said to be addressing King Alexander:

> 'O Alexander, caveas omnibus momentis tuis et operacionibus in quibus fueris operatus eas facere secundum motus, aspectus et qualitates corporum celestium; quod si ad predicta inspexeris, peticiones tue implebuntur cum effectu, et habebis quicquid in tua voluntate fueris estimatus.'
>
> ('O Alexander, be wary in every moment and action you perform! Do them according to the motions, aspects, and qualities of the celestial bodies. If you have done your research, your request will be fulfilled effectively, and you will have whatever you had reckoned in your will.') (2.3.17)

The conditional clause in the last sentence places the onus on the practitioner, and failure in any ritual can thus be attributed to inadequacy on the part of his level of preparation. A similar promise for success in magic if certain

26 Note that the science of astrology is repeatedly described as the foundation of magic; the *Picatrix* is, after all, a treatise of astral magic.

complex conditions are met occurs in several other passages in the text as well (e.g., 3.1.1 and 3.7.29).

The complexity of the discipline of magic is explicitly expressed on numerous occasions throughout the text, representing as many caveats to the reader in terms of reasons for the failure of any attempted operations. The fact that magic requires intensive study and the mastery of numerous other disciplines (Theme 4 above) leads to the claim that the *sapientes philosophi* (wise philosophers) spoke in ways 'ut nemo ipsa nisi in scienciis illustrati propter eorum difficultates necnon et profunditates intelligere valeant' (that only those enlightened in this science [i.e., magic] could understand them on account of their complexity and depth) (4.4.61). Elsewhere, the reader is warned thus: 'Et hoc quod in isto loco sumus locuti est valde obscurum et profundum. Et ideo advertas et cogites diligenter in eo' (What we said here is very obscure and profound. So pay attention, and study it diligently) (2.6.5). There are even examples of the deleterious consequences in store for a practitioner who fails to abide by the prescribed complexities:

> Nunc autem recitare proposui id quod nostro tempore accidit cuidam volenti attrahere virtutem Lune, qui hoc opus necessitate deductus confecit et fuit in quadam nocte operatus a vero tramite ipsius operacionis totaliter remotus. Et dum in una nocte actualiter operabatur, sibi apparuit homo cum una re in manu, quam in eius ore posuit; et statim clausit os, nec apparebat ipsum os umquam habuisse. Et sic per 40 horas timore maximo repletus permansit, in quarum fine fuit totaliter extinctus.
>
> > (Here I propose to relate a story that happened recently to someone who wished to draw upon the power of the Moon. He performed this ritual urged by necessity and so did it on some night totally irrelevant to the ritual. Then one night, while he was actually operating, a man appeared to him with something in his hand that he placed in the practitioners' mouth. He closed his mouth immediately, and it appeared as if he had never even had a mouth. He remained filled with the greatest terror for forty hours, at the end of which he was completely ruined.) (4.2.14)

This instance also happens to employ the first-person narrative perspective (Theme 11 below), which adds to the immediacy and potency of the warning. A similar warning occurs in the very next paragraph: '[E]t qui opera execucioni tradere nesciunt accidunt pericula et expavescenda maxima, quod quidem terribile esset narrandum' (Those who do not know how to properly perform rituals bring about the greatest perils and utmost terrors, which for my part are too awful to tell) (4.2.16). Clearly, these examples place the onus on the practitioner to perform any ritual with exactness. When juxtaposed with the lengthy and complex rituals, such as the prayers to the planetary spirits described in Book 3 — some of the prayers extend to several pages in length (e.g., 3.7.21, to Jupiter) — the repeatedly emphasized need for precision provides a ready explanation for any failure to achieve the desired effect. When such a result would come about — as it no doubt did frequently — the operator's belief in the efficacy of magic could always remain intact, since perfect precision is unattainable. Thus, the magical system described in the *Picatrix* contains this built-in mechanism of self-defence against the scepticism that can result from repeated failures.

The pursuit of knowledge that is emphasized so often in the text (Theme 6 above) implies knowledge of the complexities involved in every aspect of magical practice. In one instance where the creation of an astrological image is concerned, an assortment of factors is listed as being important to the success of the operation: the shape of the image, the reason for which it is being manufactured, as well as the material to be used. The passage continues: 'Et facias quod predicta omnia in similitudine ad invicem concordentur; et similia sint encia viribus et influenciis planete ipsi operi dominantis' (Let all these things be harmonious with one another in similitude. Similarly, let them be in the powers and influences of the planets governing that work) (2.6.1). This section culminates with an equivocal statement: 'Homines enim in quocumque tempore facientes ymagines et ignorantes predicta male faciunt' ([P]ractitioners of any era who make images and ignore these things do wrong) (2.6.1), which can be interpreted from a moral standpoint (i.e., they are doing an evil thing), or as the equivalent of 'they perform the ritual badly'. In any case, the claim of temporal universality for magic ('practitioners of any era') suggests that the statements apply as much to the ancient sages as to the current reader of the text, as if to say 'if you, reader, meet these complex requirements, you, too, can be successful at practicing magic!' It is the multiplicity of factors that leads to the complexity that in turn generates the need for diligent study.

In some instances, the complexity of the rituals manifests most in the difficulty involved in obtaining certain ingredients. This phenomenon occurs when the text suggests alternatives for material components, such as when, in a ritual 'ad amorem mulierum acquirendum' (for acquiring the love of women), we read the following recommendation: 'Et si forte propter difficultatem contigerit ut eidem in cibo vel potu nullatenus dare possis, recipe confectionem superius graduatam, et loco sanguinis superius dicti ponas sanguinem illius contra quam eris operatus' (If perchance, on account of difficulty, it happened that you could not get this chosen woman to eat or drink the mixture described above, use it, but instead of the adept's blood, use the blood of the one against whom the ritual is performed) (3.10.9). Another ritual for the same purpose includes an alternative procedure rather than an alternative ingredient:

> Quod si non eidem dare poteris in odore, facias ymaginem ceream quam superius diximus. Et agens (videlicet ille qui hoc opus fieri iussit) in eius manibus teneat ipsam, et accipiat incensi et galbani ana ʒ ii, quas propriis manibus proiciat in ignem.

> (If you cannot give her the aroma, make the wax image we mentioned above. The agent (namely, the man who ordered this work to be done) should hold it in his hands, and he should take 2 oz. each of incense and galbanum, which he should throw into a fire with his own hands.) (3.10.12)

These alternatives are likely the result of a process of trial and error arising from difficulties experienced in following the complexities of the original version, with both versions being diligently recorded.

Theme 9: Practitioner's Fully Focused Will and Devout Belief Necessary for Success in Magic

Among the prerequisites for success at the practice of magic was that the practitioner must be focused fully on the objective. Doubt of any sort was considered antithetical to magic. The requirement for a fully focused will (or, the total absence of doubt) is another manifestation of the *Picatrix*'s magical system tending to shunt blame for the failure of any given ritual on the inadequacies of the practitioner, not just because of the system's inherent complexity (as described in Theme 8), but also in relation to the practitioner's inner orientation and mental state. The *Picatrix* makes this point repeatedly: 'Illi autem qui ymagines facere querunt [...] firmiter debent credere in operibus que faciunt in ymaginibus quod illud quod faciunt erit veridicum et sine dubio' (Those who seek to make images [...] must firmly believe in the rituals they perform with the magical images such that the work they implement will be genuine and without doubt) (1.4.1); 'oportet operatorem artis magice esse credentem in suis actibus sine aliqua dubitacione operis' (a practitioner of the magical arts should be one who puts faith in his own actions without any doubt regarding the rituals) (1.4.32); 'illud quod est necessarium in dictis operibus et sine quo aliquid complere non possumus est adiungere totam voluntatem et credulitatem operi' ([w]hat is necessary to these magical operations — without which one cannot complete anything — is to link one's whole will and belief to the operation) (1.5.35); and 'Proprietates vero et exempla proposita et dicta in libris sciencie huius prophetarum, si tu ea ad opus tentativo modo deducere laborares viderentur de genere trufancium nec ad effectum per ipsum promissum deducere posses in eternum' (The properties and examples written in the books of this science of the prophets [i.e., magic] would seem fraudulent were you to apply them to a ritual half-heartedly; moreover, you would never achieve their application toward the promised effect) (3.12.2). Even when the practitioner himself is fully invested in the ritual, the mere presence of a sceptic was considered enough to ruin the procedure:

> [E]t quod aliquis nullam noticiam habeat de suis operibus nisi esset fidelis in sui amicicia et credens in opera, nec sit derisor nec incredulus

in operibus et potenciis spirituum celi et suarum potenciarum que sunt potentes in hoc mundo, et quod opera fiunt ex eis spiritibus.

> (Let no other individual know about these rituals unless he be faithful in his friendship and persuaded in the work. Let him be neither a scoffer nor a disbeliever in the works and powers of the celestial spirits, in their potencies which have effect upon this world, or in the belief that these works ensue from these spirits.) (1.5.36)

In addition, even the intent of testing the efficaciousness of a ritual was considered sufficient to ruin it: 'nec dubitent aliquid de ipsis effectibus, et quod hoc non faciunt causa tentandi aut probandi utrum sint vera anne' ([the practitioners] must not doubt a single thing concerning the rituals' effects and not perform them for the sake of testing or probing whether or not they are efficacious) (1.4.1). Here, we are in the presence of another factor that reinforces the credibility of the magical system presented in the *Picatrix*. If elaborate complexity combined with the stated need for precision led to repeated failures and thus generated doubt as to the efficacy of magic, the resulting doubt itself could subsequently be signalled as the cause of any further failures.

The *Picatrix* even goes so far as to set out a logical trap for doubters that demonstrates they are mistaken, thereby rendering the system of magic presented in the *Picatrix* immune to outside criticism. Indeed, the text states that one cannot criticize magic without knowing something about it: 'Quare scias quod hoc secretum quod in hoc nostro libro intendimus discooperire acquiri non potest nisi prius acquiratur scire. Et qui scire intendit acquirere studere debet in scienciis et eas ordinatim perscrutari' (Know, therefore, that the secret we mean to divulge in our book cannot be understood without the prerequisite knowledge. Whoever intends to understand it must focus on knowledge and pursue it exhaustively) (1.1.2); and 'nil deterius in hominibus quam scientes velle sophistice apparere et scienciam non habere. [...] [P]er hoc non attingunt scienciam sed in ea fideliter laborantes' (there is nothing worse than people who want to appear sophisticated without having the knowledge. [...] [T]hese types cannot acquire understanding, unlike those who strive for it faithfully) (1.7.1). Moreover, one cannot know anything about magic without studying it with an earnest and devout mind, which is by definition incompatible with a doubtful or sceptical outlook (as discussed in the paragraph above). Therefore, doubters can never understand magic, which in turn makes the system of beliefs impervious to sceptical critics.

Theme 10: Exhortations to Ritual Purity: Spiritual Focus vs. Rejection of the Worldly, the Impure, and the Corporeal

Ritual purity, in the context of the *Picatrix,* tends toward a rejection of the corporeal in favour of the spiritual. This phenomenon could be treated as a means toward the end of maintaining a fully focused will on the part of the practitioner (Theme 9 above), but it occurs often enough that it deserves its own treatment here. A lack of sufficient ritual purity could always be signalled as a cause for the failure of any magical operation, since the *Picatrix* presents it as every bit as important to success as the correct performance of complex rituals (Theme 8 above).

In the conclusion of a passage that contains a description of the distinction between spiritual matter (which is imperceptible to our five senses) and corporeal matter (which is composed of the four elements and is perceptible to the senses), we read the following address to the reader:

> Tu autem qui in hoc libro studere proponis considera quomodo tuam animam reducere possis ad gradus et cognicionem spirituum beatorum; quod quidem facies si in omnibus tuis operibus partem spiritualem prosequi conaberis qua cognicione eris differens a brutis.
>
> > (You who propose to study this book, however, consider how you might lead your soul back to the degree and perception of the blessed spirits. Indeed, you will succeed at this if you try to follow the spiritual part in everything you do. By understanding it, you will distinguish yourself from beasts.) (4.1.3)

The final sentence of that passage represents another instance where success at magic is something that serves to distinguish humans from animals.[27]

At the end of the first paragraph of Book 1, the reader is presented with a list of the properties of knowledge that links the purity associated with the rejection of the world with knowledge (only the first two of which are relevant here): 'Item habet tres fortitudines, quarum prima est quod facit despicere res huius mundi, secunda quod acquirit bonos mores' ([Knowledge] has three more properties: the first is that it causes humans to spurn the concerns of this world; the second is that it promotes good character [...]) (1.1.1). As we have seen above (Theme 6), the quest for knowledge is considered a good pursuit. Since it is claimed repeatedly that it is magical knowledge that is being presented throughout the book, this statement also serves to link this theme of world-rejecting purity with the good intentions of the author and, by extension, the translator (Theme 12 below). In another passage that serves a similar function, three parts to the human spirit are described: the animal,

27 See also *Picatrix,* 1.6.1, as discussed in Theme 4 above.

the natural, and the rational. We are also told: '[Q]uod si racionalis spiritus vicerit, intellectuum, bonitatum scienciarumque et in nullo alio erit amator' (If the rational spirit prevails the person will be a lover of intelligence, goodness, and knowledge) (4.5.11). The most succinct of such expressions, however, reads as follows: 'mala ex rebus corporeis, bona vero a spiritu procedunt' (evil proceeds from matter, while good proceeds from spirit) (4.1.2).

Divination, or prophecy, is one of the modalities of magic. It is presented as 'one of the faculties of the spirit' (una ex viribus spirituum), and:

> quando virtus illa in qua formantur encia est completa et munda ab omnibus superfluitatibus et immundiciis, videbit res separatas quemadmodum sunt ut encia in speculo apparencia; et eodem modo patefiunt in spiritu quando est nitidus et completus.
>
> > (When that faculty in which entities are formed is perfect and pure from all dross and impurity, one will envision discrete objects just as images in a mirror. Thus, images are revealed in the spirit when it is unpolluted and perfected.) (2.5.5)

The capacity for prophecy is then attributed to God: 'Et hoc esse non potest nisi in hominibus singularibus in quibus prophetici spiritus complete funduntur a primo encium dispositore, qui est ipse Deus [...] ipse Deus talem virtutem in eo naturaliter ponit' (This does not happen, however, except to unique individuals into whom flow perfectly prophetic spirits from the first creator of entities, God Himself [...] God Himself naturally places such a virtue into a prophet) (2.5.5), which links this theme of ritual purity with the theme relating magic to God discussed in Theme 7 above. Another passage, attributed to al-Ṭabarī, reinforces the same link, this time emphasizing the need for both spiritual and physical purity:

> Cum volueris cum aliquo planetarum loqui vel ab eo aliquid tibi necessarium petere, primo et principaliter voluntatem et credenciam tuam erga Deum mundifica, et omnino caveas ne in aliquo alio credas; deinde corpus tuum et pannos tuos ab omni sordicie mundifica.
>
> > (When you wish to speak with any of the planets or to request something necessary to you, first and foremost, purify your will and your belief in God. Take particular care lest you believe in anything else. Rid your body and your clothes of all impurity.) (3.7.1)

Concern for purity arises again in the context of a ritual to harness the power of Jupiter, where the reader is addressed directly:

> Et scias quod quanto magis humiliter et mansuete peregeris dum predictum opus feceris erit melius dum habueris voluntatem mundam et nitidam et a cunctis rebus mundi remotam, neque in mundanis cogites vel occuperis nisi in opere proposito.

> (Know that the more humbly and gently you act while you perform this ritual, the better it will be so long as you have a pure and clear will, removed from all worldly things. Do not think about or occupy yourself with worldly things except for the proposed ritual.) (3.7.22)

This passage explicitly links the need for purity with the level of focused will necessary for successful magical operations (Theme 9 above).

In a quotation attributed to Plato, the relationship between bodies and spirits is summarized thus:

> 'Corpora spiritibus sunt contraria quia ex vita unius contingit sustentacio alterius. [...] Nolite enim vestros spiritus vestris corporibus servientes ullo modo nec interficere mortuum propter vitam viventis nec interficere vivum amore defuncti.'
>
> ('Bodies are opposed to spirits since only the life of one depends on the sustenance of the other. [...] Never allow your spirits to serve your bodies nor destroy the dead on account of the life of the living nor kill the living out of a love for the dead.') (4.4.65)

This passage represents an attempt to whitewash the more nefarious rituals contained elsewhere in the book by means of a morally sound principle expressed as a universal statement.

Broadly speaking, in light of the passages quoted in this Theme, there can be said to exist a certain harmony between how the *Picatrix* presents notions of spiritual purity and the ascetic-leaning practices current in both Christianity and Islam during the period of the text's composition and transmission. This emphasis on ritual purity (which could also include bodily purity) via a focus on the spiritual at the expense of the physical reinforces the aura of legitimacy that the text aims to project, and it simultaneously serves to distract from the often heinous nature of some of the rituals that are included.

Theme 11: Narrative Devices: First-Person Perspective and Testimony

Narrative perspective contributes to the believability and coherence of the magical system presented in the *Picatrix*. In multiple instances, the text presents a first-person perspective that contributes to the tangibility and believability of the magical practices being described. These passages often take the form of the narrator interjecting statements such as, 'Dico enim tibi veraciter quod habebam amicum quendam qui hoc opus fecit ut supra' (Truly I say to you that I once had a friend who performed the ritual above) (4.2.11); or 'Et volo hic exemplum ponere quod intellexi a quodam sapiente qui in his scienciis laboravit' (I wish to put forward this example I learned from one sage who worked in these realms of knowledge) (2.1.1), followed by a recounting of how the ritual was performed successfully. In some instances, proof or verification of the effectiveness of a ritual is also related in the first person, thereby repeatedly adding to the impression of effectiveness of the contents of the book: 'Ego autem feci ymaginem illius figure in hora predicta cum qua sigillavi tam incensum quam quecumque alia sigillanda, et cum illis faciebam mirabilia quibus mirabantur omnes' (I myself have made an image of that figure [i.e., a scorpion] in this hour [i.e., when the Moon is in the second face of Scorpio]. I impressed some incense and some other imprintable things and with those I have accomplished miracles at which everyone has marvelled) (2.1.2); 'Ego enim feci sigillari trociscos de sanguine hirci secundum doctrinam istam factos, et operabantur miraculose' (I myself have made imprinted pills from the blood of a goat according to those instructions, and they worked miraculously) (2.12.39); and 'Ego autem vidi composicionem quandam ad hominem abscondendum taliter ordinatam' (I myself have seen a kind of mixture for invisibility made in this way) (4.3.2). These interjections in the first person contribute to the immediate tangibility and credibility of the content: magic is not merely something that ancient sages practiced successfully, but, more promising for a medieval reader's perspective, the text's author himself was claiming multiple instances of personal experience in achieving the desired effects of the rituals being related.

The first-person point of view extends beyond the tangible experience of rituals themselves to claims of the author's personal handling of the books that he consulted in the process of bringing together the materials that make up the *Picatrix*. Indeed, there are multiple instances where he claims to have handled the books of the ancient sages himself: 'Iohannicius [...] quendam librum Aristotelis, Grecorum domini, transtulit, quem ego aspexi; eo quod dicta sapientum suarumque estimacionum profunditates necnon et suorum intellectuum conclusiones intelligere valeas, in hoc loco proposui recitare' (Johannitius [Ḥunayn ibn Isḥāq] [...] translated a certain book by Aristotle, the teacher of the Greeks, which I myself have seen. Since you can understand the words of the wise, the depth of their intuition, and the conclusions of their minds, I propose to convey them here) (4.4.61); 'Hoc vero opus ut supra in libris Indorum invenimus scriptum; quod opus est continue in cunctis negociis apud eos' (This ritual I found as above in the books of the Indians. It is still in use among them in all matters) (4.6.13); and 'In libro vero De agricultura Caldea, quem Abubaer Abenvaxie de Caldeorum idiomate transtulit in Arabicum, multa artis nigromancie opera [...] invenimus quas in hoc loco sumus dicturi' (In the *Chaldean Agriculture*, which Abubaer Abenvaxie [Ibn Waḥshiyya] translated into Arabic from the tongue of the Chaldeans, we have found many writings on the art of magic [...] which we will now relate here) (4.7.1). These individual instances are confirmed and amplified in the passages where we hear of the monumental effort deployed in compiling the *Picatrix*:

> Et dico tibi, carissime, quod cum magno labore et studio hunc librum composui, sapientum antiquorum libros quamplurimos convolvendo, aspiciens et cogitans in opinionibus quorundam, et conclusiones verid-

icorum et effectu comprobatas scribendo, in tantum quod 224 libros predecessorum antiquorum sapientum studui de verbo ad verbum, et ex ipsis omnibus tamquam florem et lilium eorumdem hunc composui librum, vacans per sex annos continue in predictis.

> (I say to you, dearest friend, that I have composed this book with great effort and study, encompassing as large a number of books of ancient sages as possible, observing and contemplating the opinions of certain men, and writing down true conclusions and proven effects, to the extent that I had studied 224 books of wise ancient predecessors word by word. With all these, I composed this book like a flower or a lily of sorts, working on everything tirelessly for six years.) (3.5.4)[28]

And:

> Cuncta vero que hactenus in hoc libro nostro fuimus recitati ex dictis sapientum antiquorum eorundemque libris in ista sciencia et opere loquentibus extraximus ab eisdem. Qui autem hunc librum legerit ipsumque perfecte aspexerit et ea que usque nunc diximus intellexerit diligenter cognoscet et sciet laborem quem in istius libri ex diversorum librorum dictis istius sciencie tractantibus compilacione, et que sunt istorum operum radices, habuimus.

> (Everything recited so far in our book was extracted from the sayings of the ancient sages and their books concerning that science and craft. Whoever reads this book, examines it perfectly, and understands the things we have said so far will understand and thoroughly recognize the effort we invested in compiling sayings from a variety of books on that body of knowledge that are the bases of these rituals.) (4.4.1)

Together, these passages demonstrate to the reader that if the task of compiling all these materials was worth doing in the first place, it must be because the knowledge of magic, and by extension magic itself, is both worthwhile and effective. Moreover, the repeated emphasis on the author's first-hand knowledge of and privileged access to such a wide variety of sources reinforces the trustworthiness of the quotations that are strewn throughout the text.

---

28 Note that this passage is also intimated in the first paragraph of the Prologue, which appears to be from the pen of one of the book's translators: 'Sapiens enim philosophus, nobilis et honoratus Picatrix, hunc librum ex CC libris et pluribus philosophie compilavit, quem suo proprio nomine nominavit' (One wise philosopher, the noble and honoured *Picatrix*, compiled this tome from over two hundred books of philosophy and then named it after himself).

Theme 12: Author's Good Intentions[29]

The coherence and acceptability of the magical system deployed in the *Picatrix* is reinforced indirectly through explicit statements that announce the author's good intentions, as a palliative against criticism of the magical practices described therein. Indeed, the fourth section of the Prologue puts forth this appeal:

> Ego autem rogo altissimum creatorem quod iste noster liber nonnisi ad manus perveniat sapientis qui intendere possit quicquid in eo sum dicturus et tenere in bono, et quicquid operabitur ex eo ad bonum et ad Dei servicium operetur.
>
>> (I therefore pray to the highest creator that this work of ours fall into the hands of no sage lest they be capable of following everything I am about to say herein — and consider it beneficial — such that everything done through it be done for the good and in the service of God.) (Prologue.4)

Later we see the reader being addressed directly in the following manner:

> Scire autem oportet quod omnia supradicta in hoc loco non recitavi nisi ut scias et intelligas quod supradicti sapientes nonnisi in aspiciendo profunditates et secreta huius operis studebant, et ut ea simul unirent ad hoc ut ex eis possent finem optatum attingere.
>
>> (Know that I have not recited all this except for you to know and understand what only these sages studied in observing the depths and secrets of this work so that they thus may be reconciled. In this way, they were able to reach a desired end.) (4.4.60)

In other words, the aim is the perfection of the reader's knowledge, rather than enabling any potentially nefarious practices per se.

After relating a lengthy ritual drawn from a book entitled *Divisio scienciarum et panditor secretorum* ('The Division of Sciences and Revealer of Secrets') intended to draw a young woman for sexual favours, the author states: 'Hoc autem fuit unum ex mirabilibus magnis que per spacium vite mee in hac sciencia viderim. Predicta autem non recitavi nisi ut animadvertas huius sciencie mirabilia et suorum effectuum magnitudinem' ('This was one of the great wonders that I have seen in this science during my life. I have only reported these things so that you may heed the wonders of this knowledge and the magnitude of its effects) (3.5.3). The last sentence can be interpreted as

29 As far as this Theme is concerned, whether 'the author' refers to al-Qurṭubī or to the translator is immaterial, since medieval readers of the Latin text would only be aware that the text was written by a sage named 'Picatrix', whose self-stated good intentions serve to buttress belief in the magical system being presented.

providing the author with a kind of fig leaf whereby he could distance himself from the opprobrium of potential critics. With such a statement, the author could still claim benign intentions while also proclaiming a given operation's effectiveness first-hand (Theme 11 above).

Conclusion

The accumulation of evidence represented by the dozen themes discussed above yields the distinct impression that the *Picatrix* was simultaneously presenting an internally coherent system of magical operation while also defending that system in general, and the book itself in particular, from potential criticism. The end product amounts to a pro-magic apologetic, self-aware in terms of the controversial or even illegal nature of some of its contents, yet also convinced of the worthiness and value of the endeavour as a whole. It is the overall effect of the entire network of interrelated themes that achieves this end, rather than any of its constituent parts taken in isolation. The repeated reminders of each of these themes that a reader — casual or dedicated — would encounter in the book would result in a reinforcement of the belief in the efficacy of the system for those predisposed to accepting the reality of magic, while explicitly dodging critiques and excluding the critics from having any hope of disproving — or even understanding — the arcane workings presented therein. Indeed, the passages with apologetic force are so numerous that they contribute to the 'sound' of the text, without the reader even realizing their impact in each instance. The confidence and faith in the system being presented come across throughout the text. Within its self-created, hermetically sealed environment, the *Picatrix* stands not only as the principal treatise of Arabic astral magic to survive in Latin, but also as a monument of pro-magic apologetics from the Middle Ages.

Bibliography

Primary Sources

Munich Handbook, in *Forbidden Rites: A Necromancer's Manual of the Fifteenth Century*, ed. by Richard Kieckhefer (University Park, PA: The Pennsylvania State University Press, 1998), pp. 193–346

Picatrix, in *'Picatrix': Das Ziel des Weisen von Pseudo-Mağrīṭī*, German trans. from the Arabic by Hellmut Ritter and Martin Plessner (London: Warburg Institute, 1962)

Picatrix, in *Picatrix: The Latin Version of the 'Ghāyat al-Ḥakīm'*, ed. by David Pingree (London: Warburg Institute, 1986)

Picatrix, in *Picatrix: A Medieval Treatise of Astral Magic*, trans. by Dan Attrell and David Porreca (University Park, PA: The Pennsylvania State University Press, 2019)

Secondary Studies

Bakhouche, Béatrice, Frédéric Fauquier, and Brigitte Pérez-Jean, *Picatrix: Un traité de magie médiéval* (Turnhout: Brepols, 2003)

Boudet, Jean-Patrice, and Jean-Charles Coulon, 'La version arabe (*Ghāyat al-ḥakīm*) et la version latine du *Picatrix*: Points communs et divergeances', *Cahiers de recherches médiévales et humanistes*, 33 (2017), 67–101

Burnett, Charles, 'Magic in the Court of Alfonso el Sabio: The Latin Translation of the *Ghāyat al-Ḥakīm*', in *De Frédéric II à Rodolphe II: Astrologie, divination et magie dans les cours (XIIIe-XVIIIe siècle)*, ed. by Jean-Patrice Boudet, Martine Ostorero, and Paravicini Bagliani (Florence: SISMEL Edizioni del Galluzzo, 2017), pp. 37–52

de Callataÿ, Godefroid, and Sébastien Moureau, 'Again on Maslama Ibn Qāsim al-Qurṭubī, the Ikhwān al-Ṣafāʾ and Ibn Khaldūn: New Evidence from Two Manuscripts of Rutbat al-ḥakīm', *al-Qantara*, 37 (2016), 329–72

Eamon, William, *Science and the Secrets of Nature: Books of Secrets in Medieval and Early Modern Culture* (Princeton: Princeton University Press, 1994)

Fanger, Claire, 'Secrecy II: Middle Ages', in *Dictionary of Gnosis & Western Esotericism*, ed. by Wouter J. Hanegraaff (Leiden: Brill, 2006), pp. 1054–56

Fierro, Maribel, 'Bāṭinism in al-Andalus. Maslama b. Qāsim al-Qurṭubī (d. 353/964), Author of the *Rutbat al-Ḥakīm* and the *Ghāyat al-Ḥakīm* (*Picatrix*)', *Studia Islamica*, 84 (1996), 87–112

Kieckhefer, Richard, *Magic in the Middle Ages* (Cambridge: Cambridge University Press, 1990)

Ockenström, Lauri, 'Demons, Illness, and Spiritual Aids in Natural Magic and Image Magic', in *Demons and Illness from Antiquity to the Early-Modern Period*, ed. by Siam Bhayro and Catherine Rider (Leiden: Brill, 2017), pp. 291–312

Pingree, David, 'Some of the Sources of the *Ghāyat al-Ḥakīm*', *Journal of the Warburg and Courtauld Institutes*, 43 (1980), 1–15

——, 'Between the *Ghāya* and the *Picatrix* I: The Spanish Version', *Journal of the Warburg and Courtauld Institutes*, 44 (1981), 27–56

Porreca, David, 'How Hidden Was God? Revelation and Pedagogy in Ancient and Medieval Hermetic Writings', in *Histories of the Hidden God: Concealment and Revelation in Western Gnostic, Esoteric and Mystical Traditions*, ed. by April D. DeConick and Grant Adamson (Durham: Acumen, 2013), pp. 137–48

Saif, Liana, 'Between Medicine and Magic: Spiritual Aetiology and Therapeutics in Medieval Islam', in *Demons and Illness from Antiquity to the Early-Modern Period*, ed. by Siam Bhayro and Catherine Rider (Leiden: Brill, 2017), pp. 313–38

Weill-Parot, Nicolas, *Les 'images astrologiques' au Moyen Âge et à la Renaissance: Spéculations intellectuelles et pratiques magiques (XIIe-XVe siècle)* (Paris: Honoré Champion, 2002)

SOPHIE PAGE

Censorship, *maleficia*, and the Medieval Readers of the *Liber vaccae*

In this article I look at strategies of scribal censorship in surviving manuscripts of the *Liber vaccae,* a Latin translation of a late ninth-century Arabic magical-alchemical work, the *Kitāb al-nawāmīs* ('Book of Laws'). My aim is to examine some of the ways in which an ambiguous work of magic may be read and censored, especially with regard to growing anxiety about manipulations of natural substances and female workers of *maleficia* in the late Middle Ages.

The *Kitāb al-nawāmīs,* an Arabic magical text written at the end of the ninth century, is remarkable because of its extraordinary experiments to generate new, living, hybrid or mixed forms.[1] It was translated into Latin in the twelfth century and circulated under various titles: the *Liber aneguemis* (sometimes corrupted to *neumich*), a Latin transliteration of the Arabic; the *Liber institucionum activarum,* a translation of the Arabic title; the *Liber tegimenti* (or *regimenti*), from a reference in the preface; and the *Liber vaccae,* from its first experiment, which makes use of a cow. The Latin *Liber vaccae* survives in 14 full and partial manuscript copies from the period 1200–1500, and has approximately 85 experiments divided between two books.[2] The first book, the *Liber vaccae major* is presented as the revival of arts that were practised

1 On the Latin text see especially Pingree, 'From Hermes to Jābir' and 'Artificial Demons and Miracles'; Van der Lugt, 'Abominable Mixtures'; Page, *Magic in the Cloister,* and Hasse, 'Plato arabico-latinus'.

2 Four new manuscripts were identified by Maaike van der Lugt and Jean-Patrice Boudet in a paper delivered by Van der Lugt at the workshop 'Autour du Livre de la Vache du pseudo-Plato' on 14 October 2016. The number of experiments varies according to each manuscript and whether a complex experiment is counted as one or more separate activities. See Page, *Magic in the Cloister,* pp. 51–52 for a list of the experiments. *Liber vaccae major,* experiment 26 ('Capitulum aliud pulcrum'), and *Liber vaccae minor* experiments 39 ('Ad fantasma provocandum') and 40 ('Ad actionem ferri') are absent from many copies. As likely later additions to the text, experiments 39 and 40 are not discussed here. A transcription of the copy in Florence, Biblioteca Nazionale Centrale MS 2. 3. 214 was edited in *Liber Aneguemis* by Scopelliti and Chaouech.

Sophie Page (London, UK) is a historian of medieval social, cultural and intellectual history, with a focus on the history of magic, science and religion, cosmology and approaches to nature.

Mastering Nature in the Medieval Arabic and Latin Worlds: Studies in Heritage and Transfer of Arabic Science in Honour of Charles Burnett, ed. by Ann Giletti and Dag Nikolaus Hasse, CAT 4 (Turnhout: Brepols, 2023), pp. 207–229 BREPOLS PUBLISHERS 10.1484/.CAT-EB.5.134031

by past great prophets and diviners, for such purposes as gaining knowledge from God about life and death, speaking with spirits, walking on water, and receiving the reverence of animals. The second book, the *Liber vaccae minor* is focussed on more modest domestic magic or parlour tricks, experiments to produce marvellous effects with seeds, lamps, combustibles, and inks to provoke wonder and provide entertainment.

The magic of the *Liber vaccae* was understood as a natural but transgressive art by most of its medieval readers, who struggled to situate it within familiar traditions of magic. The Latin text had been largely detached from its origins in Arabic astral magic. It had a small number of similarities with alchemical processes and even fewer with ritual magic texts to conjure spirits.[3] At the same time, as medieval critics and modern scholars both acknowledged, its approach to natural substances and objects was unusual. David Pingree characterized its experiments as 'psychic magic' because (he argued) they relied on body parts and substances in which the soul was believed to reside (human semen and brains, and the hearts, brains, and gall bladders of animals). It is true that the *Liber vaccae* refers to the importance of the imperishable substance of the living soul (*substantia animae*) that can be created or generated by the zealous study of nature, and this seems to be a significant theory behind the generative and other experiments.[4] The experiments also rely on ancient theories of sexual and spontaneous generation, on the vivifying power of certain actions such as animal sacrifice or immersion in running water, and on simpler relationships of sympathy and antipathy.[5]

Maaike van der Lugt calls the procedures of the *Liber vaccae* 'organic magic', referring to their basic ingredients, 'illusionist magic' because of their theatrical quality, and 'alchemical' for their concern with transformation and generation.[6] The term 'organic magic' is useful to distinguish the *Liber vaccae*'s more complex approach to matter and generation and its extraordinary goals, from the usual scholastic understanding of natural magic texts describing the properties of stones, plants and animals.[7] Although scholastic critics of the *Liber vaccae* were divided on the question of whether its experiments would be successful, these were not usually interpreted as working with the aid of demons because of the absence of the usual indicators (invocations, images, inscribed names and figures) of rituals to conjure spirits. Instead this text's extraordinary interventions into nature, such as creating new rational and irrational animals, influencing the weather and making animals submit to the

3 On Arabic astral magic and the *Liber vaccae*, see the *Picatrix*, Book 2, Ch. 12, ed. by Pingree, pp. 88–89 and Saif, 'The Cows and the Bees', and on its relationship to alchemical traditions, Newman, *Promethean Ambitions*, pp. 177–81 and 190–91.

4 Page, *Magic in the Cloister*, pp. 63–65.

5 Page, *Magic in the Cloister*, pp. 53–54 and Van der Lugt, 'Abominable Mixtures', pp. 240–42.

6 Van der Lugt, 'Abominable Mixtures', p. 239. On 'illusionist magic' see also Kieckhefer, *Forbidden Rites*, pp. 42–68.

7 On natural magic see especially, Draelants, 'The Notion of Properties'.

practitioner, provoked discussion about how far demons could use their own natural knowledge to perform such feats.[8] In the fifteenth century, however, witchcraft literature brought the ideas of human and demonic intervention in nature together. Witches were accused of mixing natural substances to perform harmful magic with the aid of demons who were experienced in scientific pursuits.

1. The *Liber vaccae* and Censorship

The starting point for this paper is the fifteenth-century censorship of the copy of the *Liber vaccae* in Oxford, MS Corpus Christi 125, which led me to a consideration of the text's association by some medieval readers with *maleficia* (harmful magic), necromancy, and witchcraft. Corpus Christi 125 is a collection of practical recipes, medical, magical, and alchemical texts given to the library of St Augustine's Abbey, Canterbury, in the early to middle fourteenth century by the monk Thomas of Willesborough.[9] The inclusion of two recipes in code in this manuscript and a condemnation of some kinds of magic on a flyleaf suggest that the provocative nature of some of its contents was already apparent to its scribes and readers.[10] The copy of the *Liber vaccae* in the manuscript was censored, by means of physical interventions at some point between the late fourteenth century, when a monk at St Augustine's rewrote corrupted magical materials, and the mid-sixteenth century, when it was acquired by the mathematician, natural philosopher and occult practitioner John Dee.[11]

The method followed by the censor, who I think is likely to have been a fifteenth-century monk of St Augustine's, is consistent.[12] Only certain experiments were censored, and in each case only the first one or two ingredients were erased.[13] This rendered the experiments inoperable. In some cases the

8 Van der Lugt, 'Abominable Mixtures', pp. 261–64.

9 Oxford, MS Corpus Christi 125, fols 123^{v}–42^{r}. The main hand of the *Liber vaccae* is *c.* 1300. A second hand rewrites parts of this text and others at the end of the fourteenth century, probably because the pale ink on dark parchment was becoming more difficult to read as it faded. The censorship is applied to the rewritten text so must postdate the second hand. See Page, *Magic in the Cloister*, pp. 12–15.

10 Page, *Magic in the Cloister*, p. 24.

11 On John Dee's ownership, see Page, *Magic in the Cloister*, pp. 133–39.

12 I make this assumption because the censorship postdates the hand that rewrites portions of the text (see footnote 9) but is unlikely to have occurred after John Dee's acquisition due to his embrace of occult activities. The approach of careful reading and identification of challenging experiments but leaving others untouched suggests a reader who values other contents in the text and indeed the manuscript itself. The actions of this scribe may even represent a pre-emptive strike against cruder censorship as I have argued is the case for the recording of a condemnation of magic on a flyleaf of this manuscript. See Page, *Magic in the Cloister*, p. 24.

13 This approach is comparable to the erasing or striking through of spirit names to render a

ingredients are still legible under the erasure, but illegibility was not always the goal of manuscript censors. One reason the ingredients remained legible was to allow more sophisticated readers to collect the occult materials and critique them; not all souls were thought to be equally susceptible to the dangers of magic.[14] In this context I use the term 'censorship' to refer to a scribal intervention to erase content viewed as objectionable in order to make it illegible to a reader. Magic texts can of course be subject to other types of censorship, such as the omission or alteration of words, rituals, and whole experiments in the course of translation.[15]

The censorship of occult items in medieval manuscripts, that is the correcting of a text to make it more morally acceptable to the censor, is easy to identify but notoriously difficult to date. It can take the form of rubbing, scratching, scribbling or blotting out all or part of the offending material, striking it through with horizontal or diagonal lines to express the opprobrium of the censor, adding crosses or marginal comments against erroneous teachings, cutting offending folios out of a manuscript, or omitting parts of a text when making a new copy.[16] As we shall see, several copies of the *Liber vaccae* were censored.[17] In the case of Corpus Christi 125 the censorship may have been directed by a scrupulous abbot or librarian, or may express the disapproving piety of a private individual who found the volume in the abbey library.[18] Although most medieval critics of the *Liber vaccae* condemned the experiments for creating rational and irrational animals or for performing pseudo-miracles, this censor targeted experiments with morally unsuitable ingredients (a funeral cloth, human blood or body parts), and those linked to necromancy, *maleficia* or witchcraft (a black cat, a hoopoe, bat's blood, a wolf skin).[19] Other targeted experiments have provocative instruments (an

ritual magic experiment inoperable. For examples of this censorship see BL, MS Arundel 342, fol. 83ʳ (experiments for conjuring spirits and divination) and London, Wellcome, MS 517, fol. 79ᵛ (a love magic experiment).

14 See for example the note added to the image magic text entitled the *Glossulae* in this manuscript (fol. 110ᵛ): 'whoever you are who has found these <words>, I ask through Christ that you do not reveal them unless by chance to a good and benevolent man, and if you do the contrary, may your soul be imperilled and not that of the writer. Amen', quoted from Page, *Magic in the Cloister*, 146.

15 A good example of this kind of censorship is the omission of experiments promoting same-sex love and the cutting or adaptation of Islamic religious references when the Arabic magic text titled the Ghāyat al-ḥakīm, was translated into Latin. Boudet, 'L'amour et les rituels à images d'envoûtement'.

16 On censorship in manuscript culture see: Kerby-Fulton, *Books Under Suspicion*; Olsan, 'The Marginality of Charms'; Camille, 'Obscenity under Erasure'.

17 In addition, three non-extant copies referred to in medieval catalogues or book lists may have been deliberately removed and destroyed.

18 At least one librarian, Clement Canterbury (fl. 1463–1495), showed interest and possibly enthusiasm for volumes with occult contents however: Page, *Magic in the Cloister*, p. 15.

19 Oxford, MS Corpus Christi 125, fol. 125ʳ. The ingredients that are either readable under the erasure or can be recovered from other non-censored copies are: in Book 1, 35, a firefly;

image of Satan) or actions (eating something made with human blood), or involved harm to the practitioner or the harmful manipulation of others. Finally, the number of experiments with psychoactive or toxic ingredients targeted suggests that the censor may have had some medical knowledge that enabled him to identify these as harmful.

I turn now to individual censored experiments. The Corpus Christi 125 copy of the *Liber vaccae*, in common with eleven other copies, travelled with and under the title of *De proprietatibus*, a Latin translation of the *Kitāb al-khawāṣṣ* (Book of Properties) of Ibn al-Jazzār. Although the positive reception of the *Liber vaccae* was undoubtedly helped by its apparent merger with this more mainstream medical text on the use of animal parts, in Corpus Christi 125 the first censored experiment is from the *De proprietatibus*. The dead man's tooth that the *De proprietatibus* advises should be suspended over a man who is suffering stomach pain is erased. The second censored experiment, number 18 from the *Liber vaccae major*, is to make everyone who enters a house suffer epilepsy. The effect is provoked by giving the victim — or possibly willing participant since neither deceit nor force is suggested — a drug (*medicamen*) mixed into an eye salve. As part of the experiment the operator makes a large image of Satan with golden marcasite (and other unnamed materials). In this case the experiment was probably censored at an early point in its circulation, since no surviving copies include the missing ingredients. Experiment 29 explains how to make an eye lotion that will turn its user invisible. This was censored too, probably because it directs the use of the blood of a ritually beheaded hoopoe and a black cat beheaded in the hour of the dead. These animals and ritual actions were common in necromancy.[20]

Of the two books of the *Liber vaccae*, the second, on apparently more innocuous parlour tricks and small-scale experiments, attracted our censor's attention more forcefully. Ingredients are erased from the first two experiments (on how to plant flowers and fruit trees that will immediately grow from seeds to full size). The first uses blood drawn from the human body (*sanguis phlebotomia*) and ends with a warning (after the practitioner has been advised to offer the marvellous fruit to guests): 'Cave tibi ne comedas ex eo aliquid propter consientiam tuam, quia est ex sanguine humano' (Be careful not to eat it yourself on account of your conscience, because it has been made with

in Book 2, 2, hemp seed (*sedenigi*); 9, quicksilver; 13, the fat of a black snake; 15, chalk and balsam; 20, the blood of a hare and a male bird like a turtle dove; 23, *malvaviscus albus* (a flowering plant in the mallow family); 27, the blood of a bat; 33, a funeral cloth; 35, wolf skin (erased but later rewritten).

20 Oxford, MS Corpus Christi 125, fol. 135^{v}. Munich, Bayerische Staatsbibliothek, MS Clm 849, fol. 40^{v}, has an experiment for invisibility that instructs the practitioner to eviscerate a black cat, place heliotrope seeds in its eyes and bury it. He then waits until a magical plant grows on this spot and eats its seeds to turn himself invisible. See Kieckhefer, *Forbidden Rites*, p. 240. See also Oxford, Bodl. Lib., MS Ashmole 1435, fol. 25^{r}, for a similar experiment for invisibility where the buried animal is a dead dog.

human blood).[21] The ingredient erased from the second experiment is the psychoactive hemp seed. Ingredients are also erased from: experiment 9 (a wick that makes people appear black), which uses human blood, mercury, and a funeral cloth; 13 (a lamp that makes a house appear full of snakes), which uses a funeral cloth; and 15 (a wick that remains lit when placed in water) that employs sulphur.[22] Sulphur powder produces an acrid smell when burnt and the fourteenth-century French bishop and critic of magic Nicole Oresme (*c.* 1320–1382) accused magical practitioners of using it to confuse and debilitate the minds of their victims.[23]

Experiment 20 (a wick to make women dance in a frenzied way) may have attracted scrutiny because of its use of animal blood or because it causes people to lose control of their bodies. Women under its influence are said to dance exultingly, rending their clothes and acting like demoniacs.[24] Experiment 23 (a wick that allows the maker to ignite their body with fire from head to toe without feeling any pain) is potentially harmful to the body and like experiment 15 includes the use of sulphur. Experiment 27 (to see something hidden in the night) involves killing a bat and smearing its blood on the user's face. Experiment 33 (a wick that makes your home green like the forest) requires a recently used funeral cloth (*pannum exequiarium recentem*); and experiment 35 (a wick that never goes out) needs a wolf skin. Experiment 38 (to make people see each other in the form of Satan) is labelled 'fantasticum capitulum' in the margin and unsurprisingly also has erased ingredients.

Before examining the implications of this reading of the *Liber vaccae* through the lens of necromancy and *maleficia*, I will briefly survey and compare other examples of selectiveness, censorship and creative rearrangement among the fourteen known copies of this magic text. First, it is worth noting that experiments from the *Liber minor* achieved early circulation in the *De mirabilibus mundi* (Book of the Marvels of the World) which incorporated thirty of its forty-one experiments. The *De mirabilibus mundi* was a collection of experiments to produce wonders, that was probably written between 1223 and 1273 in Paris. It survives in only seven medieval manuscripts, but was also printed several times in Latin and vernacular translations.[25] The author places its experiments within a discussion of philosophical and scientific theories that emphasise the power of the imagination and the sympathetic correspondence between things. The *Liber vaccae* is treated cautiously. The excluded experiments include the first three, to make seeds grow quickly, and experiment 39, to make a lamp that will create the illusion that people

---

21 Oxford, MS Corpus Christi 125, fol. 139r.

22 I have translated *licinium* as 'wick', in the sense of a mixture (or bundle of fibres steeped in a mixture) that is supposed to be flammable and used by the practitioner in a lamp.

23 Oresme, *Tractatus de configurationibus*, II, xxxii.

24 Oxford, MS Corpus Christi 125, fol. 140v: 'scident vestimenta sua … et estiment qui vident eas quod sint demoniace ex vehementia earum et tripudii'.

25 Sannino, 'De mirabilibus mundi'.

in the audience will take the form of Satan. Selected experiments focus on lamps and fire, inks and suffumigations and have been revised so that they have more neutral, less dramatic language. For example, in the *De mirabilibus mundi* version of *Liber major* experiment 20, to make women leap and dance excitedly, there is no mention of demoniacs.[26] The *De mirabilibus mundi* adds that the experiment was proved by experience, a rather plausible claim given that the practitioner is advised to light the magic lamp in the middle of a room where singers and girls have already gathered.

Turning now to the manuscript tradition, we find three surviving copies of the *Liber vaccae* from the period 1280–1300 that bear the marks of critical scribal scrutiny.[27] As Maaike van der Lugt has pointed out, copies from this period tend to be compiled with medical works, which may be attributed to the fact that in all these copies the *Liber vaccae* is preceded without a clear break by the *De proprietatibus*.[28] In two cases scribes were clearly perturbed by the shift in the combined text from medicine to magic, and omitted all or most of the *Liber vaccae* following its prologue. A copy from Windberg Monastery, Bavaria (Munich, MS Clm 22292, fol. 70^{v}–71^{v}) contains part of the prologue and book 1, experiments 15 and 16 only (to stop rain and to make someone temporarily die when they enter a house). Experiment 16 is incomplete — the folio on which it should continue and the folio after this (the final two in the quire) have been cut out of the manuscript, so perhaps the original scribe did record a selection of experiments before they were censored. The scribe of Innsbruck, Universitäts- und Landesbibliothek Tirol, MS 489 (fol. 64^{r-v}), who was probably from southern Germany or Tyrol, kept only the first part of the prologue (up to 'bonus effectus est casus eius'), at which point its end is marked by a small cross that may signify either the pious action of the scribe in refusing to copy any more of the text or the sign that it contained materials inappropriate to a good Christian.[29]

A longer, but also cautious, selection of experiments is found in the copy of the *Liber vaccae* in Munich, Clm 615 (fol. 103^{r}–108^{v}), possibly from Germany. This selection omits the generative experiments 1–3 and experiments 18 (to induce epilepsy), 20 (to make a house of gold in which people entering look

26 Oxford, MS Corpus Christi 125, fol. 140^{v}: 'non cessabunt saltare gaudendo et scindent vestimenta sua ex gaudia ita ut admiratio fit ex eis et extiment qui vident eas et sint demoniace ex vehementia earum et tripudii'. *De mirabilibus mundi*, ed. by Sannino, p. 141: 'non cessant saltare et gaudere et insanire gaudio'.

27 A fourth copy is not discussed here because it has a complete selection of experiments up to experiment 39: Yale Medical Library, MS codex Fritz Paneth (fols 390^{v}–406^{r}), copied in Bologna *c.* 1300.

28 Maaike Van der Lugt discusses the manuscript context of the *Liber vaccae* in 'Abominable Mixtures' pp. 243–49. Her valuable table at pp. 275–76 lists manuscripts and non-extant catalogue references with brief notes on their date, shelfmark, title, the (in)completeness of the text, and other contents of the manuscript.

29 On the use of the sign of the cross to indicate erroneous teaching, see Kerby Fulton, *Books under Suspicion*, pp. xxv and 165.

gold) and 21 (to know the future) from book 1. The latter may be excluded because it requires the operator to find the teeth, nose, and bones of a dead person, but it is likely that divining the future is also under scrutiny. In an approach similar to the censor of Corpus Christi 125, the scribe of this manuscript sometimes copies the bulk of the experiment but omits a few crucial elements. So experiments 29 (how to give men the form of apes) and 30 (how to be invisible and predict the future) are rewritten in such way that the experiments cannot be properly carried out. The uncontroversial goals of the remaining selection of experiments are to create extraordinary spectacles and effects, perform weather magic, make a tree incline towards you and make a marvellous lamp so that people appear to have green faces.[30]

In the fourteenth century, copies of the *Liber vaccae* were more likely to be compiled with other magic texts, reflecting the fact that knowledge of learned magic was now disseminated more widely. These copies were probably deliberately included by scribes with occult interests rather than being accidentally inserted into medical compilations. In British Library, MS Arundel 342 (mid-fourteenth century, Italian provenance) the *Liber vaccae* is compiled with texts on natural magic, image magic, and lapidaries, but the scribe still omits experiments 16–37 (from the experiment to make someone temporarily 'die' upon entering a house to a passage on Sun worshippers). It has the same lacuna in experiment 23 (*malvaviscus*) as Corpus Christi 125, and also omits book 1, experiment 40 and a key ingredient from experiment 41.[31] There are signs of censorship throughout this manuscript, so the omissions in book 1 are probably deliberate rather than a result of the scribe working from a defective copy. The spirit names and goals of two experiments — one to conjure spirits, the other crystallomancy with the aid of a boy skryer — are erased to prevent the rituals from being carried out.[32]

It is important to note that the same magical experiments to confuse, harm and terrify people that provoked qualms in some copyists could attract others, revealing another reason for the appeal of this magic text. The German *hausbuch* Nuremberg, Germanisches Nationalmuseum, MS 3227a is a compilation of texts chosen for their usefulness in combat situations or other challenges of the knightly milieu. It was made between 1389 and 1484, and written in Latin and an East Central German dialect. It combines texts on fencing and grappling with magical experiments on military themes, such as

30 The copy breaks off in the middle of Book 2, experiment 10.

31 The flowers and leaves of the mallow plant had multiple medical uses in the Middle Ages. It is possible that this ingredient was omitted in some copies because it was difficult to identify. Alternatively, a scribe or manuscript owner with medical experience might be aware that the seeds can be poisonous if consumed in large quantities. This plant also appears in an experiment to hold fire in your hands without burning them that was included in the *De mirabilibus mundi*. See the edition by Sannino, pp. 145–46. This text is discussed below.

32 BL, MS Arundel 342, fol. 83^{r}.

conjuring an illusory army (fol. 5^v) with the aid of the demon Astaroth.[33] The excerpts from the *Liber vaccae* focus on terrifying public spectacles, unnerving domestic illusions, and physical transformations that were presumably selected for their military potential. This includes the experiments to become invisible (experiment 3), to create the appearance of battling armies in the heavens (5), to make the moon and stars appear in daytime (6), to make the moon appear split in two (7) or eclipsed (8), and to create apparitions in the air of men and beasts (9), giants (10) and huge forms (11) that will stupefy observers.

The selection then skips over weather magic to the experiments that make those who enter a house fall down as if dead (experiment 16), become terrified by a vision of the sun at night time (17) or think that they are in a house on fire or made of gold (20). Experiment 17 uses psychoactive and harmful ingredients (mandrake root and yellow arsenic) which probably were highly effective: even suffumigated in small quantities they are likely to provoke extraordinary and frightening visions. While the illusory house of gold (experiment 20) may not sound terrifying, its potential to distract or trick an enemy is shown by the evocative remark that it will capture the gaze like a hook (*domus sicut hamus capit visum*). After this experiment there are chapters to make a tree incline toward you (experiments 21–23), to understand birds and know hidden things (24, 25, 27), and to see spirits and demons (32 and 33 in reverse order). A later scribe added a new experiment at the bottom of the final folio, which fits the goals of the *Liber vaccae major* very well: to see men whose heads touch the heavens while their feet remain on earth.[34] This mingling of military and magical interests is found most famously in Conrad Keyser's popular manual *Bellifortis*, which included lamp experiments from the *Liber vaccae*, instructions to produce explosive effects from Marcus Graecus's *Liber ignium*, and experiments to raise spirits.[35]

In the fifteenth century, magic had two quite different trajectories. On the one hand, magic texts were reaching ever wider audiences through vernacular translations, and the appeal of learned magic to readers from the court to the cloister meant that many condemned texts circulated widely under the radar. Some magic texts had 'real' authors, not pseudonymous ones like Hermes or Solomon; and in Italy humanist currents of thought offered philosophical justifications for the human capacity to manipulate the universe.[36] On the other hand, magic was condemned with increasing vigour and precision in ecclesiastical sources. Inquisitors' investigation of magical practices and the development of the concept of the demonic pact widened the scope of

---

33 In other magic texts this demon is linked to love magic.

34 Nuremberg, Germanisches Nationalmuseum, MS 3227a, fol. 152^r: 'Ut videas homines qui tangent celum cum capitibus et terram pedibus'.

35 Conrad Kyeser, *Bellifortis*, ed. and trans. by Quarg.

36 On these currents in the history of magic see, for example, Weill-Parot, *Images astrologiques*, pp. 602–38.

persecution and contributed to clerical theorising about witchcraft.[37] The organic magic of the *Liber vaccae* shared features with the licit scholastic category of natural magic but also with *maleficia* or harmful magic, a category that became a key component of witchcraft mythologies.

The fifteenth-century copies of the *Liber vaccae* reflect these two diverging trajectories. They were compiled with other magic texts in manuscripts, mostly with diverse occult works and sophisticated theoretical treatises. It is therefore not surprising that some of these copyists show more interest in the complex experiments of the *Liber vaccae major* than the smaller-scale magic of the second book, and little inclination to censor or omit experiments. The copy in Vatican, MS Pal. Lat. 1892 (Southern German provenance) is compiled with texts on medicine, magical characters, astrology, chiromancy, and scholastic treatises on the body and soul.[38] The *Liber major* is complete, but the *Liber minor* stops partway through the fourth experiment in book 2, presumably because the more domestic scale of the second book was less appealing. In Cambridge, MS Corpus Christi 132 the *Liber vaccae* is compiled with a variety of short texts with religious, astrological, and medical contents. The *Liber vaccae major* is complete but it ends abruptly at book 2, experiment 10.[39] Two further fifteenth-century copies are complete, with the exception of the usual missing ingredient in experiment 18, reflecting the positive reception of learned magic in this period.[40]

Some fifteenth-century scribes showed more caution, however. Selective copying from the *Liber vaccae major* is apparent in Paris, Bibliothèque nationale de France, MS lat. 7337, which has only the experiments to influence the weather (12–15), and those to have wonderful effects on the body and nature: experiments to turn invisible, to make trees and beasts incline to the practitioner, to create inextinguishable lamps, to transform the appearance of people and to see spirits (21–33). The copy in Oxford, Bodleian Library, MS Digby 71 (fols 36^r–56^r) omits book 2, experiment 20 (to make women dance), and by omitting the title of book 1, experiment 40, turns it from 'a suffumigation that has the power to make the soul good' into a simpler experiment for 'attracting joy' and setting aside sadness.[41] The scribe of the copy of the *Liber vaccae* in Leiden, Universitatbibliotheek, MS Vossius Chym Q 60, of German provenance, censored the experiments on making rational animals.[42] The rubricated title 'Ad faciendum animal rationale' and

37 See, for example, Boureau, *Satan the Heretic*.

38 BAV, MS Pal. Lat 1892, fols 103^v–119^v.

39 Cambridge, Corpus Christi, MS 13, fols 139^r–66^r.

40 In Florence, Biblioteca Nazionale Centrale, MS 2. 3. 214 the *Liber vaccae* is compiled alongside astral magic, magic and astrology. Montpellier, Université Faculté de Médecine, MS 277 owned by a Venetian physician or scribe, is a collection of texts on the use of herbs, stones and animal parts.

41 Oxford, Bodl. Lib., MS Digby 71, fol. 36^r.

42 Leiden, Universitatbibliotheek, MS Vossius Chym Q 60, fols 52^r–64^r.

four lines have been erased, removing the direction to the practitioner to mix his own semen with an equal amount of the stone called the stone of the sun. Instructions on what to do with this mixture are also erased. There are similar erasures of the title and first five lines of the second experiment to create a rational animal using a female ape. These erasures relate to the timing of the experiment and the direction to hold the ape, perhaps because this is expressive of an inappropriate intimacy between man and beast. A folio is also cut out of the manuscript between folios 58 and 59, so that the text omits most of the experiments between 18 (to induce epilepsy) and 27 (to see something hidden in the night).

Finally, a sixteenth-century manuscript illustrates how the organic magic of the *Liber vaccae* could be a good fit for ritual magic interests because of its claims to give the practitioner extraordinary powers and access to the world of spirits.[43] The 'Experimentum de upupa' in a compilation of ritual magic written in England between 1532 and 1558 uses many of the same generative techniques and creatures (a fly like a bee, worms, a hoopoe) as the *Liber vaccae* and a related magic text, the *Liber Theysolius*.[44] In the sixteenth-century experiment, the hoopoe is to be ritually killed and its blood collected in a vessel which is put in a secret room for nine days. When the vessel is re-opened the blood will be seen to be full of worms, and it should be left for another nine days. After this the practitioner will find a fly like a bee, which has devoured all the worms; he is instructed to place it inside a sphere made of smashed nuts and fossilized stone. The next time he opens it there will be a bird with the likeness of a hoopoe, which he should then roast, collecting its fat to make an ointment. When this ointment is placed on the eyes of a person, they will be able to see and converse with spirits. This version of an Arabic generative experiment to produce an eye salve to see spirits is unlike the other necromantic rituals with which it is collected because these focus on the power of written or invoked words and characters to conjure spirits. But the hoopoe was a typical instrument of necromancy, and a note in a sixteenth-century German manuscript also makes the connection to necromancy, referring to the *Liber vaccae* as *nigromanticus*.[45]

---

43 Cambridge, Cambridge University Library, MS Additional 3544, fols 56^{r}–8^{r}, ed. Young, *The Cambridge Book of Magic*, pp. 52–54, and see discussion on pp. xxxiii–xxxiv.

44 Experiments 28 and 32 in the *Liber vaccae* and operations to compel spirits to vivify a dead body in the *Liber Theysolius* also include hoopoe blood, incubatory vessels that produce new forms, multiplying worms and ointments to see spirits. On these two texts and their relationship see Page, 'Magic and the Pursuit of Wisdom'.

45 Van der Lugt, 'Abominable Mixtures', p. 249.

2. The *Liber vaccae*, *Maleficia* and Female Practitioners of Magic

The first medieval critic to link the *Liber vaccae* with *maleficia* (harmful magic) was William of Auvergne. His interest, as Maaike van der Lugt has discussed, was in the early generative experiments that involved cross-breeding or mixtures (*conmixtiones*) of animals against the laws of nature (*contra leges natura*).[46] According to William, the offspring of these experiments (after being ritually killed and their body parts or substances extracted) were used for nefarious works and harmful magic (*nefanda opera et maleficia*). In the *Liber vaccae* the organs, body parts and blood of the extraordinary offspring are said to contain the power to create an illusion of the full moon, to bring rain, to transform people into apes, pigs or sheep, to walk the length and breadth of the earth in the blink of an eye, to see, hear and speak to spirits, to animate or make a tree fertile, to have immunity to weapons, to become invisible, and to cause serpents to spontaneously generate. Although these powers are not explicitly harmful, they were frequently associated with ambiguous female practitioners of *malefica* in literary texts, as I will demonstrate below. William's contemporary in Paris, the Dominican theologian and later inquisitor Roland of Cremona, made an explicit connection between one of these powers and female magical practitioners. Referencing the *Liber vaccae*'s description of an ointment to make the user invulnerable to weapons, he argued that if this could be made by philosophers or demons, then it might also be possible for women to make an ointment that would enable them to fly.[47]

From Late Antiquity, the term *maleficia* (literally, 'evil practices') described practices of magic deemed particularly evil. While it is undoubtedly a broad and flexible term, when linked to magic it usually had the connotation of making potions, poisons or ligatures with the intention of killing a victim, driving them mad, or provoking them to love or hatred. In the *Etymologiae* (*c.* 630), one of the lengthiest early medieval discussions of magic, Isidore of Seville stated that *maleficii* was the name given to magicians (*magi*) by the populace (*vulgus*) when their crimes were particularly serious. Isidore's characterisation of magic is rooted in ancient texts and female practitioners, notably Circe, Erichtho, the Witch of Endor, and the unnamed Massylian *sacerdos* who helped Dido.[48] Their goals were to shake up the elements, disturb human minds, and kill with the violence of their spells.[49] They are also said to summon demons, use blood and sacrifices, and work with the bodies of the dead. In the twelfth and early thirteenth centuries, *maleficia* continued to

46 William Auvergne, *De legibus*, Ch. 12. See Van der Lugt, 'Abominable Mixtures', pp. 256–57.

47 Roland of Cremona, *Summa*, II, Dist. 8. See Van der Lugt, 'Abominable Mixtures', pp. 263–64.

48 *Etymologiae*, VIII (*De ecclesia et sectis*), Ch. 8.

49 For contemporary laws mentioning *maleficii*, see Klingshirn, 'Isidore of Seville's Taxonomy'.

be associated with the trickery of demons and use of blood on the one hand, and with love and hatred on the other.[50]

This characterisation of magic was reflected in contemporary literature, particularly the romances influenced by classical sources (*romans d'antiquité*), which described female practitioners of magic (*malefica*) who drew on natural sources of power, produced extraordinary effects in nature and specialized in disturbing the minds of men. Morgan le Fay from Geoffrey of Monmouth's *Vita Merlini* of *c.* 1150 lives on an extraordinarily fertile island where she has learnt the art of healing and how to transform into a bird. Medea from Benoît de Sainte-Maure's *Le Roman de Troie* (*c.* 1155–1160) can perform weather magic, transform into a bird and subvert the flow of water. She has healing skills, giving Jason an ointment to protect his body from fire, but also conjures up snakes, prepares poisons, pours her own blood and that of animals onto the altar and puts a concoction of ingredients into the dead body of Aeson to make him appear alive. The sorceress (*sorciere*) who advises Dido in the *Roman d'Enéas*, composed in Normandy *c.* 1160, can revive the dead, predict the future, cause celestial spectacles, and make birds speak, trees walk and water flow uphill.

When twelfth- and early thirteenth-century romance authors adapted the female sorcerers of classical literature to their new literary forms and audiences, they generally retained the emphasis on harmful magic, but also used female magic workers to move the plot forward in positive ways. For example, the three sorceresses (*sorcieres*) in the anonymous *Amadas and Ydoine* (*c.* 1190–1220) help the heroine Ydoine preserve her virginity in an unwanted marriage to the Count of Nevers.[51] Their powers are described as follows:

> Qu'eles sevent de nuit voler / Par tout le mont, et de la mer / Faire les ondes estre em pais / Comme la tere, et puis après / Defors de la graine venir / Arbres, naistre, croistre et florir, / Et sevent par encantement / Resusciter la morte gent, / Des vis l'une a l'autre figure / Müer par art et par figure, / Houme faire asne devinir, / Et ceus qu'il voelent endormir / Et puis songer çou que leur plaist, / Bestes orgener en forest, / Murs remüer et trembler tours, / Et les euwes courre a rebours. / Ne puis pas dire ne conter / Le disme part, ne reconter, / Qu'eles sevent de mauvais ars: / De sage font sos et musars; / Tant ont grant sens mult m'esmervel.
>
> > (They could fly across the whole world at night, make the waves of the sea as peaceful as the land and cause trees to grow and flower straight from seeds. They could resurrect the dead by magic, transform people's

50 See for example John of Salisbury's *Polictraticus* (*c.* 1159). For examples from England and Italy, see: Rider, *Magic and Religion*, pp. 89–108; Montesano, *Classical Culture and Witchcraft*, pp. 99–131.

51 *Amadas and Ydoine* survives in only one complete manuscript, a Picard transcription of an Anglo-Norman model.

> appearances by spells and artifices and turn a man into an ass. They could put anyone they wanted to sleep and make him dream what they wanted. They could charm beasts in the forest and make walls shake, towers tremble and rivers run upstream. I could not describe even a tenth of the evil arts they knew! They turned wise men into fools and babblers: I am astonished at how much they knew.)[52]

The three sorceresses help Ydoine by using their 'evil' magic arts to enter the count's room, bypassing doors and bolts, and trick him into believing they are the three Fates. One of the ways they persuade him of their powers is by speaking at length about things that have happened to his kinsmen, peers and neighbours. This narrative of revealing hidden knowledge acquired by magic and then pretending to have other supernatural powers resembles the strategy proposed in a *Liber vaccae* experiment for invisibility. The practitioner is told to secretly enter the houses of men to watch them eating and drinking and to listen to their private conversations. When he later reveals what the men ate and said, they attribute extraordinary powers to him. At this point he can either say, 'Creator misit me cum angelis spiritualibus' (My Creator sent me with angelic spiritualities) or 'demon narravit mihi illud' (a demon told me this).[53]

The significance of these literary representations of female practitioners lies in their use of natural knowledge rather than ritual actions, in their ability to temporarily subvert the normal course of nature and in the morally ambiguous goals of their magical activities. All these characteristics (with the exception of making people love or hate one another) are found in the *Liber vaccae* too, making it unlike most learned magic texts from the Arabic, Jewish and Greco-Roman traditions circulating in Latin after the translating movement of the twelfth and thirteenth centuries, that drew more explicitly on the power of stars, celestial spirits, words invoked and inscribed, and the making of talismans.

Nicole Oresme was one of the earliest medieval thinkers to bring *maleficia* into a serious discussion of the art of magic. He links it to the *Liber vaccae* experiments and female practitioners in the classical tradition. His *Tractatus de configurationibus qualitatum et motuum* (Treatise on the Configurations of Qualities and Motions) is addressed to a general learned audience, and warns readers that magical practitioners skilled in the art of deception cause harm to themselves and others. Oresme ranges widely across the magicians' methods of deception, including their manipulation of an audience's senses and emotions using psychoactive and toxic substances, sounds and smells and their creation of dramatic shifts in light and darkness, menacing atmospheres, and tricks with mirrors to create terrifying and discombobulating illusions.[54]

52 *Amadas and Ydoine*, ll. 2023–43.
53 Oxford, MS Corpus Christi 125, fol. 158^{r}.
54 Nicole Oresme, *Tractatus de configurationibus*, II, xxvi–xxxii.

Oresme argued that magic was the art of making certain things appear which seemed to be impossible, using natural means and human artifice rather than the aid of spirits and the performance of rituals. One of the ways he thought practitioners achieved this was by using natural substances (plants, roots, seeds, semen and poisons) that had the power to alter people's minds (*mentes hominum immutare*).[55] When these substances — singly or in compounds — were consumed, anointed or suffumigated, they made people unable to recognize the correct colour, figure, motion or characteristics of things or the medium through which they were perceived.

Oresme calls this manipulation of natural substances to harmful effect *veneficia* or *maleficia*, and cites the *Liber vaccae* as a text that instructs in how to create these 'abominable mixtures and improper applications' (*mixtionibus abhominandis et applicationibus abusivis*).[56] His examples of practitioners are predominantly women from classical literature: Dipsas, an old witch and brothel keeper from Ovid's *Amores* (16 BC), Erichtho, the elderly Pharsalian sorceress from Lucan's epic poem the *Pharsalia* (61–65 AD) and the landlady in Augustine's *City of God* who turns unlucky travellers into packhorses.[57] Unsurprisingly, Oresme's characterisation of the practices of these women is extremely negative, but it emphasizes natural knowledge rather than the assistance of demons. When he mentions contemporary female practitioners, it is to state his belief that some old women can exercise the power of the evil eye, a natural ability to cause harm.

In Ovid's account Dipsas uses incantations, conjurations, natural ingredients (notably animal afterbirths to make poisons) and her own will to perform extraordinary feats of magic.[58] She can turn back a water's course, perform weather magic and create strange meteorological phenomena, such as the stars appearing to rain blood. She is rumoured to take the shape of a feathered creature at night, gives people the evil eye and can raise the dead.[59] Erichtho's primary role in Lucan's *Pharsalia* is to raise a soldier from the dead to make a prophecy about the outcome of a battle for Pompey. She vivifies the corpse by inserting into it a mixture of unpleasant natural ingredients that range from hot blood and gore to poisonous herbs, marrow, eyes, and ingredients linked to serpents, dragons, and other animals. Lucan sums up these ingredients as including 'all that Nature inauspiciously conceives and brings forth'.[60] Oresme says that practitioners of magic like Dipsas and Erichtho are *venefici* or *malefici*

55 Oresme, *Tractatus de configurationibus*, II, xxxi, ed. by Clagett, p. 356.

56 Oresme, *Tractatus de configurationibus*, II, xxxii, ed. by Clagett, pp. 358–59. Although the reference to the *Liber vaccae* only appears in three manuscript copies, Maaike Van der Lugt argues that it is an early addition by Oresme himself: Van der Lugt, 'Abominable Mixtures', p. 255.

57 Oresme, *Tractatus de configurationibus*, II, xxvi.

58 Ovid, *Amores*, I, 8, ll. 7–8.

59 Ovid, *Amores*, I, 8, ll. 1–18.

60 Lucan, *Pharsalia*, VI, ll. 669–70.

because they aim to poison the minds of men. Interestingly, Oresme's final example of this kind of magic refers to a false accusation: the story of the Christian saint Agnes, who was accused by pagan priests of being a sorceress (*maga* and *malefica*) who altered mental faculties and deranged minds after she converted pagans to Christianity, raised the Emperor's son from the dead and survived flames without being burnt. This example demonstrates, from Oresme's scientific perspective, that beholders' interpretation of extraordinary events can be wrong.

The *Liber vaccae* is a magic text in which natural ingredients are used to create things that appear to be impossible, from new living beings to celestial spectacles, visions in houses and the transformation of people into apes. Many of the recipes appear to work on the mind but others influence the body, such as the experiments to provoke epilepsy or disorderly dancing. Many of the goals of the *Liber vaccae* thus fit Oresme's characterisation of *veneficia, maleficia* and abominable mixtures. The experiments include mixtures of natural ingredients, notably animal parts, fat and blood, but also human semen and psychoactive and potentially toxic ingredients: mandrake root, hemp seeds, orpiment, mercury, and sulphur. It is perhaps because the *Liber vaccae* uses medical terms — *medicamen, alcohool* (eye salve) and *unguentum* — that Oresme emphasizes that abominable mixtures aim explicitly at the ill-health of the mind, not the health of the body.[61] This malevolent use of the occult powers of stones, plants, seeds is contrasted with uses designed to help humans live well, by practitioners like physicians, surgeons, and goldsmiths.[62] In the fifteenth century the fear of abominable mixtures applied to improper ends would have widespread currency, but it was no longer considered simply part of the natural knowledge of female practitioners. Instead it was linked to their association with demons who taught or directed the practice of harmful magic as I discuss in the next section.

3. Organic Magic and Witchcraft

The mythologies of witchcraft that started coalescing in the 1430s drew the same connections between harmful organic magic and demonic activity that were made in the Isidorean taxonomy of magicians.[63] Although the mythologies of witchcraft literature and trial accounts varied according to the different cultural and geographical contexts in which they were produced, there are some commonalities concerning the use of natural substances. Many fifteenth-century witchcraft theorists gave material texture to the reality of witchcraft by referring to substances encountered in daily life

61 Oresme, *Tractatus de configurationibus*, II, xxxi.

62 Oresme, *Tractatus de configurationibus*, II, xxxi.

63 Kieckhefer, 'Mythologies of Witchcraft in the Fifteenth Century'.

such as blood, plants, and animal fats (used for cooking, candles, medicinal salves, waterproofing clothes, and cosmetic purposes), but placing them in contexts that were horrific (killing babies), devastating (destroying crops) or fantastical (flying on a stick). Their descriptions of 'abominable mixtures' built on the stereotype of illiterate practitioners of 'common magic' in rural communities who worked closely with natural materials, especially herbs and animal parts. These activities were now interpreted more forcefully as an unnatural manipulation of nature.[64]

Witches were not usually thought to follow the instructions of complex magic texts or to have the power to conjure spirits to do their bidding. Their 'abominable mixtures' were efficacious either because demons (with God's permission) were the true agents behind them, or because they had taught witches about the unpleasant secrets concealed in post-lapsarian nature. Some writers clearly preferred the idea of extraordinary but natural properties to the idea that demons subverted the normal laws of nature directly, but others thought that these mixtures were a ruse by demons to give witches an illusory sense of their own agency. Even writers like Jean Vincent, the Prior of Les Moustiers — who in 1475 argued in a tract against magic that there were extraordinary powers in the stones, herbs and liquids that witches mixed into potions, powders, and ointments to confuse minds (*mentes humanae perturbare*), transform bodies, and harm their neighbours — thought that demons were the most important active agent behind this *maleficia*.[65]

Witches were thought to use their 'abominable mixtures' to harm people directly, to influence the weather or fertility of crops, and to enable extraordinary feats like flight or transformation. The potions, powders, and ointments to kill or otherwise influence people included parts and substances of human bodies, plants, and animals, especially those that were poisonous or had associations with organic magic and necromancy. For example, the *Errores gazariorum, seu illorum qui scopam vel baculum equitare probantur*, an anonymous text written before 1437 by an inquisitor in the Val d'Aosta, describes a deadly ointment made from children's fat and poisonous animals such as snakes, toads, lizards, and spiders.[66] This author also described how a powder made with similar ingredients was thrown into the air on foggy days to kill people and infect them with diseases; and a ritual in which a cat's skin was filled with various fruits, then soaked, dried, pulverized, and thrown into the wind to render agricultural lands infertile.[67]

The idea that witches' mixtures brought down storms and hail on their neighbours' crops and animals is common in sources from the Alpine region,

64 Rider, *Magic and Religion*, pp. 61–69; Kieckhefer, *European Witch Trials*, p. 56.

65 Jean Vincent, *Liber adversus magicas artes*, Chs 10–11, ed. by Hansen, p. 230.

66 *Errores gazariorum*, 5, ed. by Ostorero, Paravicini Bagliani, and Utz Tremp, p. 291.

67 *Errores gazariorum*, 6 and 8, ed. by Ostorero, Paravicini Bagliani, and Utz Tremp, pp. 292 and 294.

the heartland of early witch hunting. For example, in the Zurich canon Felix Hemmerlin's *De nobilitate et rusticitate dialogus* (1444–1450) a nobleman accuses peasant women of mixing together the most poisonous and unclean things (*res materiales venonissimas et immunissimas*) which, together with plants, roots, and the aid of demons were cooked and placed in the sun until the smoke rising from their pots thickened in the sky and produced devastating storms and hail.[68] The *History of the Case, State, and Condition of the Waldensian Heretics (Recollectio casus, status et condicionis Valdensium ydolatrarum)* (1460), an anonymous text linked to the witchcraft trials at Arras, describes a similar action of harmful magic: casting powder into the wind to cause agricultural destruction. The author offers various alternative explanations of its efficacy, leaving the reader to decide whether it was made by the practitioners of *vauderie* (diabolic magic), caused by natural storms, or stirred up by demons.[69]

Finally, mixtures of natural substances, especially in ointments, were thought to enable extraordinary feats, such as flying to Sabbath meetings, becoming invisible, transforming into animals or creating new beings with human forms but insubstantial bodies. The idea that witches flew to their gatherings by rubbing ointment onto sticks and other implements is common in witchcraft literature, though the ingredients of the ointments are rarely described.[70] The *Errores gazariorum* referred to above describes an ointment made from the fat of murdered children and other unnamed ingredients. Johannes Hartlieb's 1456 *Book of All Forbidden Arts* (*Buch aller verbotenen Künste*) calls it a 'horse-brass ointment' (*unguentum pharelis*) and states that it was made from seven plants picked or dug up on special days and mixed with the blood of a bird and animal fat.[71] As a court physician to the dukes of Bavaria and author of a compendium on herbs, Hartlieb probably had particular authority to pronounce on the inappropriate use of natural substances.

There are accounts of witches using ointments to turn themselves invisible and transform into wolves in lay writings and trials conducted by secular authorities in the Alpine region. The chronicle of Hans Fründ, written in Lucerne between 1428 and 1430 and dealing with witchcraft in the Valais, relates how witches used different plants to become invisible.[72] Contemporary trials in this region describe witches transforming themselves into wolves by applying over their whole bodies an ointment made from bones ground into powder.[73] Although Martine Ostorero rightly points to the likely origins of

68 Felix Hemmerlin, *De nobilitate*, Ch. 32, ed. by Hansen, p. 110.

69 *Arras Witch Treatises*, trans. by Gow, Desjardins, and Pageau, Section 4, p. 49.

70 Narratives of women using flying ointments can be found as early as the thirteenth century: Montesano, *Classical Culture and Witchcraft*, p. 112.

71 Johann Hartlieb, *Book on all the Forbidden Arts*, trans. by Kieckhefer, p. 38.

72 Hans Fründ, *Rapport sur la chasse aux sorciers*, ed. by Ostorero, Paravicini Bagliani, and Utz Tremp, p. 36.

73 Ammann-Doubliez, 'La première chasse aux sorciers en Valais', pp. 86–89.

these stories in local folklore, they also resonated with literary traditions for educated authors and readers.[74] In his dialogue on the powers and credibility of witches, the legal scholar Ulrich Molitoris's *De lamiis et phitonicis mulieribus* (1489) cites the stories of Circe giving Ulysses's companions a magic potion (*pocula malefica*) to transform them into wolves, bears, lions, tigers, and pigs, Apuleius's character Lucius turning into an ass after taking a potion, and Augustine's story of the landlady transforming an unwitting man into her packhorse with some doctored cheese.[75] Ultimately, however, Ulrich dismisses these stories, arguing that demons transformed appearances by influencing people's sense organs, imaginative powers or humours.[76]

The phenomenon of changeling children is part of the same medieval fascination with unnatural generation that attracted commentaries on the generative experiments in the *Liber vaccae*. In the trial documents for Jubert of Bavaria recounted by the secular judge Claude Tholosan (1437), the accused is said to work with devils to make mixtures from the bodies of babies, unbaptised children, semen, and menstrual blood. Ephemeral changelings — unlucky children who are demons (*pueros infelices qui sunt demones*) — take the place of those who have been stolen to trick the parents.[77] Jubert's practices of necromancy and *maleficia* included making poisons taken from animals (basilisks, toads, snakes, spiders, and scorpions) and herbs collected with the invocation of his devils to infuse them with power.

Although direct influences cannot be demonstrated, the *Liber vaccae*'s 'organic magic' provides a plausible context for one strand of influence on the unpleasant 'abominable mixtures' with extraordinary effects that are common motifs in witchcraft literature. Moreover, we know that some readers of the *Liber vaccae* — scholars like Roland of Cremona and Nicole Oresme and the censor of Corpus Christi 125 — made connections between its organic magic, dangerous female power and natural knowledge.

In conclusion, this preliminary study is intended to be a small contribution to scholarship on this fascinating magic text, building on Maaike van der Lugt's work on its scholastic reception and looking forward to a future edition of the text. The *Liber vaccae* and its more mainstream counterparts in the medical and *experimenta* traditions encouraged the idea of the impressionability and permeability of human bodies susceptible to the influence of suffumigations, ointments, comestibles, and suspensions. At the end of the Middle Ages and in a post-plague context there was less idealisation of post-lapsarian nature as God's good creation. But uncertainty remained about the extent to which

74 Ostorero, 'The Concept of the Witches' Sabbath', p. 5. For examples of clerical writers hostile to magic citing examples from classical literature, see Montesano, *Classical Culture and Witchcraft*, pp. 113–17.

75 Ulrich Molitoris, *De lamiis*, Ch. 3, trans. by Kieckhefer, pp. 107–08. The choice of animals follows the account of Boethius in *De consolatione*, also cited by Ulrich in Ch. 3.

76 Ulrich Molitoris, *De lamiis*, Ch. 4, trans. by Kieckhefer, pp. 138–41.

77 Claude Tholosan, *Record of the trial of Jubert of Bavaria*, ed. by Hansen, p. 542.

witches and demons could really undermine natural laws, especially those linked to the human body and its capacities.

Organic magic and its scribes and censors must be placed in a complex context alongside the idea of extraordinary properties and processes in nature; the contemporary scientific understanding of psychoactive and poisonous properties; the practices of 'common magic' and clerical critiques of it; literary and scholastic representations of female sorceresses and demons; and the strong associations of particular substances and animals with harmful magic and the binding of demons. Although the strands of learned magic, literary witches and theological critique are present throughout the three centuries covered by my article, the tension between positive conceptions of natural magic and unease at 'abominable mixtures' grew stronger in the fifteenth century with an increasing construction of witchcraft mythologies and the beginning of sustained persecutions of witches in some parts of Europe.

Bibliography

Manuscripts

Cambridge, Corpus Christi, MS 132
Cambridge, Cambridge University Library, MS Additional 3544
Città della Vaticano, Biblioteca Apostolica Vaticana, MS Pal. Lat. 1892
Innsbruck, Universitäts- und Landesbibliothek Tirol, MS 489
Leiden, Universitatbibliotheek, MS Vossius Chym Q 60
London, British Library, MS Arundel 342
Munich, Bayerische Staatsbibliothek, MS Clm 615
Munich, Bayerische Staatsbibliothek, MS Clm 22292
Nuremberg, Germanisches Nationalmuseum, MS 3227a
Paris, Bibliothèque nationale de France, MS fonds latin 7337
Oxford, Bodleian Library, MS Ashmole 1435

Primary Sources

Amadas et Ydoine, ed. by John R. Reinhard (Paris: Honoré Champion, 1974)
Amadas and Ydoine, trans. by Ross G. Arthur (New York: Garland, 1993)
The Cambridge Book of Magic, ed. and trans. by Francis Young (Cambridge: Texts in Early Modern Magic, 2015)
The Arras Witch Treatises: Johannes Tinctor's 'Invectives contre la secte de vauderie' and the 'Recollectio casus, status et condicionis Valdensium ydolatrarum' by the Anonymous of Arras (1460), trans. by Andrew Colin Gow, Robert B. Desjardins, and François V. Pageau (University Park, PA: Penn State Press, 2017)
Errores gazariorum, in *L'imaginaire du sabbat: Édition critique des textes les plus anciens (1430 c. –1440 c.)*, ed. by Martine Ostorero, Agostino Paravicini Bagliani, and Kathrin Utz Tremp (Lausanne: Université de Lausanne, 1999), pp. 278–99
Hans Fründ, *Rapport sur la chasse aux sorciers et aux sorcières menée dès 1428 dans le diocèse de Sion*, in *L'imaginaire du sabbat: Édition critique des textes les plus anciens (1430 c.-1440 c.)*, ed. by Martine Ostorero, Agostino Paravicini Bagliani, and Kathrin Utz Tremp (Lausanne: Université de Lausanne, 1999), pp. 30–51
Johannes Hartlieb, *Book on all the Forbidden Arts, Heresy and Magic*, in *Hazards of the Dark Arts: Advice for Medieval Princes on Witchcraft and Magic*, trans. by Richard Kieckhefer (University Park, PA: Penn State Press, 2017), pp. 21–92
Felix Hemmerlin, *De nobilitate*, in *Quellen und Untersuchungen zur Geschichte des Hexenwahns und der Hexenverfolgung im Mittelalter*, ed. by Joseph Hansen (Bonn: C. Georgi, 1900)
Conrad Kyeser aus Eichstätt, *Bellifortis*, ed. and trans. by Götz Quarg (Düsseldorf: VDI-Verlag, 1967)
Liber Aneguemis. Un antico testo ermetico tra alchimia pratica, esoterismo e magia nera, ed. by Paolo Scopelliti and Abdessattar Chaouech (Milan: Mimesis Edizioni, 2006)

De mirabilibus mundi, ed. by Antonella Sannino (Florence: SISMEL – Edizioni del Galluzzo, 2011)

Ulrich Molitoris, *De lamiis et phitonicis mulieribus*, in *Hazards of the Dark Arts: Advice for Medieval Princes on Witchcraft and Magic*, trans. by Richard Kieckhefer (University Park, PA: Penn State Press, 2017), pp. 93–153

Nicole Oresme, *Tractatus de configurationibus qualitatum et motuum*, ed. and trans. by Marshall Clagett (Madison, WI: University of Wisconsin Press, 1968)

Picatrix: The Latin version of the Ghāyat al-ḥakīm, ed. by David Pingree (London: Warburg Institute, 1986)

Picatrix: A Medieval Treatise on Astral Magic, trans. by Dan Attrell and David Porreca (University Park, PA: Penn State Press, 2019)

Jean Vincent, *Liber adversus Magicas Artes et eos qui dicunt artibus eisdem nullam inesse efficaciam*, in *Quellen und Untersuchungen zur Geschichte des Hexenwahns und der Hexenverfolgung im Mittelalter*, ed. by Joseph Hansen (Bonn: C. Georgi, 1900)

Claude Tholosan, *Ut magorum et maleficiorum errores manifesti ignorantibus fiant*, in *L'imaginaire du sabbat: Édition critique des textes les plus anciens (1430 c.-1440 c.)*, ed. by Martine Ostorero, Agostino Paravicini Bagliani, and Kathrin Utz Tremp (Lausanne: Université de Lausanne, 1999), pp. 362–415

Secondary Studies

Ammann-Doubliez, Chantal, 'La première chasse aux sorciers en Valais (1428–1436)', in *L'imaginaire du sabbat: Édition critique des textes les plus anciens (1430 c.-1440 c.)*, ed. by Martine Ostorero, Agostino Paravicini Bagliani, and Kathrin Utz Tremp (Lausanne: Université de Lausanne, 1999), pp. 86–89

Boudet, Jean-Patrice, 'L'amour et les rituels à images d'envoûtement dans le 'Picatrix' latin', in *Images et magie: Picatrix entre Orient et Occident*, ed. by Jean-Patrice Boudet, Anna Caiozzo, and Nicolas Weill-Parot (Paris: Honoré Champion, 2011), pp. 149–62

Boureau, Alain, *Satan the Heretic*, trans. by Theresa L. Fagan (Chicago: University of Chicago Press, 2006)

Camille, Michael, 'Obscenity under Erasure: Censorship in Medieval Illuminated Manuscripts', in *Obscenity: Social Control and Artistic Creation in the European Middle Ages*, ed. by Jan M. Ziolkowski (Leiden: Brill, 1998)

Draelants, Isabelle, 'The Notion of Properties: Tensions between Scientia and Ars in Medieval Natural Philosophy and Magic', in *The Routledge History of Medieval Magic*, ed. by Sophie Page and Catherine Rider (London: Routledge, 2019), pp. 169–86

Hasse, Dag Nikolaus, 'Plato arabico-latinus: Philosophy – Wisdom Literature – Occult Sciences', in *The Platonic Tradition in the Middle Ages: A Doxographic Approach*, ed. by Stephen Gersh and Maarten J. F. M. Hoenen (Berlin: De Gruyter, 2002), pp. 31–65

Kerby-Fulton, Kathryn, *Books Under Suspicion: Censorship and Tolerance of Revelatory Writing in Late Medieval England* (Notre Dame: University of Notre Dame, 2006)
Kieckhefer, Richard, *European Witch Trials: Their Foundations in Popular and Learned Culture, 1300–1500* (Los Angeles: University of California Press, 1976)
——, *Forbidden Rites: A Necromancer's Manual of the Fifteenth Century* (University Park, PA: Penn State Press, 1997)
——, 'Mythologies of Witchcraft in the Fifteenth Century', *Magic, Ritual and Witchcraft*, 1 (2006), 79–107
Klingshirn, William E., 'Isidore of Seville's Taxonomy of Magicians and Diviners', in *Traditio*, 58 (2003), 59–90
Montesano, Marina, *Classical Culture and Witchcraft in Medieval and Renaissance Italy* (London: Palgrave Macmillan, 2018)
Newman, William R., *Promethean Ambitions: Alchemy and the Quest to Perfect Nature* (Chicago: University of Chicago Press, 2004)
Olsan, Lea, 'The Marginality of Charms in Medieval England' in *The Power of Words: Studies on Charms and Charming in Europe*, ed. by James Kapaló, Eva Pócs, and William Francis Ryan (Budapest: Central European University Press, 2013)
Ostorero, Martine, 'The Concept of the Witches' Sabbath in the Alpine Region (1430–1440), Text and Context', in *Demons, Spirits, Witches 3: Witchcraft Mythologies and Persecutions*, ed. by Gabor Klaniczay and Eva Pócs (Budapest: CEU Press, 2008)
Page, Sophie, 'Magic and the Pursuit of Wisdom: The "Familiar" Spirit in the *Liber Theysolius*', in *La corónica. A Journal of Medieval Hispanic Languages Literatures and Cultures*, 36 (2007), 41–70
——, *Magic in the Cloister* (University Park, PA: Penn State Press, 2013)
Pingree, David, 'From Hermes to Jābir and the *Book of the Cow*', in *Magic and the Classical Tradition*, ed. by Charles Burnett and William Francis Ryan (London: The Warburg Institute, 2006), pp. 19–28
——, 'Artificial Demons and Miracles', in *Démons et merveilles d'Orient*, ed. by Rika Gyselen and Pierfrancesco Callieri (Bures-sur-Yvette: Groupe pour l'Étude de la Civilisation du Moyen-Orient, 2001), pp. 109–22
Rider, Catherine, *Magic and Religion in Medieval England* (London: Reaktion, 2012)
Saif, Liana, 'The Cows and the Bees: Arabic Sources and Parallels for Pseudo-Plato's *Liber vaccae* (*Kitāb al-Nawāmīs*)', *Journal of the Warburg and Courtauld Institutes*, 6 (2017), 1–47
Van der Lugt, Maaike, '"Abominable Mixtures": The *Liber vaccae* in the Medieval West, or the Dangers and Attractions of Natural Magic', *Traditio*, 64 (2009), 229–77
Weill-Parot, Nicolas, *Les 'images astrologiques' au moyen âge et à la Renaissance: Spéculations intellectuelles et pratiques magiques (XIIe-XVe siècle), Sciences, Techniques et Civilisations du Moyen Age à l'Aube des Lumières, 6* (Paris: Honoré Champion, 2002)

KOENRAAD VAN CLEEMPOEL

The Transmission of Materialized Knowledge

A Medieval Saphea with Islamic Projections, Re-engraved in the Renaissance

A European *saphea* — a single plate astrolabe with a universal projection — that recently surfaced at an auction in London,[1] and which is now preserved at the History of Science Museum in Oxford (inv. 14645), is a unique testimony of intellectual and artisanal transmission from the Middle Ages to the Renaissance (hereafter, the 'Oxford *Saphea*'). It was probably made *c.* 1450 in France and includes an important projection of eleventh-century Islamic origin; and it was re-engraved and completed during the Renaissance. This is the only example we have of a medieval astrolabe with the universal projection known as *saphea* in combination with the ecliptic projected on one arc. What makes it even more remarkable is the fact that the astrolabe resurfaced around 1600 in the workshop of the Louvain mathematician Adrian Zeelst, who completed it with additional scales; and, as shown below, Zeelst used the original instrument as the basis for an illustrated treatise on the astrolabe (Liège, 1602) and for the construction of at least one other astrolabe.

The unusual story of this astrolabe spans an intellectual arc of almost 450 years and links early medieval Toledo with late Renaissance Louvain. The knowledge embodied in the instrument was recognized, analysed, and even completed on the same instrument centuries after it was conceived.

* I acknowledge the valuable comments and suggestions on the manuscript by Dr Stephen Johnston of the History of Science Museum in Oxford and by Prof. Em. David King.

1 Bonhams catalogue, *Scientific and Mechanical Musical Instruments and Cameras*: 'An important single sheet European universal astrolabe, not signed or dated, engraved by two hands, the first working in the mid- to late 15th century, the second in the third quarter (or slightly later) of the 16th century', lot 189.

Koenraad Van Cleempoel (Hasselt, Belgium) is a specialist in medieval scientific instruments including astrolabes, as well as conceptual perspectives on architecture and design.

Mastering Nature in the Medieval Arabic and Latin Worlds: Studies in Heritage and Transfer of Arabic Science in Honour of Charles Burnett, ed. by Ann Giletti and Dag Nikolaus Hasse, CAT 4 (Turnhout: Brepols, 2023), pp. 231–252 BREPOLS PUBLISHERS 10.1484/.CAT-EB.5.134032

The Oxford *Saphea* is derived from the better-known planispheric astrolabe.[2] Based on a stereographic projection, the planispheric astrolabe represents 'earth' and 'heaven' in two dimensions on top of each other, usually constructed as circular brass instruments. The heavenly part, called 'rete', shows a selection of stars and the zodiac, which can rotate over a latitude plate. It thus evokes the apparent movement of stars (including the sun) over the horizon of one particular latitude. Both these latitude plates and the rete are contained in a circular body, called 'mater' with a raised, graduated rim and a 'throne' with a ring to hang the instrument vertically. This allows one to measure the altitude of a celestial body over the horizon by means of two sighting vanes mounted on an 'alidade' that moves over a graduated rim of 360°.

The astronomical and mathematical understanding for building astrolabes was available in Antiquity, but so far no such instruments have surfaced from that period. The knowledge was passed on to Muslim scholars producing the earliest known astrolabe *c.* 920, soon followed by examples from the Latin West in the eleventh century. The construction of these sophisticated instruments mostly coincides with centres of knowledge production, such as court environments, monasteries and early universities.

Conventional planispheric astrolabes thus feature a stereographic projection of one particular latitude engraved on one side of a plate. The geographical reach of the instrument could be expanded by including different plates with projections for different latitudes. Some astrolabes could include up to seven plates engraved on both sides with projections for 14 latitudes. Designing a projection for making an astrolabe 'universal' — and thus independent of a large set of plates with projections for particular latitudes — was both practical and an intellectual challenge. It was practical in that it would save the artisans the labour of engraving different plates, and the user would be able to make observations with one projection in all latitudes. Mostly, however, it was an intellectual challenge, as it asked for the most advanced knowledge in mathematics and astronomy available at the time. Artisanal craftsmanship and technical skill to engrave these sophisticated projections on brass plates was also necessary.

The Oxford *Saphea* described here is an exceptional example of such a universal astrolabe. As there is no mater and rete, it is also called a 'single-plate astrolabe'.

2 For a clear understanding of an astrolabe and its anatomy, John North's article of 1974 is still a reliable reference, completed recently by David King's historic survey of 2019 (with the most extensive bibliography available on the subject), also elaborating on the relationship between east and west: North, 'The Astrolabe'; King, 'The Astrolabe'.

Figure 8.1. Oxford *Saphea*, *c.* 1450 (probably France) and *c.* 1580 (Louvain or Liège), Side A. Oxford, History of Science Museum (inv. 14645). Reproduced with permission. Engraved with a universal stereographic projection equipped with an arc-shaped ecliptic, a cursor and a radial arm.

1. Description of the Oxford *Saphea*

The circular instrument is 29 cm in diameter and has an integral two-lobed throne with suspension ring. Both sides of the plate are engraved and equipped with sliding rules, an arc, cursors, moveable pointers and one alidade.

Side A (Fig. 8.1)

On the rim is one scale of 360° with divisions for single degrees, numbered three times in different styles: (1) outermost for degrees in gothic numerals, clockwise for every 10° starting from the south (on the top when the throne is upwards); (2) on the inside of the scale for the hours in Roman numerals

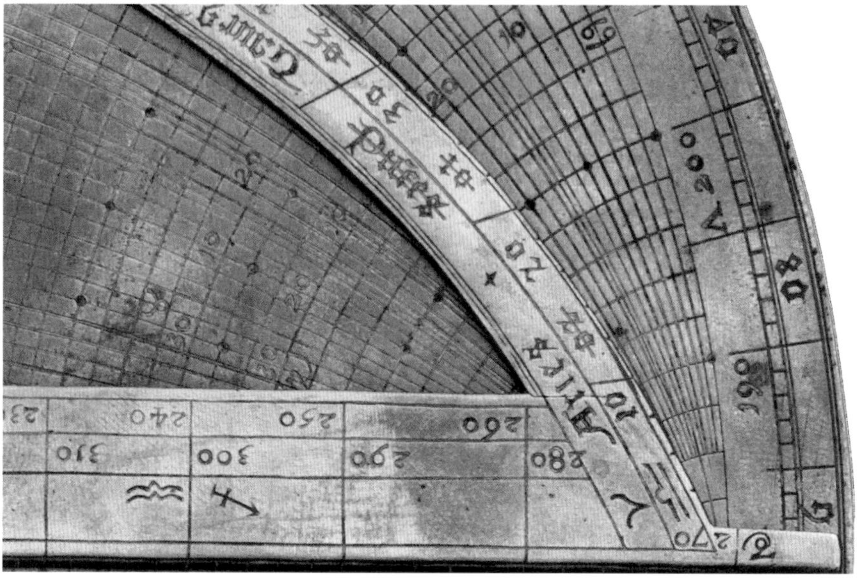

Figure 8.2. Oxford *Saphea*, *c.* 1450 and *c.* 1580, Side A (detail). Oxford, History of Science Museum (inv. 14645). Reproduced with permission. Detail of the arc with crossed-through numbers turned up-side-down for reading in the reverse direction, corresponding with crossed-through names of the zodiac.

every 15°, numbered twice I–XII; and (3) alongside for 360° in italic numbers, anticlockwise for every 10° starting from the east (on the left when the throne is upwards). The rest of the surface is engraved with a very precise universal stereographic projection with the colure of the solstices as the plane of the projection. There are half ecliptic arcs along the circles of declination at 58° and 62° on either side of the equator, with each zodiacal sign represented by its sign, divided in 30 degrees, numbered every 10° in italic numerals. The poles are situated on the east-west axis, and not as was common on Renaissance instruments on the north-south axis. Through the centre runs a straight line at an angle of 23°30' to the equator representing the ecliptic, with subdivision for every 30° and with each sign of the zodiac indicated by its symbol. This is mirrored by a faintly engraved line left without graduations or symbols. The hours are marked in both hemispheres at a declination of 35°. On top of the projection is a rotating arc attached to a diametrical rule with, on top, a cursor and a radial arm. The arc is engraved with a projected ecliptic in medieval letters and numbers. It is graduated in 180°, numbered six times 0–30° for every 10°: 10–20-30 and also in crossed-through numbers turned up-side-down for reading in the reverse direction, corresponding with the crossed-through names of the zodiac (Fig. 8.2). The solstices are indicated near the centre of the arc, the equinoctial signs towards the extremes. The arc is connected to the centre with a straight diametrical arm engraved with

Figure 8.3.
Oxford *Saphea*, *c.* 1450 and *c.* 1580, Side B. Oxford, History of Science Museum (inv. 14645). Reproduced with permission.
Eccentric calendar-zodiac scales, shadow squares, various sets of scales and projections.

a degree scale in Roman numerals reading from the centre to the right (with the arc pointing downward) from 0–90° and below in the reverse direction from 100° to 270° and continuing above again from 280° to 360°. The twelve signs of the zodiac are also engraved, this time with the equinoctial signs at the centre and the solsticial at the ends. Attached to the centre and mounted on top of the arc is a rotating radial arm (labelled *index*) with a scale for 90° in italic numbers, reaching with its tip the scale on the edge of the universal projection. Finally, on top of this rule, is a *brachiolum* arm with a sharp pointer.

Side B (Fig. 8.3)

This side has an eccentric zodiac calendar scale surrounding a latitude plate combined with a shadow square, over which rotates an alidade with two pointers in place of a rete. Starting from the circumference moving inwards are the following:

1. Alongside the rim a scale for 360° with divisions for single degrees and an extra mark for every 5° running towards the edge of the plate. The scale is numbered for three purposes (a and b outermost of the scale, c on the inside):
   a. For degrees, numbered four times 10–90° for every 10° in gothic numerals;
   b. For equal hours, indicated with Roman numerals for every 15°, numbered twice I–XII;
   c. For the zodiac with degrees numbered twelve times 10–20-30 in gothic numerals and with the names of the constellations also engraved in gothic letters.
2. Eccentric to this scale is a calendrical scale, divided into 365 units for single days, each month indicated with gothic letters and its days 10–20-30 with gothic numbers. February has 28 days, and the first point of Aries (vernal equinox) corresponds with 11.8 March.
3. The upper half (if the throne is pointing upwards) of the central surface is engraved with a stereographic projection for (calculated value) latitude 51°, the almucantars engraved for every 2°, numbered twice every 10°, along azimuths for the four cardinal directions. The numbers are in italic letters. Part of the same stereographic projection, but also running over the lower half, are concentric circles representing the tropics of Cancer and Capricorn and in between the equator.
4. On top of the almucantars is a projected ecliptic, with a scale graduated for single degrees with every 5th and 10th degree marked with a longer line. The twelve constellations are indicated with their symbols.
5. On the right-hand side (if the throne is pointing upwards), and again on top of the almucantar grid, is a small set of unequal arcs with italic numbers, numbered 1–6 in one direction and 7–11 in the opposite direction, used with the scales on the alidade. Note that the arcs spring from the intersection of the circles of the unequal hour diagram and an arc which is just inside the equator.
6. From the north point of the horizon radiate arcs for the twelve astrological houses. They are numbered counter-clockwise in Roman numerals, and their engraved arcs are marked with small hachures, like pecked arcs.
7. Under the horizon are two sets of scales:
   a. Two shadow squares, labelled with engraved letters in italic letters *Latus rectum* and *Latus versum*. There are two scales: outermost for twelve units numbered in gothic numbers in groups of four with each set divided to fifths to give a total of 60 units; and innermost a second shadow square of ten, numbered in five groups of two (2–4-6–8-10) with each group subdivided per ten units, giving a total of 50 units.

b. Between the tropic of Cancer and the tropic of Capricorn is a set of twelve dotted arcs numbered clockwise 1–12 in italic numbers for the unequal hours.

On top of this side is a rotating counter-changed alidade with a scale of 10–14 on one arm and a zodiac scale on the other. There are also fixed sighting vanes at both ends. Just off-centre from the alidade are two *brachiola* with two hinges and sharp pointers.

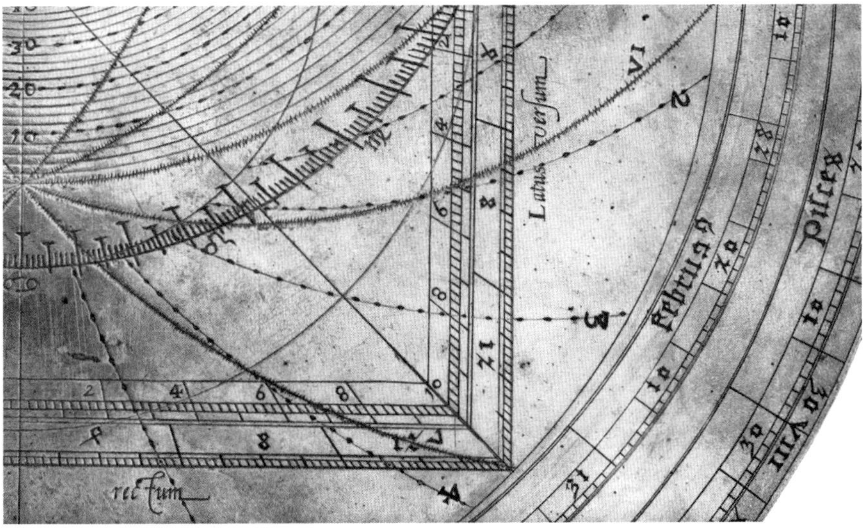

Figure 8.4. Oxford *Saphea*, *c.* 1450 and *c.* 1580, Side B (detail). Oxford, History of Science Museum (inv. 14645). Reproduced with permission. Detail showing the two different stages of engraving.

2. Phase 1: The Original Medieval Astrolabe

This *saphea* clearly shows the hand of two different engravers: on side A, the universal projection, the graduated rim and the superimposed arc are medieval, as are the eccentric calendar-zodiac scales and the shadow square of twelve on side B. (Fig. 8.4) The rest was engraved around 1600 in Louvain or Liège, as discussed below. Dating the initial, medieval phase is challenging — indeed, dating medieval astrolabes in general remains difficult as reliable parameters are not readily available.[3] The auction catalogue dates the earliest phase as 'mid to late 15th century'. This feels right on stylistic grounds, and its design is closely related to a mid-fourteenth century manuscript of a Paris-based astronomer, as discussed below.

3 Poulle, 'Peut-on dater les astrolabes médiévaux?'.

A. Saphaea *Arzachelis: European Afterlife of an Islamic Invention*

Knowledge on astrolabes in the Latin West was much indebted to the development of instruments in the Muslim world and to Arabic treatises on the astrolabe.[4] This is particularly the case for astrolabes with more advanced features, such as the universal projection.

The necessary competences to design and build such complex instruments were available in eleventh-century Toledo at the court of al-Maʾmūn (1037–1075). Two astronomers there developed universal projections for astrolabes: ʿAlī ibn Khalaf and Abū Isḥāq Ibrāhīm al-Zarqālī, also known as Azarquiel or Arzachel (flourished between 1048/49, date of the instrument for al-Maʾmūn, and 1087, date of the observations in Cordoba). Azarchel's name derives from the Arabic al-Zarqālī al-Naqqāsh. He was also named al-Naqqāsh, the engraver, but the Latin names derive from the name al-Zarqālī.[5]

Both projections were described and illustrated in the *Libros del saber*, c. 1277, of King Alfonso X, respectively as *lámina universal* or *orizon universal* (universal plate) and *azafea* (*saphea*). The manuscripts include very elegant and accurate coloured parchment drawings of the variations in universal projections.[6] The basic concept is that the plane of projection is the colure of solstices which also forms the outer circle of the plate.

The projection of Azarchel, or 'the tablet of al-Zarqālī', became known in the Latin west as *saphea*.[7] Azarchel describes his astrolabe, the *saphea*, in a treatise of which two versions were translated in the Middle Ages:[8] a longer one, with 100 chapters, translated into Castilian at the court of Alfonso X, and a shorter one, with 61 chapters, translated into Hebrew by Profacius Judaeus (Jacob ben Makir ibn Tibbon), and, in 1263, 'from Arabic into Latin' by Profacius and John of Brexia.[9] William the Englishman composed a treatise on Azarchel's astrolabe in 1231 with the help of Yehuda ben Mose, but the exact dependence of William upon Azarchel is still unclear.[10] In the sixteenth century, new treatises on Azarchel's instrument were written by Gemma Frisius, Jacob Ziegler, Johannes Schöner, Andreas Stöberl, and Juan de Rojas. Despite the fact that the astronomical and mathematical principles remained known among authors of astronomical treatises in the Latin West,

---

4 King, *In Synchrony with the Heavens*, II, p. 556: 'We also need to be aware that astronomical instrumentation in medieval Europe was largely indebted to Islamic instrumentation. Indeed, for most technical innovations in European instrumentation up to *ca.* 1600 we can identify Islamic precedents, although this does not, of course, always imply direct transmission or exclude independent initiative'.

5 Kennedy, *Studies in the Islamic Exact Sciences*, p. 502; Puig, 'Zarqālī', pp. 1258–60.

6 Fernández Fernández, 'Astrolabes on Parchment', pp. 287–310.

7 Zinner, *Deutsche und Niederländische Astronomische Instrumente*, pp. 145–49.

8 For this paragraph, I am grateful for advice from Dag Nikolaus Hasse.

9 Azarchel, *Tractat de l'assafea*, and introduction by Millàs Vallicrosa, pp. xxii and 152.

10 Introduction by Millàs Vallicrosa to Azarchel, *Tractat de l'assafea*, pp. xxxviii–xliii; Poulle, 'Un instrument astronomique', pp. 492–93.

prior to the discovery of the Oxford *Saphea* only two medieval instruments with a *saphea* projection were known:

1. Anonymous, *c.* 1400 (Oxford, History of Science Museum, inv. 52869). The universal projection is engraved inside the mater but has lost its moveable parts. Another intriguing feature on this astrolabe is the rim of the rete with cut-out teeth, like a gearwork.

2. The Marcin Bylica astrolabe, attributed to Hans Dorn in Cracow, 1486 (Cracow, Jagiellonian University, Copernicus Treasury). The reverse of this unusually large astrolabe (45 cm) shows the *saphea.*[11]

This scarcity of European instruments with universal projections changed in the sixteenth century; first when a treatise attributed to Regiomontanus was published in 1534 on the use of an instrument with the same kind of projection as the *saphea*: *Problemata XXIX saphaeae nobilis instrumenti astronomici*. Then Gemma Frisius (1508–1555) in Louvain elevated the projection as a standard feature on astrolabes, usually engraved on the reverse of the mater. He explained this in his treatise on the astrolabe, published posthumously by his son Cornelius Gemma.[12] The title 'Catholic Astrolabe' refers precisely to this universal projection that could be used from pole to pole, but, ironically, Frisius failed to mention its eleventh-century Islamic origin in Toledo. The Louvain-based instrument maker Gualterus Arsenius, who signed as nephew (*nepos*) of Gemma Frisius, especially favoured it. Over 20 recorded astrolabes of his workshop include the universal projection on the reverse. The earliest one is dated 1554, two years prior to Frisius's treatise.[13] Petrus Ab Aggere was a Louvain-trained mathematician and instrument maker who travelled with the court of Felipe II from Brussels to Madrid, Toledo, and finally El Escorial to start making astrolabes around 1570 with projections of Ibn Khalaf and Azarchel.[14] They even feature the

11 Described in exhibition catalogue Levenson ed., *Circa 1492*, cat. no. 121, pp. 222–23.

12 Reinerus Gemma Frisius, Cornelius Gemma, Jan Steels, and Joannes Grapheus, *De astrolabo catholico liber quo latissime patentis instrumenti multiplex vsus explicatur, & quicquid vspiam rerum mathematicarum tradi possit continetur*.

13 Surprisingly the three known astrolabes of Gerard Mercator of *c.* 1545 do not include the universal projection, but they excel in many other aspects: Turner, 'Three Astrolabes'. The first Louvain astrolabe with a *saphea* grid dates from 1554 and is signed *Authore Gemma Frisio et exaratus a Gualtero Arsenio Louanij 1554*, with a diameter of 330 mm (Mexico City). In fact, this astrolabe forms a pair with another astrolabe of the same diameter with an orthographic universal projection on the back (Madrid). The pair thus shows two variations of universal projections, possibly made as a royal gift on the occasion of the wedding of King Felipe II to Queen Mary I on July 25, 1554. One can only speculate that his Spanish roots, and having King Alfonso X as his early predecessor, may have inspired the Louvain mathematicians. Both astrolabes are discussed in Van Cleempoel, *Catalogue Raisonné of Scientific Instruments*, pp. 35–37 and 87–90.

14 Van Cleempoel, 'Migration of "Materialised Knowledge"'.

complicated rete pattern with lace-work of the latitude coordinates, as illustrated in the *Libros del saber*.[15]

This fascination among Renaissance mathematicians with Islamic universal projection was not limited to Louvain and El Escorial. John Blagrave in London published in 1585 his 'Mathematical Jewel'.[16] Although the 'jewel' was a universal astrolabe as invented in eleventh-century Toledo, the subtitle claims Blagrave as the inventor: 'The most part newly found out by the author'. A unique astrolabe by Charles Whitwell of 1595 turned Blagrave's paper treatise into a refined and elaborate brass astrolabe, including the stereographic grid on the rete.[17] It is not entirely clear how the transmission of knowledge of the *saphea* moved from Toledo *c.* 1050 to London *c.* 1580, but in what follows we see at least one important and early intermediate step that helps to frame the *Saphea* which is the subject of this essay.

B. Al-āsī *or 'Myrtle'-Shaped Retes*

The Oxford *Saphea* features an arc-shaped ecliptic that is unique among European medieval astrolabes. Standard astrolabes are based on a northern stereographic projection with the northern sky being projected in the plane of the celestial equator with the south pole as the pole of the projection. A strange effect of this projection is apparent on the rete: the northern signs of the ecliptic are squeezed together on a small arc and the southern signs spread out on an arc twice the size. Therefore virtually all surviving astrolabes have retes with stars extending towards the equator and an ecliptic circle tangent to the tropics. There is, however, a handful of astrolabes with a projected ecliptic with the full diameter of the stereographic projection. Despite operating as a projection on the solsticial colure, the *saphea* projection can also be interpreted as a regular astrolabe projection onto the equator — but reaching only as far as the equator, so having room only for the northern part of the ecliptic as a rete. The ecliptic is then shaped as an arc; also known as 'myrtle shaped'. The term 'myrtle' for this characteristic comes from the Arabic *al-asṭurlāb al-āsī* as described in a manuscript by al-Bīrūnī (973-post 1050) and Abū ʿAlī al-Ḥasan al-Marrākushī or simply al-Ḥasan al-Marrākushī (fl. 1275–1282), meaning 'the astrolabe with an ecliptic shaped like a myrtle-leaf on its rete'.[18] There are two important Islamic astrolabes with such exceptional retes:

1. One was made by ʿAlī ibn Ibrāhīm al-Jazzār (or al-Ḥarrār) in Taza, Morocco in 728 AH/ 1327–1328 AD (Oxford, History of Science Museum, inv. 50853).

15 Moreno, 'A Recently Discovered Sixteenth-Century Spanish Astrolabe'.

16 Blagrave, *Mathematical Jewel*.

17 Now preserved in Florence at Museo Galileo, inv. 1095. See Turner, *Elizabethan Instrument Makers*.

18 Illustrated in Michel, *Traité de l'astrolabe*, pp. 69–70; quoted in Calvo, 'Ibn Bāso's Universal Plate', p. 67 n. 11.

This rete features a double, or symmetrical myrtle shaped ecliptic, with vernal and autumn points touching the equator (outer rim of the plate) and the summer-winter points at opposite positions near the centre.

2. The other was made by Ibn al-Sarrāj in Aleppo in 1328 (Athens, Benaki Museum, inv. 13178), labelled by David King as the most sophisticated astrolabe of the Middle Ages and the beginning of the Renaissance.[19] Even more complex than the previous astrolabe, the myrtle-shaped ecliptic is here folded over so that summer and winter solstices coincide.

The aforementioned instruments by Petrus Ab Aggere (Spain, *c.* 1580) and John Blagrave (London, *c.* 1580) have rete designs identical to those of Ibn al-Sarrāj, again illustrating that Renaissance instruments are embedded in an Islamic tradition of instrumentation. Henri Michel's *Traité de l'astrolabe* (1976) includes a rete with double-arc myrtle ecliptic (p. 72, fig. 52) associated with the Jesuit of Brussels origin, Odo van Maelcote (1572–1615). Van Maelcote designed the *astrolabium aequinoctiale* for both northern and southern latitudes, providing the Jesuits with an instrument to operate in both hemispheres, by reconciling septentrional-meridional polarities — both astronomically and metaphorically.[20] Michel does not indicate the whereabouts of van Maelcote's astrolabe, but the style of engraving and the overall design of the rete indicate Petrus Ab Aggere as a possible manufacturer.

A remarkable witness of the knowledge transition from east to west comes from an unpretentious and seemingly crude astrolabe with Latin inscriptions, thought to have been made in Sicily *c.* 1300 (Fig. 8.5).[21] With its humble size of 59 mm it is the smallest astrolabe in the collection of the History of Science Museum in Oxford (inv. 40829). R. T. Gunther calculated the latitude as 38° and therefore attributed it to Sicilian origin; however, David King corrected this calculation to 24°, but maintains, nevertheless, that the origin is Sicily. The rete on this miniature astrolabe is of the rotated half myrtle variety — exactly the same as that on the newly discovered Oxford *Saphea*. The northern projection of the northern ecliptic and the southern projection of the southern ecliptic are engraved on a single, miniature arc of a circle. The graduation is double and can be read in two directions, again similar to the arc on the Oxford *Saphea*.

19 King, 'A Quintuply-Universal Astrolabe Dated 1328/29' in King, *In Synchrony with the Heavens*, II, pp. 694–700, at 694, cf. pp. 561 and 670.

20 Odo van Maelcote, *Astrolabium aequinoctiale*.

21 First mentioned in Gunther, *The Astrolabes of the World*, II, pp. 319–20; and described more in detail by David King, who also proposed the earlier dating of *c.* 1300 'based on a feeling that they are not 15th century, nor are they 14th century': King, 'A Medieval Italian Testimonial to an Early Islamic Tradition of Non-Standard Astrolabes', in King, *In Synchrony with the Heavens*, II, pp. 553–74, at 564.

Figure 8.5. ‘Sicilian’ Astrolabe, *c.* 1300. Oxford, History of Science Museum (inv. 40829). Reproduced with permission. This miniature arc-shaped ecliptic shows the same type of projection as on the newly discovered Oxford *Saphea*. The graduation is double and can be read in two directions, again similar to the arc on the Oxford *Saphea*.

Prior to 2016 it was claimed that this small astrolabe was ‘the sole surviving *medieval* European instrument with a half myrtle with double graduation.’[22] The Oxford *Saphea* presented in this paper challenges this assumption, and provides further evidence of how Latin instrument makers were indebted to their Islamic predecessors.

22 King, *In Synchrony with the Heavens*, II, p. 570.

C. *Jean de Lignères and His* Circulus Mobilis

An interesting crossing point between the Toledo projections and the arc-shaped ecliptic comes from mid-fourteenth-century Paris. Jean de Ligneres's astronomical work includes treatises on three instruments: the *saphea*, the *equatorium*, and the *directorium*.[23] The description of *saphea* in his *Descriptiones que sunt in facie instrumenti notificate* introduces an important addition to al-Zarqālī's design: a kind of rete which he calls *circulus mobilis*. It consists of an ensemble of three elements attached to each other: (1) a graduated semicircle running along the outer edge of the plate; (2) inside this an arc-shaped ecliptic also reaching to the edge of the plate; and (3) a straight bar or graduated rule with a hole in the centre to fix to the *saphea* plate. This combination can rotate around the centre. On top is another rotatable, graduated rule.[24]

This *circulus mobilis* of Jean de Lignères seems to be the earliest reference in the Latin West to an arc-shaped ecliptic, and it is tempting to associate his manuscript with the Oxford *Saphea*, despite the fact that there is no immediate evidence for making this connection. An interesting sixteenth-century manuscript in Munich[25] called *De Astrolabio universalis et Saphea sine cursore* describes and illustrates another type of *rete* which is, in fact, identical to the one on the Oxford *Saphea* (Fig. 8.6): the arc-shaped ecliptic (called *zodiacus mobilis*) with full diameter is still there, but the semicircle has disappeared; and the rotatable, graduated ruler is replaced by a ruler (labelled *index*) half its size and a moveable arm with an adjustable pointer (labelled *brachiolum*). The straight graduated rule is identified as *regula horizontalis*. This unpublished autograph deserves further research but it is already clear that it can be linked to the influential Louvain mathematician and instrument maker Adrian Zeelst, who was, in fact, the author who engraved the second phase of the Oxford *Saphea*.

3. Phase 2: The Renaissance Additions

The particular style of engraving of the Louvain-trained mathematician and instrument maker Adrian Zeelst makes it possible to attribute to him the additions labelled on the medieval Oxford *Saphea*. This is further supported by a treatise that he published at the court of the prince-bishop of Liège in 1602 with illustrations that are virtually identical to the instrument.

23 Poulle, 'John of Lignères', p. 124.

24 Emmanuel Poulle made a reconstruction drawing in an article in which he traces the influence of the *saphea* in the Latin West: Poulle, 'Un instrument astronomique', p. 509.

25 Munich, Bayerische Staatsbibliothek, MS lat. 25026, fols 13^{r}–16^{v} and 20^{r}–22^{v}.

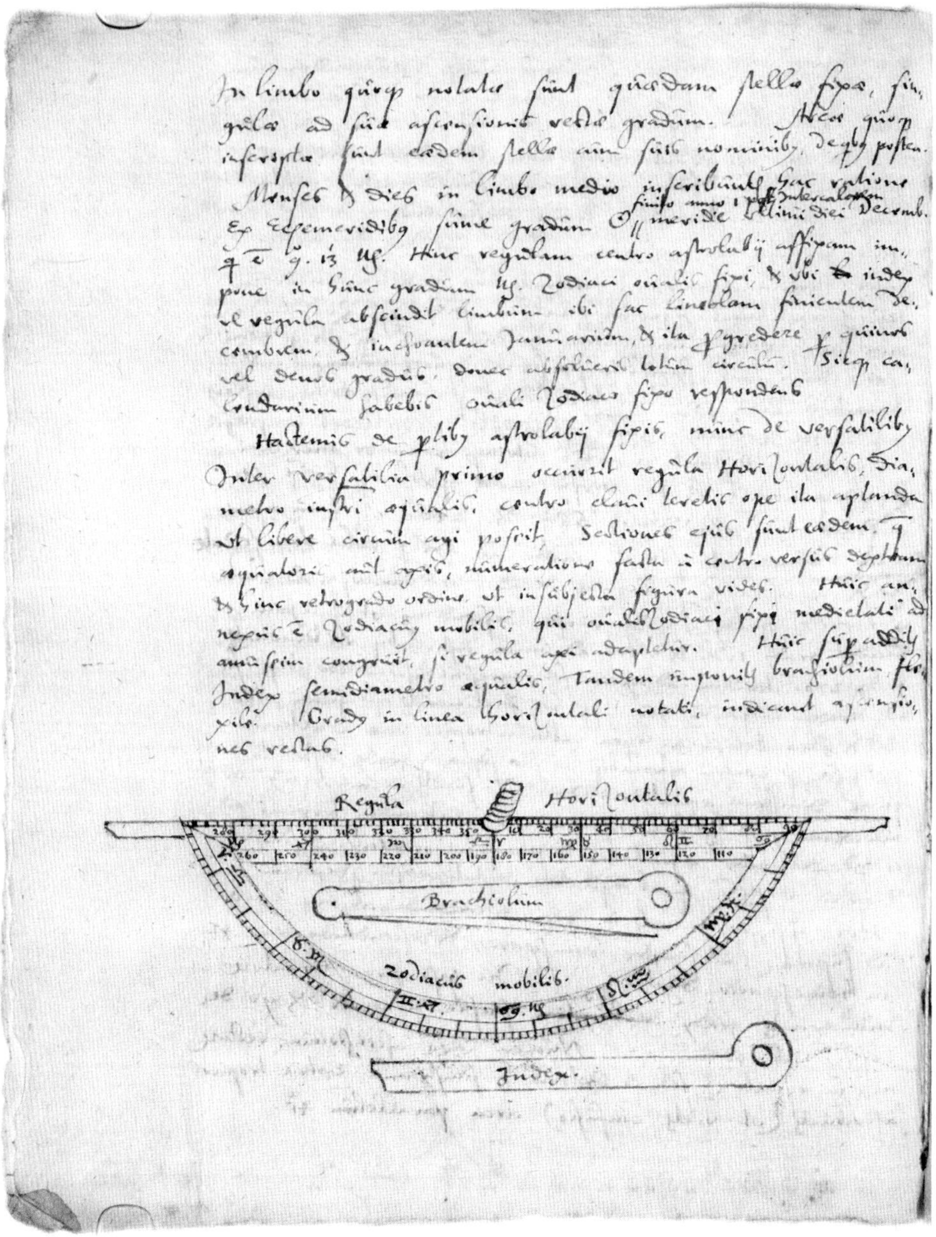

Figure 8.6. *De Astrolabio universalis et Saphea sine cursore*, Munich, Bayerische Staatsbibliothek, MS lat. 25026 (16th c.), fol. 14^v^. Reproduced with permission. The manuscript describes a rete corresponding to the zodiac on the Oxford *Saphea*; the arc-shaped ecliptic is called *zodiacus mobilis*.

A. Adrian Zeelst and the Court of Prince-Bishop Ernst of Bavaria in Liège

The panorama of Renaissance instrument making workshops can be grouped in two categories or profiles. There was a group of makers that concentrated on producing a large number of instruments of technical accuracy and great esthetical refinement, but who did not necessarily add new personal innovations. Erasmus Habermel (1583–1606) in Prague and Gualterus Arsenius (*c.* 1530-*c.* 1580) in Louvain would seem to fall into this category. Their instruments were sought after by wealthy and important patrons all over Europe. Beautiful instruments such as astrolabes and armillary spheres were used as diplomatic gifts and appear in royal inventories. Then there was a category of instrument makers who would nowadays be labelled 'innovative', transforming their research and their own ideas into brass objects. This group of makers includes names like Gerard Mercator and Adrian Zeelst. Their instruments — and in some cases their extant manuscripts — witness their invention of new and original scales, projections or additional mathematical and astronomical information. Zeelst's extant oeuvre may consist of fewer instruments when compared to other Renaissance masters, but all his instruments are of great refinement and include interesting new additions. Zeelst was active in Louvain as a mathematician and instrument maker between *c.* 1572 and *c.* 1595, and then moved to the scientific court of Prince-Bishop Ernst of Bavaria in Liège.

Several instruments and manuscripts witness this rich period in which Zeelst developed new inventions of various types of instrument. His work is closely related to Mercator, both intellectually and materially, but it is more likely than not that they never met in person, as Mercator had left Louvain already in 1556. Like Mercator, Zeelst had a particular style of engraving — for example the very dense italic letters and the peculiar shape of several signs of the zodiac — making it possible to attribute to him several unsigned instruments, including the Oxford *Saphea*. It also makes sense on a more conceptual, intellectual level: the Oxford *Saphea* was a vehicle, or beholder of a very interesting, yet hermetic and advanced tradition of universal projections with Islamic origin. The fact that it surfaced around 1600 in the context of Zeelst can be no coincidence. Few other instrument makers in Europe would have had the capacity to understand the contents of the instrument and the skill to start engraving a second layer of projections and scales on top of it. The attribution of this second phase of the Oxford *Saphea* to Zeelst is based not only on the style of engraving but also on illustrations published in a treatise of 1602 at the court of Ernst of Bavaria in Liège.

In 1602 Adrian Zeelst illustrated a printed treatise on the universal astrolabe which he wrote in collaboration with Gerard Stempel of Gouda, Ernst of Bavaria's court mathematician: *Utriusque astrolabii tam particularis quam universalis fabrica et usus*. The book was dedicated to Ernst of Bavaria, who commissioned it. All the observations for it were carried out in 1599 in Liège, at the court. In the preface, which is dated November 1601, the authors explain that they wrote the book at the palace, at the expense of the prince-bishop.[26]

The resemblance between the set of engravings and the Oxford *Saphea* is striking and set a unique standard in the image-instrument relationship in the Renaissance. The *saphea* projection, with its exceptional arc-shaped zodiac projection, is identical. (Fig. 8.7) Note, also, how the poles are rotated 90° and how small construction and design details are identical. The arc-shaped ecliptic and straight rule are engraved with the same set of scales organized and labelled in exactly the same fashion. This also goes for the reverse: the graphic design of the stereographic projection in combination with the astrological houses; and the shadow square and the fixed zodiac. Note in particular the decimal division of the shadow square which Zeelst labels as *latus rectum* and *latus versum*; the letters are even engraved in the same style and the labels are positioned on the same location.

It is evident that the medieval instrument was the basis of this treatise and that Zeelst completed the instrument by engraving additional scales on it. It was not the first time that Zeelst pushed his burin in an astrolabe that was made by someone else. We know of at least one astrolabe by Gualterus Arsenius which he partly erased and then re-engraved with a more complex hour scale than the original.[27] We also know that Zeelst made a large astrolabe with a *saphea* on the reverse — now preserved in Cologne — that is also indebted to the medieval *saphea*.

B. The Cologne Astrolabe (Fig. 8.8)

In the Kölnisches Stadtmuseum (inv. 1.184) is an exceptionally large astrolabe with a diameter of 460 mm that is attributable to Adrian Zeelst.[28] The design of the elaborate rete follows the pattern of Zeelst's earlier astrolabes with the typical decorative additions on the doubled strapwork. Inside the mater is a large *quadratum nauticum*, a typical feature of the Louvain astrolabes by Mercator and Arsenius that were produced in Louvain, where Zeelst must have learned the craft of instrument making. But it is the universal projection on the back that relates this spectacular astrolabe to the Oxford *Saphea*. The projected ecliptic on an arc extending towards the edge of the plate, in combination with the still present rule and *brachiolum*, is identical to the illustration of the 1602 treatise. Even the engraved stars are positioned identically. We are left to guess how the transmission occurred: was the re-engraved medieval astrolabe first, followed by the treatise and then the Cologne astrolabe? Perhaps the

26 Stempel and Zeelst, *Utriusque astrolabii tam particularis quam universalis fabrica et usus*, fol. 3^{v}.

27 St Petersburg, Musei Lomonosov, IC 3101; illustrated image available in Van Cleempoel, *Catalogue Raisonné of Scientific Instruments*, pp. 144–45.

28 First described in Dieckhoff, 'Cosmographia planisphaeria'. Regarding the attribution to Zeelst and the relationship to his other astrolabes see Van Cleempoel, *Catalogue Raisonné of Scientific Instruments*, pp. 61–70.

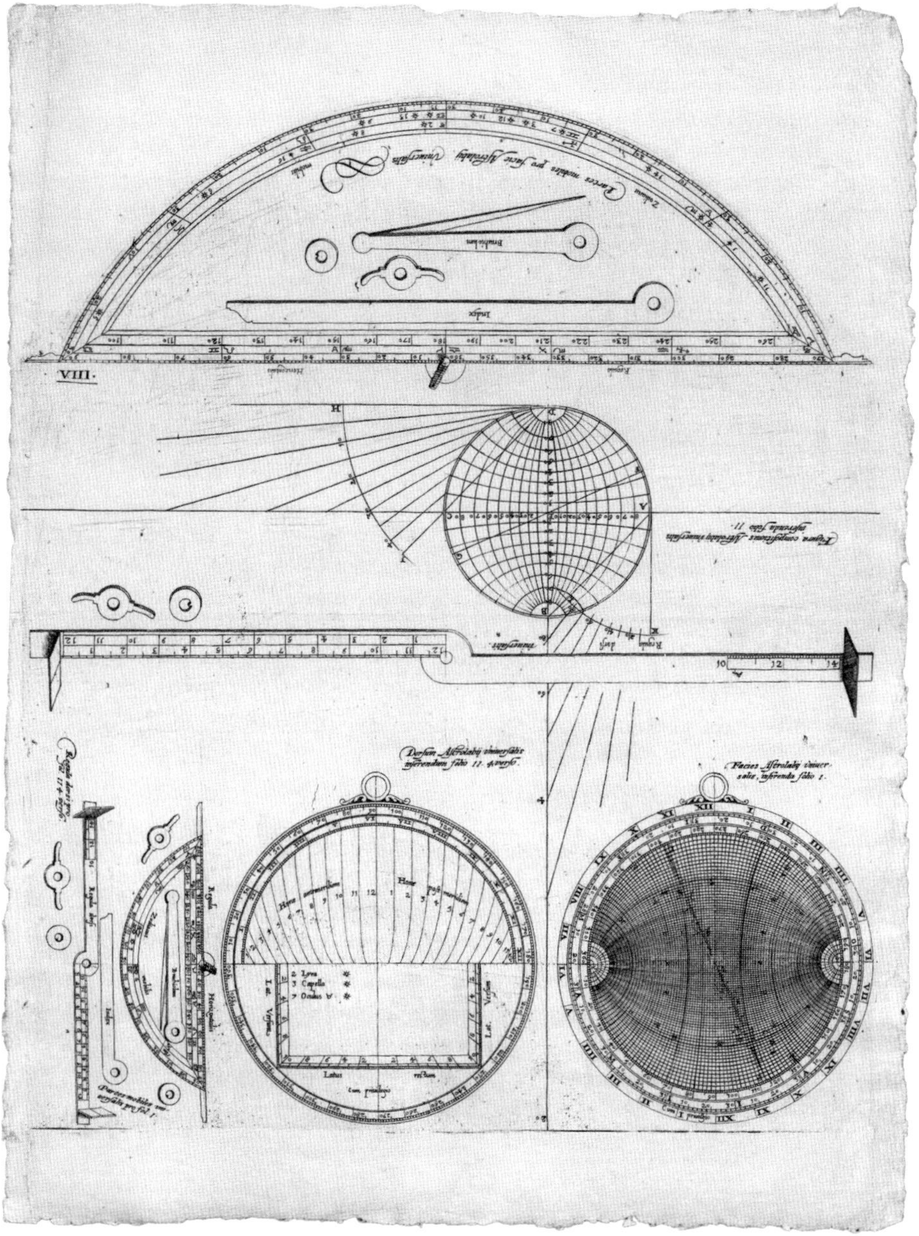

Figure 8.7. Adrian Zeelst and Gerard Stempel, *Utriusque astrolabii tam particularis quam universalis fabrica et usus* (Liège: Ouwerx, 1602), unnumbered fol. Oxford, History of Science Museum (inv. 55945). Reproduced with permission. The lower-right design shows the same type of projection on the astrolabe's reverse as on the Oxford *Saphea*, as well as the same arc-shaped zodiac.

Figure 8.8. Astrolabe by Adrian Zeelst, *c.* 1590 (Cologne or Louvain). Cologne, Kölnisches Stadtmuseum (inv. 1.184). Photo: © Rheinisches Bildarchiv Köln. Zeelst's exceptionally large astrolabe with a universal stereographic projection on the reverse including the arc-shaped zodiac, brachiolum and index, as represented in the engraving.

astrolabe was made on the occasion of the publication of the treatise of 1602 as an official gift for the prince-bishop who had commissioned the treatise.

C. *Zeelst's Additions on the Medieval* Saphea

On side A, Zeelst completed the medieval *saphea* with the following engravings:

1. On the outside of the 360° scale are the gothic numerals 0–360 for every 10° in a clockwise direction starting from the top. Zeelst added two systems on the inside of the 360° scale: (1) for the hours in Roman numerals every 15°, numbered twice I–XII; and (2) alongside for 360° in italic numbers, anticlockwise for every 10° starting from the east (on the left when the throne is upwards).
2. The projection of the *saphea* is medieval but all the engraved numbers on its surface are by Zeelst, including the symbols of the zodiac on the ecliptic.

The arc with the projected zodiac is medieval but Zeelst attached this to the straight rule which he also engraved with a zodiacal scale, italic numbers and the signs of the zodiac. It is interesting to see how he imitated the gothic letters on the place where the newly added brass attached to the medieval part (Aries in particular, Fig. 8.2). The graduated ruler or *index* with its italic numbers is also an addition of Zeelst, as well as the *brachiolum*.

On side B are more sixteenth-century additions:

1. The stereographic projection for (calculated value) latitude 51° occupying a large part of the upper half (almucantars) but with the concentric circles of the tropics and equator also below the horizon.
2. The projected ecliptic, with a scale graduated for single degrees with every 5th and 10th degree marked with a longer line and with the twelve constellations indicated with their symbols.
3. On the medieval set of unequal hour lines on the right-hand side on top of the almucantar, Zeelst has added the hours 7–12 to supplement the typically medieval pattern of 5–1.
4. The arcs for the twelve astrological houses, numbered counter-clockwise in Roman numerals and their engraved arcs are marked with small hachures, like pecked arcs.
5. The lower half has two sets of scales:
    a. A shadow square of ten, numbered in five groups of two (2–4-6–8-10) with each group subdivided into ten units, giving a total of fifty units, named in italic letters *Latus rectum* and *Latus versum*, located on the inside of the larger medieval hour scale.

b. Between the tropic of Cancer and the tropic of Capricorn he engraved a set of twelve dotted arcs numbered clockwise 1–12 in italic numbers for calculating the unequal hours.

The alidade with fixed sights and two *brachiola* are also additions by Zeelst.

It is the layout and graphic design of this side that are almost identical to the illustration in the Stempel-Zeelst treatise of 1602.

4. Conclusion: Artisanal Knowledge and 'Superior Artisans'

The remarkable Oxford *Saphea* makes a strong case for artisanal knowledge, where the physical instrument, with all its historical layers, becomes a primary historical source. More often than not, scientific instruments relate to a body of knowledge embedded in manuscripts and published treatises. In this case, however, it seems that a Renaissance treatise was inspired by a medieval instrument, which a Renaissance mathematician further completed. The Oxford *Saphea* thus precedes a treatise on which more traditional historical research would have focused strongly. It is a clear — and beautiful — example of how advanced and innovative knowledge lies embedded in a brass object.

The high quality and sense of detailing of the instrument also show the advanced skill of its makers — despite the time span of 150 years — who were experts in refined craftsmanship and theoretical astronomical knowledge. They were 'superior artisans', blurring the boundaries between artisan-scholar, handworker-theorist or, eventually, practice-theory.[29] The instrument is also a witness of tacit knowledge, beyond the knowledge contained in the treatises. Like a palimpsest, the instrument shows the various layers of transmission and accumulation of astronomical and mathematical knowledge, spanning almost five centuries: the projections were conceived in Toledo *c.* 1080; the instrument was made in France *c.* 1450; and it was eventually completed in Liège by a Louvain-trained instrument maker and mathematician around 1600.

29 Initially this approach was developed in Zilsel, 'Sociological Roots of Science', and then further elaborated in Long, *Artisan/Practitioners and the Rise of the New Sciences*. It also echoes the concept of 'Instrument Epistemology' where a materialistic conception of knowledge is argued, or even that 'instruments are not in the intellectual basement, but they occupy the same floor as our greatest theoretical contributions to understand the world': Baird, *Thing Knowledge*, p. xvii. Derek Price advocated for a proper status of instruments *per se* as carriers of a unique kind of knowledge: 'It is unfortunate that so many historians of science and virtually all of the philosophers of science are born-again theoreticians instead of bench scientists': Price, 'Philosophical Mechanism and Mechanical Philosophy', p. 75.

Bibliography

Manuscript

Munich, Bayerische Staatsbibliothek, MS lat. 25026

Primary Sources

Azarchel (ʿAlī ibn Khalaf and Abū Isḥāq Ibrāhīm al-Zarqālī), *Tractat de l'assafea* (*Don Profeit Tibbon Tractat de l'assafea d'Azarquiel*), Hebrew trans. by Jacob ben Makir ibn Tibbon, ed. and intro. by Josep Maria Millàs Vallicrosa (Barcelona: Editorial Alpha, 1933)

Blagrave, John, *The Mathematical Jewel* (London: Walter Venge, 1585)

Reinerus Gemma Frisius, Cornelius Gemma, Jan Steels, and Joannes Grapheus, *De astrolabo catholico liber quo latissime patentis instrumenti multiplex vsus explicatur, & quicquid vspiam rerum mathematicarum tradi possit continetur* (Antwerp, Ioannis Steels, 1556)

Stempel, Gerard, and Adrian Zeelst, *Utriusque astrolabii tam particularis quam universalis fabrica et usus* (Liège: Ouwerx, 1602)

van Maelcote, Odo, *Astrolabium aequinoctiale* (Brussels: Rutgerum Velpium, 1607)

Secondary Studies

Baird, Davis, *Thing Knowledge: A Philosophy of Scientific Instruments* (Berkeley: University of California Press, 2004)

Bonhams catalogue, *Scientific and Mechanical Musical Instruments and Cameras* (London, 4 November 2014)

Calvo, Emilia, 'Ibn Bāso's Universal Plate and Its Influence on European Astronomy', *Scientiarum Historia*, 18 (1992), 61–70

Dieckhoff, Reiner, 'Cosmographia planisphaeria. Ein Arsenius-Astrolabium des späten 16. Jahrhunderts im Kölnischen Stadtmuseum', *Kölner Museums-Bulletin. Berichte und Forschungen aus den Museen der Stadt Köln*, 2 (1990), 23–44

Fernández Fernández, Laura, 'Astrolabes on Parchment: The Astrolabes Depicted in Alfonso X's *Libro del saber de astrologia* and their Relationship to Contemporary Instruments', *Medieval Encounters*, 22 (2017), 287–310

Gunther, Robert Theodore, *The Astrolabes of the World*, 2 vols (Oxford: Oxford University Press, 1932; repr. London: Holland Press, 1976), II: 319–32

Kennedy, Edward Stuart, *Studies in the Islamic Exact Sciences* (Beirut: American University of Beirut, 1983)

King, David, *In Synchrony with the Heavens: Studies in Astronomical Timekeeping and Instrumentation in Medieval Islamic Civilization*, 2 vols, Islamic Philosophy, Theology and Science, 55 (Leiden: Brill, 2004–2005)

——, 'The Astrolabe: What It Is and What It Is Not (A Supplement to the Standard Literature)' (2018) <https://www.researchgate.net/publication/331327866_The_astrolabe_-_what_it_is_what_it_is_not> [accessed 19 April 2023]

Levenson, Jay A., ed., *Circa 1492: Art in the Age of Exploration* (New Haven: Yale University Press, 1991)

Long, Pamela O., *Artisan/Practitioners and the Rise of the New Sciences, 1400–1600* (Corvallis: Oregon State University Press, 2011)

Michel, Henri, *Traité de l'astrolabe* (Paris: Gauthier-Villars, 1947)

Moreno, Roberto, David King, and Koenraad Van Cleempoel, 'A Recently Discovered Sixteenth-Century Spanish Astrolabe', *Annals of Science*, 59 (2002), 331–62

North, John D., 'The Astrolabe', *The Scientific American*, 230 (1974), 96–107

Poulle, Emmanuel, 'Peut-on dater les astrolabes médiévaux?', *Revue d'Histoire des Sciences et de leurs applications*, 9 (1956), 301–22

——, 'Un instrument astronomique dans l'occident latin, la "saphea"', *Studi medievali*, Series 3, 10 (1969), 491–510

——, 'John of Lignères, or Johannes de Lineriis', in *Complete Dictionary of Scientific Biography*, ed. by Charles Coulston Gillispie (Detroit: Charles Scribner's Sons, 2008), pp. 123–28

Price, Derek, 'Philosophical Mechanism and Mechanical Philosophy: Some Notes Towards a Philosophy of Scientific Instruments', *Annali dell' Istituto e Museo di Storia della Scienza di Firenze*, 5 (1980), 75–85

Puig, Roser, 'Zarqālī: Abū Isḥāq Ibrāhīm ibn Yaḥyā al-Naqqāsh al-Tujībī al-Zarqālī', in *Biographical Encyclopedia of Astronomers*, 2 vols, Springer Reference, ed. by Thomas A. Hockey (New York: Springer, 2007)

Turner, Gerard l'Estrange, 'The Three Astrolabes of Gerard Mercator', *Annals of Science*, 51 (1994), 329–53

——, *Elizabethan Instrument Makers: The Origins of the London Trade in Precision Instrument Making* (Oxford: Oxford University Press, 2000)

Van Cleempoel, Koenraad, *A Catalogue Raisonné of Scientific Instruments from the Louvain School, 1530 to 1600* (Turnhout: Brepols, 2002)

——, 'The Migration of "Materialised Knowledge" from Flanders to Spain in the Person of the Sixteenth-Century Flemish Instrument Maker Pertus Ab Aggere', in *Silent Messengers: The Circulation of Material Objects of Knowledge in the Early Modern Low Countries*, ed. by Sven Dupré and Christoph Herbert Lüthy (Berlin: LIT Verlag, 2011), pp. 69–88

Zilsel, Edgar, 'The Sociological Roots of Science', *American Journal of Sociology*, 47 (1942), 544–62

Zinner, Ernst, *Deutsche und Niederländische Astronomische Instrumente des 11.-18. Jahrhunderts*, 2nd edn (Munich: Beck, 1967)

Bibliography of Works by Charles S. F. Burnett

Sections

I. Books and Articles over 100 Pages Long

II. Edited Volumes

III. Articles and Pamphlets arranged Thematically, and in Chronological Order within each Topic

- **A. The Arabic-Latin Translators**
- **B. Natural Science and Philosophy**
- **C. Arithmetic and Geometry**
- **D. Astronomy and Astrology**
- **E. Medicine and Psychology**
- **F. Magic and Divination**
- **G. Anglo-Norman Science and Learning in the Twelfth Century**
- **H. Peter Abelard and the French Schools**
- **I. Music**
- **J. Contacts between the West and the Far East**
- **K. Miscellaneous**

IV. Editions of Arabic and Latin (and some Greek and Hebrew) texts in the books and articles catalogued

* = Article reprinted in *Magic and Divination in the Middle Ages: Texts and Techniques in the Islamic and Christian Worlds*, Variorum Collected Studies Series, C557 (Aldershot: Variorum, 1996)

‡ = Article reprinted in *Arabic into Latin in the Middle Ages: The Translators and their Intellectual and Social Context*, Variorum Collected Studies Series, CS939 (Farnham: Ashgate, 2009)

§ = Article reprinted in *Numerals and Arithmetic in the Middle Ages*, Variorum Collected Studies Series, CS967 (Farnham: Ashgate Variorum, 2010)

I. Books and Articles over 100 Pages Long

1. Hermann of Carinthia, *De essentiis*, ed., trans. and commentary (Leiden: Brill, 1982)
2. 'A Checklist of the Manuscripts Containing Writings of Peter Abelard and Heloise and Other Works Closely Associated with Abelard and his School', with David Luscombe and Julia Barrow, *Revue d'histoire des textes*, 14–15 (1984–1985), 183–302
3. Pseudo-Bede, *De mundi celestis terrestrisque constitutione: A Treatise on the Universe and the Soul*, ed., trans. and commentary, Warburg Institute Surveys and Texts, 10 (London: Warburg Institute, 1985)
4. 'Zādānfarrūkh al-Andarzaghar on Anniversary Horoscopes', with Ahmed al-Hamdi, *Zeitschrift für Geschichte der arabisch-islamischen Wissenschaften*, 7 (1991–1992), 294–398
5. Al-Kindī, *Iudicia, The Two Latin Versions*, edition (London: privately printed, 1993)
6. Abū Maʿšar, *The Abbreviation of the Introduction to Astrology, together with the Medieval Latin translation of Adelard of Bath*, ed. and trans. with Michio Yano and Keiji Yamamoto (Leiden: Brill, 1994)
7. *Jesuit Plays on Japan and English Recusancy*, with Masahiro Takenaka, Renaissance Monographs, 21 (Tokyo: Renaissance Institute, Sophia University, 1995)
8. '*Algorismi vel helcep decentior est diligentia*: The Arithmetic of Adelard of Bath and His Circle', in *Mathematische Probleme im Mittelalter: Der lateinische und arabische Sprachbereich*, ed. by Menso Folkerts (Wiesbaden: Harrassowitz, 1996), pp. 221–331
9. *Magic and Divination in the Middle Ages: Texts and Techniques in the Islamic and Christian Worlds*, Variorum Collected Studies Series, C557 (Aldershot: Variorum, 1996)
10. *The 'Liber Aristotilis' of Hugo of Santalla*, ed. with David Pingree, Warburg Institute Surveys and Texts, 26 (London: Warburg Institute, 1997)
11. *The Introduction of Arabic Learning into England*, The Panizzi Lectures, 1996 (London: British Library, 1997)
12. Abū Maʿšar, *The Abbreviation of the Introduction to Astrology*, trans. by Charles Burnett, annotations by C. Burnett, G. Tobyn, G. Cornelius, and V. Wells (ARHAT Publications, 1997)
13. *Adelard of Bath, Conversations with His Nephew: 'On the Same and the Different', 'Questions on Natural Science'*, and '*On Birds*', ed. and trans. with collaboration of Italo Ronca, Pedro Mantas-España, and Baudouin van den Abeele (Cambridge: Cambridge University Press, 1998)
14. *Scientific Weather Forecasting in the Middle Ages: The Writings of Al-Kindī. Studies, Editions, and Translations of the Arabic, Hebrew and Latin Texts*, with Gerrit Bos (London: Kegan Paul International, 2000)
15. *Abū Maʿšar on Historical Astrology, The Book of Religions and Dynasties (On the Great Conjunctions)*, ed. and trans. with Keiji Yamamoto, 2 vols, Islamic Philosophy, Theology, and Science, 33–34 (Leiden: Brill, 2000)

16. Hermes Trismegistus, *Astrologica et divinatoria*, ed. with Gerrit Bos, Thérèse Charmasson, Paul Kunitzsch, Fabrizio Lelli, and Paolo Lucentini, Hermes Latinus, IV/4, Corpus Christianorum Continuatio Mediaevalis, 144C (Turnhout: Brepols, 2001)

17. Al-Qabīṣī (Alcabitius), *The Introduction to Astrology*, Arabic and Latin editions and English trans. with Keiji Yamamoto and Michio Yano, Warburg Institute Studies and Texts, 2 (London: Warburg Institute, 2004)

18. *Abbreviatio Petri Abaelardi Expositionis in Hexameron*, edition, in *Expositio in Hexameron, Abbreviatio Petri Abelaerdi Expositionis in Hexameron*, ed. by Mary Romig, David Luscombe and Charles Burnett, Petri Abaelardi Opera Theologica, V, Corpus Christianorum Continuatio Mediaevalis, 15 (Turnhout: Brepols, 2004)

19. '*Sefer ha-Middot*: A Mid-Twelfth-Century Text on Arithmetic and Geometry Attributed to Abraham Ibn Ezra', with Tony Lévy, *Aleph*, 6 (2006), 57–238

20. *Ibn Baklarish's Book of Simples: Medical Remedies between Three Faiths in Twelfth-Century Spain*, edition, Studies in the Arcadian Library, 3 (London: Arcadian Library, 2008)

21. *Arabic into Latin in the Middle Ages: The Translators and their Intellectual and Social Context*, Variorum Collected Studies Series, CS939 (Farnham: Ashgate, 2009)

22. Raymond de Marseille, *Opera omnia: Traité d'astrolabe, Liber cursuum planetarum*, ed. with Marie-Thérèse d'Alverny and Emmanuel Poulle, Sources d'histoire médiévale, 40 (Paris: CNRS, 2009)

23. *Numerals and Arithmetic in the Middle Ages*, Variorum Collected Studies Series, CS967 (Farnham: Ashgate Variorum, 2010)

24. *Textes médiévaux de scapulomancie*, ed. by Stefano Rapisarda, with collaboration of Charles Burnett, trans. by Benoît Grévin, Marco Miano, and Stéphanie Vlavianos, Textes littéraires du Moyen Âge, 43 Serie Divinatoria, 6 (Paris: Classiques Garnier, 2017)

25. Maimonides' *On Coitus*, Arabic ed. and English trans. by Gerrit Bos, medieval Hebrew translation ed. by Gerrit Bos, medieval Latin translation ed. by Charles Burnett, and Slavonic trans. by William Francis Ryan and Moshe Taube, Medical Works of Moses Maimonides, 11 (Leiden: Brill, 2019)

26. *The Great Introduction to Astrology by Abū Maʿšar*, Arabic ed. and trans. with Keiji Yamamoto, ed. of Greek version by David Pingree, Islamic Philosophy, Theology and Science, 106, 2 vols (Leiden: Brill, 2019)

27. 'A Newly Discovered Treatise by Abraham ibn Ezra and Two Treatises Attributed to al-Kindī in a Latin Translation by Henry Bate', with Shlomo Sela, Carlos Steel, C. Philipp E. Nothaft, and David Juste, *Mediterranea: International Journal on the Transfer of Knowledge*, 5 (2020), 193–305

28. *Thābit ibn Qurra 'On Talismans' and Pseudo-Ptolemy 'On Images 1–9' together with the 'Liber prestigiorum Thebidis' of Adelard of Bath*, with Gideon Bohak, Micrologus Library, 106 (Florence: SISMEL–Edizioni del Galluzzo, 2021)

II. Edited Volumes

29. *Adelard of Bath: An English Scientist and Arabist of the Early Twelfth Century*, ed., Warburg Institute Surveys and Texts, 14 (London: Warburg Institute, 1987)

30. *The Second Sense: Studies in Hearing and Musical Judgement from Antiquity to the Seventeenth Century*, ed. with Michael Fend and Penelope M. Gouk, Warburg Institute Surveys and Texts, 22 (London: Warburg Institute, 1991)

31. *Glosses and Commentaries on Aristotelian Logical Texts: The Syriac, Arabic and Medieval Latin Traditions*, ed., Warburg Institute Surveys and Texts, 23 (London: Warburg Institute, 1993)

32. *Constantine the African and ʿAlī ibn al-ʿAbbās al-Maǧūsī: The 'Pantegni' and Related Texts*, ed. with Danielle Jacquart (Leiden: Brill, 1994)

33–36. Edition of four volumes of collected articles of Marie-Thérèse d'Alverny for Variorum reprints: *Études sur le symbolisme de la Sagesse et sur l'iconographie*, 1993, *La connaissance de l'Islam dans l'Occident médiéval*, 1994, *La transmission des textes philosophiques et scientifiques au moyen âge*, 1994, and *Pensée médiévale en Occident*, 1995

37. *Hildegard of Bingen: The Context of Her Thought and Art*, ed. with Peter Dronke, Warburg Institute Colloquia, 4 (London: Warburg Institute, 1998)

38. *Islam and the Italian Renaissance*, ed. with Anna Contadini, Warburg Institute Colloquia, 5 (London: Warburg Institute, 1999)

39. *Studies in the History of the Exact Sciences in Honour of David Pingree*, ed. with Jan Pieter Hogendijk, Kim Plofker, and Michio Yano, Islamic Philosophy, Theology, and Science, 54 (Leiden: Brill, 2004)

40. *Britannia Latina: Latin in the Culture of Great Britain from the Middle Ages to the Twentieth Century*, ed. with Nicholas Mann, Warburg Institute Colloquia, 8 (London: Warburg Institute, 2005)

41. *Scientia in Margine: Études sur les marginalia dans les manuscrits scientifiques du moyen âge à la Renaissance*, ed. with Danielle Jacquart (Geneva: Droz, 2005)

42. *Hebrew Medical Astrology: David Ben Yom Tov, Kelal Qatan. Original Hebrew Text, Medieval Latin Translation, Modern English Translation*, ed. with Gerrit Bos and Tzvi Langermann, Transactions of the American Philosophical Society, 95, Part 5 (Philadelphia: American Philosophical Society, 2005)

43. *Magic and the Classical Tradition*, ed. with William Francis Ryan, Warburg Institute Colloquia, 7 (London: Warburg Institute, 2006)

44. *Continuities and Disruptions between the Middle Ages and the Renaissance*, ed. with José Meirinhos and Jacqueline Hamesse, Textes et études du Moyen Âge, 48 (Louvain-la-Neuve: Brepols, 2008)

45. *The Winding Courses of the Stars: Essays in Ancient Astrology*, ed. with Dorian Gieseler Greenbaum, *Culture and Cosmos*, 11, nos 1 and 2 (2008)

46. *Astro-Medicine: Astrology and Medicine, East and West*, ed. with Anna Akasoy and Ronit Yoeli-Tlalim, Micrologus Library, 25 (Florence: SISMEL–Edizioni del Galluzzo, 2008)

47. *The Word in Medieval Logic, Theology and Psychology*, ed. with Tetsuro Shimizu, Rencontres de philosophie médiévale, 14 (Turnhout: Brepols, 2009)

48. *Ancient and Medieval Alchemy*, ed. with Bink Hallum, *Ambix*, 56/1 (2009)

49. *Between Orient and Occident: Transformation of Knowledge*, ed. with Benno van Dalen, *Annals of Science*, 68/4 (2011)

50. *Islam and Tibet: Interactions along the Musk Routes*, ed. with Anna Akasoy and Ronit Yoeli-Tlalim (Farnham: Ashgate, 2011)

51. *Ptolemy's Geography in the Renaissance*, ed. with Zur Shalev, Warburg Institute Colloquia, 17 (London: Warburg Institute, 2011)

52. *Ritual Healing: Magic, Ritual and Medical Therapy from Antiquity until the Early Modern Period*, ed. with Ildikó Csepregi, Micrologus' Library, 48 (Florence: SISMEL–Edizioni del Galluzzo, 2012)

53. *Medieval Arabic Thought, Essays in Honour of Fritz Zimmermann*, ed. with Rotraud Hansberger and Muhammad Afifi al-Akiti, Warburg Institute Studies and Texts, 4 (London: Warburg Institute, 2012)

54. *Greek into Latin from Antiquity until the Nineteenth Century*, ed. with John Glucker, Warburg Institute Colloquia, 18 (London: Warburg Institute, 2012)

55. *Rashīd al-Dīn: Agent and Mediator of Cultural Exchanges in Ilkhanid Iran*, ed. with Anna Akasoy and Ronit Yoeli-Tlalim, Warburg Institute Colloquia, 24 (London: Warburg Institute, 2013)

56. *Time, Astronomy, and Calendars in the Jewish Tradition*, ed. with Sacha Stern, Time, Astronomy, and Calendars, 3 (Leiden: Brill, 2014)

57. *Mapping Knowledge: Cross-Pollination in Late Antiquity and the Middle Ages*, ed. with Pedro Mantas-España, Arabica Veritas, 1 (Córdoba – London: Oriens Academic–Córdoba Near Eastern Research Unit – Warburg Institute, 2014)

58. *From Māshā'allāh to Kepler: Theory and Practice in Medieval and Renaissance Astrology*, ed. with Dorian Gieseler Greenbaum (Ceredigion, Wales: Sophia Centre Press, 2015)

59. Ex Oriente lux: *Translating Words, Scripts and Styles in Medieval Mediterranean Society*, ed. with Pedro Mantas-España, Arabica Veritas, 2 (Córdoba – London: University of Córdoba Press–Córdoba Near Eastern Research Unit – Warburg Institute, 2016)

60. *The Teaching and Learning of Arabic in Early Modern Europe*, ed. with Jan Loop and Alastair Hamilton, History of Oriental Studies, 3 (Leiden: Brill, 2017)

61. *Astrolabes in Medieval Cultures*, ed. with Josefina Rodríguez-Arribas, Silke Ackermann, and Ryan Szpiech (Leiden: Brill, 2019), edited reprint of *Medieval Encounters*, 23 (2017)

62. *Spreading Knowledge in a Changing World*, ed. with Pedro Mantas-España, Arabica Veritas, 3 (Córdoba – London: University of Córdoba Press–Córdoba Near Eastern Research Unit – Warburg Institute, 2019)

63. *Pregnancy and Childbirth in the Premodern World: European and Middle Eastern Cultures, from Late Antiquity to the Renaissance*, ed. with Costanza Gislon Dopfel and Alessandra Foscati, Cursor Mundi, 36 (Turnhout: Brepols, 2019)
64. *Ptolemy's Science of the Stars in the Middle Ages*, ed. with David Juste, Benno van Dalen, and Dag Nikolaus Hasse, Ptolemaeus Arabus et Latinus Studies, 1 (Turnhout: Brepols, 2020)
65. *Falconry in the Mediterranean Context during the Pre-Modern Era*, ed. with Baudouin van den Abeele, Bibliotheca Cynegetica, 9 (Geneva: Droz, 2021)
66. *Why Translate Science? Documents from Antiquity to the 16th Century in the Historical West (Bactria to the Atlantic)*, ed. with Dimitri Gutas and Uwe Vagelpohl (Leiden: Brill, 2022)
67. *Mark of Toledo: Intellectual Context and Debates between Christians and Muslims in Early Thirteenth-Century Iberia*, ed. with Pedro Mantas-España, Arabica Veritas, 4 (Córdoba – London: University of Córdoba Press–Córdoba Near Eastern Research Unit – Warburg Institute, 2022)
68. *A Cultural History of Chemistry in the Middle Ages*, ed. with Sébastien Moureau, A Cultural History of Chemistry, 6 vols (London: Bloomsbury Academic, 2022), II

III. Articles and Pamphlets arranged Thematically, and in Chronological Order within each Topic

A. The Arabic-Latin Translators

1. 'A Group of Arabic-Latin Translators Working in Northern Spain in the Mid-Twelfth Century', *Journal of the Royal Asiatic Society*, 109 (1977), 62–108
2. 'Arabic into Latin in Twelfth-Century Spain: The Works of Hermann of Carinthia', *Mittellateinisches Jahrbuch*, 13 (1978), 100–34
3. 'The Impact of Arabic Science on Western Civilisation in the Middle Ages', *Bulletin of the British Association of Orientalists*, 11 (1979–1980), 40–51
4. 'Plato of Tivoli' and 'Translations: Western European', in *Dictionary of the Middle Ages*, ed. by Joseph R. Strayer, 13 vols (New York: Scribner, 1982–1989), IX (1987): 704–05, and XII (1989): 136–42
5. 'Some Comments on the Translating of Works from Arabic into Latin in the Mid-Twelfth Century', in *Orientalische Kultur und europäisches Mittelalter*, ed. by Albert Zimmerman, Ingrid Craemer-Ruegenberg, and Gudrun Vuillemin-Diem, Miscellanea Mediaevalia, 17 (Berlin: De Gruyter, 1985), pp. 161–71
6. 'Literal Translation and Intelligent Adaptation amongst the Arabic-Latin Translators of the First Half of the Twelfth Century', in *La diffusione delle scienze islamiche nel Medio Evo Europeo*, ed. by Biancamaria Scarcia Amoretti (Rome: Accademia Nazionale dei Lincei, 1987), pp. 9–28

7. 'Hermann of Carinthia', in *A History of Twelfth-century Western Philosophy*, ed. by Peter Dronke (Cambridge: Cambridge University Press, 1988), pp. 386–406

*8. 'The Translating Activity in Medieval Spain', in *The Legacy of Muslim Spain*, ed. by Salma Khadra Jayyusi, Handbuch der Orientalistik, 12 (Leiden: Brill, 1992), pp. 1036–58

9. 'Cultural Contacts Between Christians and Muslims in Northern Spain in the Middle Ages', *Bulletin of the Confraternity of Saint James*, 42 (1992), 22–25

10. Introduction to Catalogue *Arabic Science and Medicine: A Collection of Manuscripts and Early Printed Books Illustrating the Spread and Influence of Arabic Learning in the Middle Ages and the Renaissance* (London: Bernard Quaritch, 1993), pp. 3–6

‡11. 'Michael Scot and the Transmission of Scientific Culture from Toledo to Bologna via the Court of Frederick II Hohenstaufen', in *Le scienze alla corte di Federico II / Sciences at the Court of Frederick II, Micrologus*, 2 (1994), 101–26

12. '*Magister Iohannes Hispanus*: Towards the Identity of a Toledan Translator', in *Comprendre et maîtriser la nature au moyen âge: Mélanges d'histoire des sciences offerts à Guy Beaujouan*, Hautes études médiévales et modernes, 73 (Geneva: Droz, 1994), pp. 425–36

13. 'The Institutional Context of Arabic-Latin Translations of the Middle Ages: A Reassessment of the "School of Toledo"', in *Vocabulary of Teaching and Research between the Middle Ages and Renaissance*, ed. by Olga Weijers, CIVICIMA, Études sur le vocabulaire intellectuel du moyen âge, 8 (Turnhout: Brepols, 1995), pp. 214–35

‡14. 'Master Theodore, Frederick II's Philosopher', in *Federico II e le nuove culture, Atti del XXXI Convegno storico internazionale, Todi, 9–12 ottobre 1994* (Spoleto: Centro Italiano di Studi sull'Alto Medioevo, 1995), pp. 225–85

‡15. '"Magister Iohannes Hispalensis et Limiensis" and Qusṭā ibn Lūqā's *De differentia spiritus et animae*: A Portuguese Contribution to the Arts Curriculum?', *Mediaevalia, Textos e estudos*, 7–8 (1995), 221–67

16. 'The Works of Petrus Alfonsi: Questions of Authenticity', *Medium Ævum*, 66 (1997), 42–79; Spanish version: 'Las obras de Pedro Alfonso: Problemas de autenticidad', in *Estudios sobre Pedro Alfonso de Huesca*, ed. by María Jesús Lacarra (Huesca: Instituto de Estudios Altoaragoneses, 1996), pp. 313–48

17. 'Translating from Arabic into Latin in the Middle Ages: Theory, Practice, and Criticism', in *Éditer, traduire, interpreter: Essais de methodologie philosophique*, ed. by Steve G. Lofts and Philipp W. Rosemann, Philosophes médiévaux, 36 (Louvain-la-Neuve: Éditions de l'Institut Supérieur de Philosophie, 1997), pp. 55–78

18. 'Tolède: Le réveil des Latins', *Les Cahiers de science et vie* (February 1998), 24–29

19. 'The Second Revelation of Arabic Philosophy and Science', in *Islam and the Italian Renaissance*, ed. by Charles Burnett and Anna Contadini, Warburg Institute Colloquia, 5 (London: Warburg Institute, 1999), pp. 185–98

20. 'Dialectic and Mathematics according to Aḥmad ibn Yūsuf: A Model for Gerard of Cremona's Programme of Translation and Teaching?', in *Langage, sciences, philosophie au xiie siècle*, ed. by Joël Biard, Sic et Non (Paris: Vrin, 1999), pp. 83–92

21. 'Learned Knowledge of Arabic Poetry, Rhymed Prose, and Didactic Verse from Petrus Alfonsi to Petrarch', in *Poetry and Philosophy in the Middle Ages. A Festschrift for Peter Dronke*, ed. by John Marenbon, Mittellateinische Studien und Texte, 29 (Leiden: Brill, 2000), pp. 29–62

‡22. 'Antioch as a Link between Arabic and Latin Culture in the Twelfth and Thirteenth Centuries', in *Occident et Proche-Orient: Contacts scientifiques au temps des croisades*, ed. by Isabelle Draelants, Anne Tihon, and Baudouin van den Abeele (Louvain-la-Neuve: Brepols, 2000), pp. 1–78

‡23. 'The Coherence of the Arabic-Latin Translation Program in Toledo in the Twelfth Century', *Science in Context*, 14 (2001), 249–88; revision of *The Coherence of the Arabic-Latin Translation Programme in Toledo in the Twelfth Century*, Max-Planck-Institut für Wissenschaftsgeschichte, Preprint 78 (1997)

24. 'The Strategy of Revision in the Arabic-Latin Translations from Toledo: The Case of Abū Maʿshar's *On the Great Conjunctions*', in *Les traducteurs au travail: Leurs manuscrits et leurs méthodes*, ed. by Jacqueline Hamesse, Textes et études du Moyen Âge, 18 (Louvain-la-Neuve: Brepols, 2001), pp. 51–113 and 529–40 (plates)

‡25. 'John of Seville and John of Spain: A *mise au point*', *Bulletin de philosophie médiévale*, 44 (2002), 59–78

26. 'The Translation of Arabic Science into Latin: A Case of Alienation of Intellectual Property?', *Bulletin of the Royal Institute for Inter-Faith Studies* (Amman), 4 (2002), 145–57

27. 'Hugo of Santalla', 'John of Seville', 'Picatrix', and 'Translations – Scientific, Philosophical and Literary (Arabic)', in *Medieval Iberia: An Encyclopedia*, ed. by E. Michael Gerli (New York – London: Routledge, 2003), pp. 401, 446, 650 and 801–04

28. 'The Transmission of Arabic Astronomy via Antioch and Pisa in the Second Quarter of the Twelfth Century', in *The Enterprise of Science in Islam: New Perspectives*, ed. by Jan P. Hogendijk and Abdelhamid I. Sabra, Dibner Institute studies in the history of science and technology (Cambridge, MA – London: MIT Press, 2003), pp. 23–51

29. 'Myth and Astronomy in the Frescoes at Sant'Abbondio in Cremona', with Marika Leino, *Journal of the Warburg and Courtauld Institutes*, 66 (2003), 273–88

30. 'Translation and Transmission of Greek and Islamic Science to Latin Christendom', in *The Cambridge History of Science*, multiple vols (Cambridge: Cambridge University Press, 2003-), II (2013): *Medieval Science*, ed. by David C. Lindberg and Michael H. Shank, 341–64

31. 'Translation from Arabic into Latin in the Middle Ages' and 'Aristotle in Translation in Medieval Europe', in *Übersetzung: ein internationales Handbuch zur Übersetzungsforschung / Translation: An International Encyclopedia of Translation Studies / Traduction: Encyclopédie internationale de la recherche sur la traduction*, ed. by Harald Kittel, Armin Paul Frank, Norbert Greiner, Theo Hermans, Werner Koller, José Lambert, and Fritz Paul, together with Juliane House and Brigitte Schultze, 3 vols (Berlin – New York: De Gruyter Mouton, 2004–2011), II (2007): 1231–37 and 1308–10

32. 'The Translation of Arabic Works on Logic into Latin in the Middle Ages and the Renaissance', in *Handbook of the History of Logic*, ed. by Dov M. Gabbay and John Woods, multiple vols (Amsterdam – London: Elsevier North Holland, 2004-), I (2004): 597–606

33. 'Adelard of Bath', 'Alfred of Shareshill', 'Daniel of Morley', 'Petrus Alfonsi', 'Robert of Ketton', and 'Roger of Hereford', *Oxford Dictionary of National Biography* (Oxford: Oxford University Press, 2004-), print and online

34. 'Abū Maʿshar', 'Adelard of Bath', 'Alfred of Sareschel', 'Arabic Numerals', 'Gerard of Cremona', 'Hermann of Carinthia', 'Hugh of Santalla', 'John of Seville', 'Mark of Toledo', 'Michael Scot', 'Thābit ibn Qurra', 'Translation Norms and Practices', in *Medieval Science, Technology, and Medicine: An Encyclopedia*, ed. by Thomas F Glick, Steven J. Livesey, and Faith Wallis (New York – London: Routledge, 2005), 4–5, 5–6, 26–27, 39–40, 191–92, 220–21, 231–32, 292–93, 327–28, 344–45, 472–73 and 486–88

35. 'The Decline of Poetry in the Translations from Arabic and Greek into Latin in the Twelfth Century', in *Poesía latina medieval (siglos V–XV). Actas del IV Congreso del 'Internationales Mittellateinerkomitee', Santiago de Compostela, 12–15 de septiembre de 2002*, ed. by Manuel C. Díaz y Díaz and José M. Díaz de Bustamante (Florence: SISMEL–Edizioni del Galluzzo, 2005), pp. 1069–75

36. 'Arabic into Latin: The Reception of Arabic Philosophy into Western Europe', in *The Cambridge Companion to Arabic Philosophy*, ed. by Peter Adamson and Richard C. Taylor (Cambridge: Cambridge University Press, 2005), pp. 370–404

37. 'Humanism and Orientalism in the Translations from Arabic into Latin in the Middle Ages', in *Wissen über Grenzen: Arabisches Wissen und lateinisches Mittelalter*, ed. by Andreas Speer and Lydia Wegener, Miscellanea Mediaevalia, 33 (Berlin – New York: De Gruyter, 2006), pp. 22–31

38. 'Stephen, the Disciple of Philosophy, and the Exchange of Medical Learning in Antioch', *Crusades*, 5 (2006), 113–29

39. 'The "Translation" of Diagrams and Illustrations from Arabic into Latin', in *Arab Painting: Text and Image in Illustrated Arabic Manuscripts*, ed. by Anna Contadini (Leiden: Brill 2007), pp. 161–76

40. 'Abū Maʿshar', 'Adelard of Bath', 'Alfonso the Wise', 'Astrology', 'Gerard of Cremona', and 'Hermann of Carinthia', in *Encyclopaedia of Islam, THREE*, ed. by Gudrun Krämer, Denis Matringe, John Abdallah Nawas, and Everett K. Rowson, 3rd ed. (Leiden: Brill, 2007-), print and online

41. 'Scientific Translations from Arabic: The Question of Revision', in *Science Translated: Latin and Vernacular Translations of Scientific Treatises in Medieval Europe*, ed. by Michèle Goyens, Pieter De Leemans, and An Smets, Mediaevalia Lovaniensia, Series 1, Studia 40 (Leuven: Leuven University Press, 2008), pp. 11–34

42. 'Arabic Philosophical Works Translated into Latin', in *The Cambridge History of Medieval Philosophy*, ed. by Robert Pasnau and Christina van Dyke, 2 vols (Cambridge: Cambridge University Press, 2010), II: 814–22; revised version of the table accompanying 'Arabic into Latin: The Reception of Arabic Philosophy into Western Europe' in *The Cambridge Companion to Arabic Philosophy*, no. 36 above)

43. 'The Arabic and Latin Tradition of Ptolemy's Almagest', and 'The Theoretical Arguments for Astrology in al-Farabi, al-Kindi and Abu Maʿshar', in *Transmission of Sciences: Greek, Syriac, Arabic and Latin*, ed. by H. Kobayashi and M. Kato (Tokyo, 2010), pp. 1–8 and 45–57

44. 'Communities of Learning in Twelfth-Century Toledo', in *Communities of Learning: Networks and the Shaping of Intellectual Identity in Europe, 1100–1500*, ed. by Constant J. Mews and John N. Crossley (Turnhout: Brepols, 2011), pp. 9–18

45. 'Plato Amongst the Arabic-Latin Translators of the Twelfth Century', in *Il Timeo: Esegesi greche, arabe, latine*, ed. by Francesco Celia and Angela Ulacco, Greco, arabo, latino: Le vie del sapere, 2 (Pisa: Pisa University Press, 2012), pp. 269–306

46. 'Revisiting the 1552–1550 and 1562 Aristotle-Averroes Editions', in *Renaissance Averroism and Its Aftermath: Arabic Philosophy in Early Modern Europe*, ed. by Anna Akasoy and Guido Giglioni, International Archives of the History of Ideas, 211 (Dordrecht: Springer, 2012), pp. 55–64

47. 'Petrus Alfonsi and Adelard of Bath Revisited', in *Petrus Alfonsi and his Dialogus: Background, Context, Reception*, ed. by Carmen Cardelle de Hartmann and Philipp Roelli, Micrologus Library, 66 (Florence: SISMEL–Edizioni del Galluzzo, 2014), pp. 77–91

48. 'The Roads of Córdoba and Seville in the Transmission of Arabic Science in Western Europe', in *Mapping Knowledge: Cross-Pollination in Late Antiquity and the Middle Ages*, ed. by Charles Burnett and Pedro Mantas-España, Arabica Veritas, 1 (Córdoba – London: Oriens Academic–Córdoba Near Eastern Research Unit – Warburg Institute, 2014), pp. 143–52

49. 'The Transmission of Science and Philosophy', in *The Cambridge World History*, 9 vols (Cambridge: Cambridge University Press, 2015), V: *Expanding Webs of Exchange and Conflict 500 CE–1500 CE*, ed. by Benjamin Z. Kedar and Merry E. Wiesner-Hanks, 339–58

50. 'Manuscripts of Latin Translations of Scientific Texts from Arabic', in *Taxonomies of Knowledge: Information and Order in Medieval Manuscripts*, ed. by Emily Steiner and Lynn Ransom (Philadelphia, The Schoenberg

Institute for Manuscript Studies, 2015), pp. 80–89; originally published digitally in *Digital Proceedings of the Lawrence J. Schoenberg Symposium on Manuscript Studies in the Digital Age*, vol. 1, issue 1 (2009), Article 1

51. 'A Mid-Seventeenth-Century View of the History of Arabic Scholarship in England. Gerard Langbaine's Notes under the Ascending Node', *Rivista di Storia e Letteratura Religiosa*, 51 (2015), 585–605

52. 'The Translator as an Authority', in *Translation and Authority: Authorities in Translation*, ed. by Pieter De Leemans and Michèle Goyens, The Medieval Translator, 16 (Turnhout: Brepols, 2016), pp. 53–67

53. 'Michael Scot', in *Grundriss der Geschichte der Philosophie: Die Philosophie des Mittelalters*, ed. by Alexander Brungs, Vilem Mudroch, and Peter Schulthess, multiple vols (Basel: Schwabe Verlag, 2017-), IV (2017): 13. Jahrhundert, 135–36

54. Foreword, in *Knowledge in Translation: Global Patterns of Scientific Exchange 1000–1800 CE*, ed. by Patrick Manning and Abigail Owen (Pittsburgh: University of Pittsburgh Press, 2018)

55. '"Arabica Veritas": Europeans' Search for "Truth" in Arabic Scientific and Philosophical Literature of the Middle Ages', in *The Diffusion of the Islamic Sciences in the Western World*, Micrologus, 28 (2020), 69–86

56. 'Imagined and Real Libraries in the Case of Medieval Latin Translators from Greek and Arabic', in *Die Bibliothek – The Library – La Bibliothèque: Denkräume und Wissensordnungen*, ed. by Andreas Speer and Lars Reuke, Miscellanea Mediaevalia, 41 (Berlin: De Gruyter, 2020), pp. 735–45

57. 'Robert of Ketton and Mark of Toledo and the Rise and Development of the Literal Translation of the Qur'an', in *The Iberian Qur'an from the Middle Ages to Modern Times*, ed. by Mercedes García-Arenal and Gerard A. Wiegers (Berlin: De Gruyter, 2022), pp. 49–68

58. 'Mark of Toledo's Rendering of the Declaration of Faith of the Almohads', in *Mark of Toledo: Intellectual Context and Debates between Christians and Muslims in Early Thirteenth-Century Iberia*, ed. by Charles Burnett and Pedro Mantas-España, Arabica Veritas, 4 (Córdoba – London: University of Córdoba Press–Córdoba Near Eastern Research Unit – Warburg Institute, 2022), pp. 39–50

59. 'The Statements of Medieval Latin Translators on Why and How They Translate Works on Science and Philosophy from Arabic', in *Why Translate Science? Documents from Antiquity to the 16th Century in the Historical West (Bactria to the Atlantic)*, ed. by Dimitri Gutas, Charles Burnett, and Uwe Vagelpohl (Leiden: Brill, 2022), pp. 445–87

60. 'Latin Translators from Greek in the Twelfth Century on Why and How They Translate', in *Why Translate Science? Documents from Antiquity to the 16th Century in the Historical West (Bactria to the Atlantic)*, ed. by Dimitri Gutas, Charles Burnett, and Uwe Vagelpohl (Leiden: Brill, 2022), pp. 488–524

B. Natural Science and Philosophy

1. 'A Note on the Origins of the Third Vatican Mythographer', *Journal of the Warburg and Courtauld Institutes*, 44 (1981), 160–66
2. 'High Altitude Mountaineering 1600 years ago', *Alpine Journal* (1983), p. 127
3. 'Science and Islam', *News and Events*, publication of the York Religious Education Centre (Autumn, 1983)
4. 'Scientific Speculations', in *A History of Twelfth-Century Western Philosophy*, ed. by Peter Dronke (Cambridge: Cambridge University Press, 1988), pp. 151–76
5. 'Innovations in the Classification of the Sciences in the Twelfth Century', in *Knowledge and the Sciences in Medieval Philosophy*, 3 vols (Helsinki: Yliopistopaino, 1990), II: ed. by Simo Knuuttila, Reijo Työrinoja, and Sten Ebbesen, 25–42
6. 'Report on Current Work on Islamic Philosophy and Science', *Bulletin de philosophie médiévale*, 35 (1993), 9–17
7. 'Aristotle and Averroes on Method in the Middle Ages and Renaissance: The "Oxford Gloss" to the *Physics* and Pietro d'Afeltro's *Expositio Proemii Averroys*', with Andrew Mendelsohn, in *Method and Order in Renaissance Philosophy of Nature: The Aristotle Commentary Tradition*, ed. by Daniel A. Di Liscia, Eckhard Kessler, and Charlotte Methuen (Aldershot: Ashgate, 1997), pp. 53–111
8. 'The Latin and Arabic Influences on the Vocabulary Concerning Demonstrative Argument in the Versions of Euclid's *Elements* Associated with Adelard of Bath', in *Aux origines du lexique philosophique européen: L'influence de la latinitas*, ed. by Jacqueline Hamesse, Textes et études du Moyen Âge, 8 (Louvain-la-Neuve: Fédération internationale des instituts d'études médiévales, 1997), pp. 117–35
9. 'Petrarch and Averroes: An Episode in the History of Poetics', in *The Medieval Mind: Hispanic Studies in Honour of Alan Deyermond*, ed. by Ian Richard Macpherson and Ralph J. Penny (Woodbridge: Tamesis, 1997), pp. 49–56
10. 'Encounters with Rāzī the Philosopher: Constantine the African, Petrus Alfonsi and Ramón Martí', in *Pensamiento hispano medieval: Homenaje a Horacio Santiago-Otero*, ed. by José María Soto Rábanos (Madrid: Consejo Superior de Investigaciones Científicas, 1998), pp. 973–92
11. '*Proverbia Senece et versus Ebrardi super eadem*: MS Cambridge, Gonville and Caius College, 122/59, pp. 19–29', with Barry Taylor and Martin J. Duffell, *Euphrosyne* 26 (1998), 357–78
12. 'The "Sons of Averroes with the Emperor Frederick" and the Transmission of the Philosophical Works by Ibn Rushd', in *Averroes and the Aristotelian Tradition: Sources, Constitution and Reception of the Philosophy of Ibn Rushd (1126–1198)*, ed. by Gerhard Endress and Jan A. Aertsen with the assistance of Klaus Braun (Leiden: Brill, 1999), pp. 259–99

13. 'A Note on The Origins of the *Physica Vaticana* and *Metaphysica media*', in *Tradition et traduction: Les textes philosophiques et scientifiques grecs au moyen âge latin. Hommage à Fernand Bossier*, ed. by Rita Beyers, Jozef Brams, Dirk Sacré, and Koenraad Verrycken, Ancient and Medieval Philosophy, Series 1, 25 (Leuven: Leuven University Press, 1999), pp. 59–68

14. 'Al-Kindī in the Renaissance', in *Sapientiam amemus: Humanismus und Aristotelismus in der Renaissance. Festschrift für Eckhard Keßler zum 60. Geburtstag*, ed. by Paul Richard Blum, Constance Blackwell, and Charles H. Lohr (Munich: Fink, 1999), pp. 13–30

15. 'Abū Muḥammad ʿAbdallāh ibn Rushd (Averroes junior), *On Whether the Active Intellect Unites with the Material Intellect whilst it is Clothed with the Body*: A Critical Edition of the Three Extant Medieval Versions, together with an English Translation', with Mauro Zonta, *Archives d'histoire doctrinale et littéraire du moyen âge*, 67 (2000), 295–335

16. 'The Two Faces of Averroes in the Renaissance', in *Al-Ufq al-kawnī li-fiqr Ibn Rushd*, Proceedings of a conference held in Marrakesh, 12–15 December, 1998, ed. by M. Masbahi (Marrakesh, Manshūrāt al-jamiyya al-falsafiyya al-maghribiyya, 2001), pp. 87–94

‡17. 'Physics before the *Physics*: Early Translations from Arabic of Texts concerning Nature in MSS British Library, Additional 22719 and Cotton Galba E IV', in *Medioevo*, 27 (2002), 53–109

18. 'Filosofía natural, secretos y magia', in *Historia de la ciencia y de la técnica en la Corona de Castilla*, ed. by Luis García Ballester, 4 vols (Valladolid: Junta de Castilla y León, 2002), I: 95–144

19. 'The Twelfth-Century Renaissance', in *The Cambridge History of Science*, multiple vols (Cambridge: Cambridge University Press, 2003-), II (2013): *Medieval Science*, ed. by David C. Lindberg and Michael H. Shank, 365–84

20. 'The Blend of Latin and Arabic Sources in the Metaphysics of Adelard of Bath, Hermann of Carinthia, and Gundisalvus', in *Metaphysics in the Twelfth Century: On the Relationship among Philosophy, Science and Theology*, ed. by Matthias Lutz-Bachmann, Alexander Fidora, and Andreas Niederberger, Textes et études du moyen âge, 19 (Turnhout: Brepols, 2004), pp. 41–65

21. 'Does the Sea Breathe, Boil or Bloat? A Textual Problem in Abū Maʿshar's Explanation of Tides', in *Mélanges offerts a Hossam Elkhadem par ses amis et ses élèves*, ed. by Frank Daelemans (Brussels: Archives et bibliothèques de Belgique, 2007), pp. 73–79

22. 'Science in the World of Islam', in *The World of 1607: Special Exhibition. Artifacts of the Jamestown Era from Around the World, Jamestown Settlement, Williamsburg, Virginia, April 2007-April 2008* (Williamsburg, VA: Jamestown-Yorktown Foundation, 2007), pp. 203–10

23. '*Experimentum* and *Ratio* in the Salernitan *Summa de saporibus et odoribus*', in *Expertus sum: L'expérience par les sens dans la philosophie naturelle médiévale, XIIe-XIVe siècles*, ed. by Thomas Bénatouïl and Isabelle Draelants, Micrologus' Library, 40 (Florence: SISMEL–Edizioni del Galluzzo, 2011), pp. 337–58

24. 'The Five Senses in Ramon Llull's *Liber contemplationis in Deum*', in *Gottes Schau und Weltbetrachtung: Interpretationen zum 'Liber Contemplationis' des Raimundus Lullus*, ed. by Fernando Domínguez Reboiras, Viola Tenge-Wolf, and Peter Walter, Subsidia Lulliana, 4 (Turnhout: Brepols, 2011), pp. 181–208

25. 'Coniunctio-Continuatio', *Mots médiévaux offerts a Ruedi Imbach*, ed. by Iñigo Atucha, Dragos Calma, Catherine König-Pralong, and Irene Zavattero, Textes et études du Moyen Âge, 57 (Porto: Fédération internationale des instituts d'études médiévales, 2011), pp. 185–98

26. 'Two Approaches to Natural Science in Toledo of the Twelfth Century', in *Christlicher Norden, Muslimischer Süden: Ansprüche und Wirklichkeiten von Christen, Juden und Muslimen auf der Iberischen Halbinsel im Hoch- und Spätmittelalter*, ed. by Matthias M. Tischler and Alexander Fidora, Erudiri Sapientia, 7 (Münster: Aschendorff, 2011), pp. 69–80

27. 'Adelard of Bath', 'Arabic Texts: Natural Philosophy', and 'Hermes Trismegistus', in Henrik Lagerlund, *Encyclopedia of Medieval Philosophy* (Dordrecht: Springer, 2011), pp. 24–26, 88–92 and 470–71

28. 'Le *Picatrix* à l'Institut Warburg: Histoire d'une recherche et d'une publication', in *Images et magie:* Picatrix *entre Orient et Occident*, ed. by Jean-Patrice Boudet, Anna Caiozzo, and Nicholas Weill-Parot, Sciences, techniques et civilisations du Moyen Âge à l'aube des Lumières, 13 (Paris: Honoré Champion, 2011), pp. 25–38

29. 'Illustrations and Diagrams in Arabic and Latin Scientific Works', in *Proceedings of the Symposium for International Comparative Study on Illustrated Books of Fortunetelling* (Tokyo: Keio University EIRI Joint Study Group on Illustrated Books, 2012), pp. 5–11

30. 'Reading the Sciences', in *The European Book in the Twelfth Century*, ed. by Erik Kwakkel and Rodney M. Thomson (Cambridge: Cambridge University Press, 2018), pp. 259–76

C. Arithmetic and Geometry

1. 'Abacus', with William Francis Ryan, in *Instruments of Science: An Historical Encyclopedia*, ed. by Robert Bud and Deborah Jean Warner (London – Washington, DC: Science Museum – National Museum of American History-Smithsonian Institution, 1998), pp. 5–7

§ 2. 'Why We Read Arabic Numerals Backwards', in *Ancient and Medieval Traditions in the Exact Sciences: Essays in Memory of Wilbur Knorr*, ed. by Patrick Suppes, Julius M. Moravcsik and Henry Mendell (Stanford, CA: Center for the Study of Language and Information, 2000), pp. 197–202

§ 3. 'Latin Alphanumerical Notation, and Annotation in Italian, in the Twelfth Century: MS London, British Library, Harley 5402', in *Sic itur ad astra: Studien zur Geschichte der Mathematik und Naturwissenschaften: Festschrift für den Arabisten Paul Kunitzsch zum 70. Geburtstag*, ed. by Menso Folkerts and Richard Lorch (Wiesbaden: Harrassowitz, 2000), pp. 76–90

§ 4. 'The Abacus at Echternach in ca. 1000 A.D.', *SCIAMUS*, 3 (2002), 91–108

§ 5. 'Learning Indian Arithmetic in the Early Thirteenth Century', *Boletín de la Asociación Matemática Venezolana*, 9 (2002), 15–26

§ 6. 'Indian Numerals in the Mediterranean Basin in the Twelfth Century, with Special Reference to the "Eastern Forms"', in *From China to Paris: 2000 Years Transmission of Mathematical Ideas*, ed. by Yvonne Dold-Samplonius, Joseph W. Dauben, Menso Folkerts, and Benno van Dalen (Stuttgart: Steiner, 2002), pp. 237–88

7. 'Euclid and al-Farabi in MS Vatican, Reg. Lat. 1268', in *Words, Texts and Concepts Cruising the Mediterranean Sea: Studies on the Sources, Contents and Influences of Islamic Civilization and Arabic Philosophy and Science: Dedicated to Gerhard Endress on His Sixty-fifth Birthday*, ed. by Rüdiger Arnzen, and Jörn Thielmann, Orientalia Lovaniensia Analecta, 139 (Leuven: Peeters, 2004), pp. 411–36

§ 8. 'Abbon de Fleury, *abaci doctor*', in *Abbon de Fleury: Philosophie, sciences et comput autour de l'an mil*, ed. by Barbara Obrist, Oriens-Occidens, 6 (Paris: Centre d'histoire des sciences et des philosophies arabes et médiévales, 2004), pp. 129–39

§ 9. 'Fibonacci's "Method of the Indians"', *Leonardo Fibonacci: Matematica e società nel Mediterraneo nel secolo XIII*, ed. by Menso Folkerts, 2 vols, Bollettino di Storia delle scienze matematiche, 23.2 and 24.1 (Pisa – Rome: Istituti editoriali e poligrafici internazionali, 2005), I: 87–97

§ 10. 'The Use of Arabic Numerals among the Three Language Cultures of Norman Sicily', *Art and Form in Norman Sicily*, ed. by David Knipp, *Römisches Jahrbuch der Biblioteca Hertziana*, 35, 2003–2004 (published 2005), 37–48

11. 'The Semantics of Indian Numerals in Arabic, Greek and Latin', *Journal of Indian Philosophy*, 34 (2006), 15–30

§ 12. 'The Toledan *Regule* (*Liber Alchorismi*, Part II): A Twelfth-Century Arithmetical Miscellany', with Ji-Wei Zhao and Kurt Lampe, *SCIAMUS*, 8 (2007), 141–231

13. 'Ten or forty? A confusing numerical symbol in the Middle Ages', in *Mathematics Celestial and Terrestrial: Festchrift for Menso Folkerts zum 65. Geburtstag*, ed. by Joseph W. Dauben, Stefan Kirschner, Andreas Kühne, Paul Kunitzsch, and Richard P. Lorch, Acta Historica Leopoldina, 54 (Halle Saale – Stuttgart: Deutsche Akademie der Naturforscher Leopoldina – Wissenschaftliche Verlagsgesellschaft, 2008), pp. 81–89

14. 'Learning to Write Numerals in the Middle Ages', in *Teaching Writing, Learning to Write: Proceedings of the XVIth Colloquium of the Comité International de Paléographie Latine*, ed. by Pamela Robinson (London: King's College London-Centre for Late Antique and Medieval Studies, 2010), pp. 233–40

15. 'The Geometry of the *Liber Ysagogarum Alchorismi*', *Sudhoffs Archiv*, 97 (2013), 143–73

16. 'The Palaeography of Numerals', in *The Oxford Handbook of Latin Palaeography*, ed. by Frank T. Coulson and Robert G. Babcock (Oxford: Oxford University Press, 2020), pp. 25–36

D. Astronomy and Astrology

1. 'A New Source for Dominicus Gundissalinus's Account of the Science of the Stars?', *Annals of Science*, 47 (1990), 361–74
2. 'Al-Kindī on Judicial Astrology: "The Forty Chapters"', *Arabic Sciences and Philosophy*, 3 (1993), 77–117
3. 'Advertising the New Science of the Stars *circa* 1120–50', in *Le XIIe siècle: Mutations et renouveau en France dans la première moitié du XIIe siècle*, ed. by Françoise Gasparri (Paris: Le Léopard d'or, 1994), pp. 147–57
4. 'Astrology', in *Medieval Latin: An Introduction and Bibliographical Guide*, ed. by Frank Anthony Carl Mantello and Arthur George Rigg (Washington, DC: Catholic University of America Press, 1996), pp. 369–82
5. '*Imperium, Ecclesia Romana* and the Last Days in William Scot's Astrological Prophecy of ca. 1266–73', in *Forschungen zur Reichs-, Papst- und Landesgeschichte: Peter Herde zum 65. Geburtstag*, ed. by Karl Borchardt and Enno Bünz, 2 vols (Stuttgart: Hiersemann, 1998), I: 347–60
6. 'Hildegard of Bingen and the Science of the Stars', in *Hildegard of Bingen: The Context of Her Thought and Art*, ed. by Charles Burnett Peter Dronke, Warburg Institute Colloquia, 4 (London: Warburg Institute, 1998), pp. 111–20
7. 'The *Sortes regis Amalrici*: An Arabic Divinatory Work in the Latin Kingdom of Jerusalem?', *Scripta Mediterranea*, 19–20 (1998–1999), 229–37

‡8. 'King Ptolemy and Alchandreus the Philosopher: The Earliest Texts on the Astrolabe and Arabic Astrology at Fleury, Micy and Chartres', *Annales of Science*, 55 (1998), 329–68; Addendum, *Annales of Science*, 57 (2000), 187

9. '*Partim de suo et partim de alieno*: Bartholomew of Parma, the Astrological Texts in MS Bernkastel-Kues, Hospitalsbibliothek 209, and Michael Scot', in *Seventh Centenary of the Teaching of Astronomy in Bologna 1297–1997*, ed. by Pierluigi Battistini, Musei e archivi dello Studio Bolognese, 8 (Bologna: CLUEB, 2001), pp. 38–76
10. 'Bartholomeus Parmensis, *Tractatus spere*, pars tercia', in *Seventh Centenary of the Teaching of Astronomy in Bologna 1297–1997*, Pierluigi Battistini, Musei e archivi dello Studio Bolognese, 8 (Bologna: CLUEB, 2001), pp. 151–212
11. 'The Certitude of Astrology: The Scientific Methodology of al-Qabīṣī and Abū Maʿshar', in *Early Science and Medicine*, 7 (2002), 198–213
12. '"Albumasar in Sadan" in the Twelfth Century', in *Ratio et Superstitio: Essays in Honor of Graziella Federici Vescovini*, ed. by Giancarlo Marchetti, Orsola Rignani, and Valeria Sorge, Textes et études du Moyen Âge, 24 (Louvain-le-Neuve: Fédération internationale des instituts d'études médiévales, 2003), pp. 59–67
13. 'Lunar Astrology. The Varieties of Texts Using Lunar Mansions, with Emphasis on *Jafar Indus*', in *Il sole e la luna / The Sun and the Moon*, *Micrologus*, 12 (2004), 43–133

14. 'Arabic and Latin Astrology Compared in the Twelfth Century: Firmicus, Adelard of Bath and "Doctor Elmirethi" ("Aristoteles Milesius")', in *Studies in the History of the Exact Sciences in Honour of David Pingree*, ed. by Charles Burnett, Jan P. Hogendijk, Kim Plofker, and Michio Yano, Islamic Philosophy, Theology, and Science, 54 (Leiden: Brill, 2004), pp. 247–63

15. 'Weather Forecasting in the Arabic World', in *Magic and Divination in Early Islam*, ed. by Emilie Savage-Smith (Aldershot: Ashgate Variorum, 2004), pp. 201–10

16. 'A Hermetic Programme of Astrology and Divination in mid-Twelfth-Century Aragon: The Hidden Preface in the *Liber novem iudicum*', in *Magic and the Classical Tradition*, ed. by Charles Burnett and William Francis Ryan, Warburg Institute Colloquia, 7 (London: Warburg Institute, 2006), pp. 99–118

17. 'The Astrological Categorization of Religions in Abū Maʿshar, the *De vetula* and Roger Bacon', in *Language of Religion, Language of the People: Medieval Judaism, Christianity and Islam*, ed. by Michael Richter, Ernst Bremer, Jörg Jarnut, and David J. Wasserstein, MittelalterStudien, 11 (Munich: Fink, 2007), pp. 127–38

18. 'Adelard of Bath', 'Petrus Alfonsi', 'Raymond of Marseilles', and 'Roger of Hereford', in *The Biographical Encyclopedia of Astronomers*, ed. by Thomas Hockey (New York: Springer, 2007), pp. 27–28, 46–47, 1804–05 and 1856

19. 'Astrology, Astronomy and Magic as the Motivation for the Scientific Renaissance of the Twelfth Century', in *The Imaginal Cosmos: Astrology, Divination and the Sacred*, ed. by Angela Voss and Jean Hinson Lall (Canterbury: University of Kent, 2007), pp. 55–61

20. Contribution to *The Medieval Imagination: Illuminated Manuscripts from Cambridge, Australia and New Zealand*, ed. by Bronwyn Stocks and Nigel Morgan (South Yarra, Vic.: Macmillan Art, 2008), pp. 214–17

21. 'Why Study Ptolemy's *Almagest*? The Evidence of MS Melbourne, State Library of Victoria, Sinclair 224', *The La Trobe Journal*, 81 (Autumn 2008), 126–43

22. 'Weather Forecasting, Lunar Mansions and a Disputed Attribution: The *Tractatus pluviarum et aeris mutationis* and *Epitome totius astrologiae* of "Iohannes Hispalensis"', in *Islamic Thought in the Middle Ages: Studies in Text, Transmission and Translation, in Honour of Hans Daiber*, ed. by Anna Akasoy and Wim Raven, Islamic Philosophy, Theology, and Science, 75 (Leiden: Brill, 2008), pp. 219–65

23. 'Aristotle as an Authority on Judicial Astrology', in *Florilegium Mediaevale: Études offertes à Jacqueline Hamesse à l'occasion de son éméritat*, ed. by José Meirinhos and Olga Weijers, Textes et études du Moyen Âge, 50 (Louvain-la-Neuve: Fédération internationale des instituts d'études médiévales, 2009), pp. 41–62

24. 'Abū Maʿshar (AD 787–886) and His Major Texts on Astrology', in *Kayd: Studies in History of Mathematics, Astronomy and Astrology in Memory of David Pingree*, ed. by Gherardo Gnoli and Antonio Panaino, Serie orientale Roma, 102 (Rome: Istituto Italiano per l'Africa e l'Oriente, 2009), pp. 17–29

25. 'Postscript' to reprint of J. M. Millás Vallicrosa, 'Pedro Alfonso's Contribution to Astronomy', *Aleph*, 10 (2010), 166–68

26. 'Hebrew and Latin Astrology in the Twelfth Century: The Example of the Location of Pain', in *Stars, Spirits, Signs: Towards a History of Astrology 1100–1800. Studies in History and Philosophy of Science*, ed. by Lauren Kassell and Robert Ralley, Studies in History and Philosophy of Science Part C: Studies in History and Philosophy of Biological and Biomedical Sciences, 41/2 (2010), pp. 70–75

27. '"Ptolemaeus in Almagesto dixit": The Transformation of Ptolemy's *Almagest* in its Transmission via Arabic into Latin', in *Transformationen antiker Wissenschaften*, ed. by Georg Toepfer and Hartmut Böhme (Berlin: De Gruyter, 2010), pp. 115–40

28. 'Albumazar, *De magnis coniunctionibus*', in *Early Medicine, from the Body to the Stars*, ed. by Gérald d'Andiran (Basel: Schwabe, 2010), p. 455

29. 'Al-Qabīṣī's *Introduction to Astrology*: From Courtly Entertainment to University Textbook', in *Studies in the History of Culture and Science: A Tribute to Gad Freudenthal*, ed. by Resianne Fontaine, Ruth Glasner, Reimund Leicht, and Giuseppe Veltri, Studies in Jewish History and Culture, 30 (Leiden: Brill, 2011), pp. 43–69

30. '*De meliore homine*. 'Umar ibn al-Farrukhān al-Ṭabarī on Interrogations: A Fourth Translation by Salio of Padua?', in *Adorare caelestia, gubernare terrena: Atti del colloquio internazionale in onore di Paolo Lucentini (Napoli, 6–7 novembre 2007)*, ed. by Pasquale Arfé, Irene Caiazzo, and Antonella Sannino, Instrumenta Patristica et Mediaevalia, 58 (Turnhout: Brepols, 2011), pp. 295–325

31. 'The Astrologer's Advice in Matters of War', *Mantova e il Rinscimento italiano: Studi in onore di David S. Chambers*, ed. by Philippa Jackson and Guido Rebecchini (Mantua: Sometti, 2011), pp. 251–58

32. 'New Manuscripts of *On the astrolabe* by Raymond of Marseilles', with Irene Caiazzo, *Scriptorium*, 65 (2011), 338–49

33. 'John of Gmunden's Astrological Library', in *Johannes von Gmunden: zwischen Astronomie und Astrologie*, ed. by Rudolf Simek and Manuela Klein, Studia medievalia septentrionalia, 22 (Vienna: Fassbaender, 2012), pp. 55–71

34. 'Astrological Translations in Byzantium', in *Le Symposium International: Le livre, la Roumanie, l'Europe (4ème édition, 20–23 September 2011)*, Troisiéme section – Latinité orientale, ed. by Martin Hauser, Ioana Feodorov, Nicholas V. Sekunda, and Adrian George Dumitru (Bucarest: Biblioteca Bucureştilor, 2012), pp. 178–83

35. 'Teaching the Science of the Stars in Prague University in the Early Fifteenth Century: Master Johannes Borotin', *Aither*, 8, international issue 2 (2012), 9–50

36. 'Doctors versus Astrologers: Medical and Astrological Prognosis Compared', in *Die mantischen Künste und die Epistemologie prognostischer Wissenschaften im Mittelalter*, ed. by Alexander Fidora (Cologne: Böhlau Verlag, 2013), pp. 101–11

37. 'Stephen of Messina and the Translation of Astrological Texts from Greek in the Time of Manfred', in *Translating at the Court: Bartholomew of Messina and Cultural Life at the Court of Manfred, King of Sicily*, ed. by Pieter De Leemans, Mediaevalia Lovaniensia, Series 1, Studia 45 (Leuven: Leuven University Press, 2014), pp. 123–32

38. 'The Interpretation of a Horoscope Cast by Abraham the Jew in Béziers for a Child Born on 29 November (*recte* October) 1135: An Essay in Understanding a Medieval Astrologer', with Helena Avelar de Carvalho, *Culture and Cosmos*, 18 (2014), 19–40

39. 'Introducing Astrology: Michael Scot's *Liber introductorius* and Other Introductions', in *Astrologers and their Clients in Medieval and Early Modern Europe*, ed. by Wiebke Deimann and David Juste, Beihefte zum Archiv für Kulturgeschichte, 73 (Cologne: Böhlau Verlag, 2015), pp. 17–27

40. '*Amitegni*: A Newly-Discovered Text on Astrological Judgements', *Bruniana & Campanelliana*, 21 (2015), 653–61

41. 'On Judging and Doing in Arabic and Latin Texts on Astrology and Divination', in *The Impact of Arabic Sciences in Europe and Asia, Micrologus*, 24 (2016), 3–11

42. 'A New Catalogue of Medieval Translations into Latin of Texts on Astronomy and Astrology', in *Medieval Textual Cultures: Agents of Transmission, Translation and Transformation*, ed. by Faith Wallis and Robert Wisnovsky, Judaism, Christianity, and Islam – Tension, Transmission, Transformation, 6 (Berlin: De Gruyter, 2016), pp. 63–76

43. 'Béziers as an Astronomical Center for Jews and Christians in the Mid-Twelfth Century', *Aleph* 17/2 (2017), 197–219

44. 'Harmonic and Acoustic Theory: Latin and Arabic Ideas of Sympathetic Vibration as the Causes of Effects between Heaven and Earth', in *Sing Aloud Harmonious Spheres: Renaissance Conceptions of Cosmic Harmony*, ed. by Jacomien Prins and Maude Vanhaelen (New York: Routledge, 2018), pp. 31–43

45. 'Richard de Fournival and the *Speculum Astronomiae*', in *Richard de Fournival et les sciences au xiiie siècle*, ed. by Joëlle Ducos and Christopher Lucken, Micrologus Library, 88 (Florence: SISMEL–Edizioni del Galluzzo, 2018), pp. 339–48

46. 'Agency and Effect in the Astrology of Abū Maʿshar of Balkh (Albumasar)', *Oriens*, 47 (2019), 348–64

47. 'Astrology for the Doctor in a Work Addressed to Robert, Earl of Leicester', in *De l'homme, de la nature et du monde. Mélanges d'histoire des sciences médiévales offerts à Danielle Jacquart*, Hautes études médiévales et modernes, 113 (Geneva: Droz, 2019), pp. 179–96

48. 'The Doctrine on Kings and Empires in Abu Maʿshar's *Book on Religions and Dynasties* and its Application in the Medieval West', in *Political Astrology in the Mediterranean Area from the Middle Ages to the Renaissance*, ed. by Marienza Benedetto, Pasquale Arfé, and Pasquale Porro, *Quaestio*, 19 (2019), pp. 15–31

49. 'From *Astronomica* to *Exotica*: Jacob Golius's Edition of al-Farghānī's *On the Science of the Stars* in Comparison with the Earlier Versions', in *Scholarship between Europe and the Levant. Essays in Honour of Alastair Hamilton*, ed. by Jan Loop and Jill Kraye, History of Oriental Studies, 8 (Leiden: Brill, 2020), pp. 60–86

50. 'Cleaning up the Latin Language in Mid-Sixteenth-Century Basel: Antonius Stuppa's Purgation of Albohazen's *De iudiciis astrorum*', in *Centres and Peripheries in the History of Philosophical Thought / Centri e periferie nella storia del pensiero filosofico: Essays in Honour of Loris Sturlese*, ed. by Nadia Bray, Diana Di Segni, Fiorella Retucci, and Elisa Rubino, Rencontres de philosophie médiévales, 24 (Turnhout: Brepols, 2021), pp. 69–82

51. 'Astral Sciences. Traditions and Practices in the Medieval Western Christian World', in *Prognostication in the Medieval World: A Handbook*, ed. by Matthias Heiduk, Klaus Herbers, and Hans-Christian Lehner (Berlin: De Gruyter, 2021), pp. 485–501

52. 'Weather Forecasting. Traditions and Practices in the Medieval Islamic World', in *Prognostication in the Medieval World: A Handbook*, ed. by Matthias Heiduk, Klaus Herbers, and Hans-Christian Lehner (Berlin: De Gruyter, 2021), pp. 689–95

53. 'The Astrological *Liber novem iudicum*: A Kind of Encyclopedia?', in *Speculum Arabicum: Intersecting Perspectives on Medieval Encyclopaedism*, ed. by Godefroid de Callataÿ, Mattia Cavagna, and Baudouin van den Abeele (Louvain-la-Neuve: Publications de l'Institut d'Études médiévales, 2021), pp. 83–96

54. 'The History of the Text: From Abū Ma'shar to Georgius Fendulus', and 'Translation of the Text and Description of the Illustrations' (with Kristen Lippincott), in *Liber Astrologiae Abū Ma'shar Treatise*, ed. by Izabela Liśkiewicz and Mónica Miró (Barcelona: M. Moleiro, 2023), pp. 83–103 and 105–264

E. Medicine and Psychology

1. 'Astrology and Medicine in the Middle Ages', *Bulletin of the Society for the Social History of Medicine*, 37 (1985), 16–18

2. 'Hermann of Carinthia's Attitude Towards his Arabic Sources, in Particular in Respect to Theories of the Human Soul', in *L'Homme et son univers au moyen âge*, ed. by Christian Wenin, 2 vols, Philosophes médiévaux, 26–27 (Louvain: Institut supérieur de philosophie, 1986), I: 306–22

3. 'The Planets and the Development of the Embryo', in *The Human Embryo: Aristotle and the Arabic and European Traditions*, ed. by Gordon Reginald Dunstan (Exeter: University of Exeter Press, 1990), pp. 95–112

4. 'The Superiority of Taste', *Journal of the Warburg and Courtauld Institutes*, 54 (1991), 230–38

5. 'The Chapter on the Spirits in the *Pantegni* of Constantine the African', in *Constantine the African and ʿAlī ibn al-ʿAbbās al-Maǧūsī: The* Pantegni *and Related Texts*, ed. by Charles Burnett and Danielle Jacquart, Studies in Ancient Medicine, 10 (Leiden: Brill, 1994), pp. 99–120
6. 'Appendix: Constantine the African, *De oblivione*', in Gerrit Bos, 'Ibn al-Ǧazzār's *Risāla fī l-nisyān* and Constantine's *Liber de oblivione*', in *Constantine the African and ʿAlī ibn al-ʿAbbās al-Maǧūsī: The* Pantegni *and Related Texts*, ed. by Charles Burnett and Danielle Jacquart, Studies in Ancient Medicine, 10 (Leiden: Brill, 1994), pp. 224–32
7. '*Sapores sunt octo*: The Medieval Latin Terminology for the Eight Flavours', in *I cinque sensi / The Five Senses, Micrologus*, 10 (2002), 99–112
8. '*Verba Ypocratis preponderanda omnium generum metallis*. Hippocrates on the Nature of Man in Salerno and Montecassino, with an Edition of the Chapter on the Elements in the *Pantegni*', in *La Scuola Medica Salernitana: Gli autori e i testi*, ed. by Danielle Jacquart and Agostino Paravicini Bagliani (Florence: SISMEL–Edizioni del Galluzzo, 2007), pp. 59–92
9. 'The Latin Versions of Maimonides' *On Sexual Intercourse (De coitu)*', in *Between Text and Patient: The Medical Enterprise in Medieval and Early Modern Europe*, ed. by Florence Eliza Glaze and Brian K. Nance, Micrologus Library, 39 (Florence: SISMEL–Edizioni del Galluzzo, 2011), pp. 467–80
10. 'Simon of Genoa's Use of the *Breviarium* of Stephen, the Disciple of Philosophy', in *Simon of Genoa's Medical Lexicon*, ed. by Barbara Zipser (Warsaw: De Gruyter, 2013), pp. 67–78
11. 'The Legend of Constantine the African', in *The Medieval Legends of Philosophers and Scholars, Micrologus*, 21 (2013), 277–94
12. 'The Synonyma Literature in the Twelfth and Thirteenth Centuries', in *Globalization of Knowledge in the Post-Antique Mediterranean, 700–1500*, ed. by Sonja Brentjes and Jürgen Renn (London: Routledge, 2016), pp. 131–39
13. 'Preface' in *Pregnancy and Childbirth in the Premodern World*, ed. by Costanza Gislon Dopfel, Alessandra Foscati, and Charles Burnett, Cursor Mundi, 36 (Turnhout: Brepols, 2019), pp. xi–xii

F. Magic and Divination

*1. 'The Legend of the Three Hermes and Abū Maʿshar's *Kitāb al-Ulūf* in the Latin Middle Ages', *Journal of the Warburg and Courtauld Institutes*, 39 (1976), 231–34
*2. 'What is the *Experimentarius* of Bernardus Silvestris? A Preliminary Survey of the Material', *Archives d'histoire doctrinale et littéraire du moyen âge*, 44 (1977), 79–125
*3. 'Hermann of Carinthia and the *Kitāb al-Istamāṭīs*: Further Evidence for the Transmission of Hermetic Magic', *Journal of the Warburg and Courtauld Institutes*, 44 (1981), 167–69
*4. 'Scandinavian Runes in a Latin Magical Treatise', *Speculum*, 58 (1983), 419–29

*5. 'Arabic Divinatory Texts and Celtic Folklore: A Comment on the Theory and Practice of Scapulimancy in Western Europe', *Cambridge Medieval Celtic Studies*, 6 (1983), 31–42

*6. 'An Apocryphal Letter from the Arabic Philosopher al-Kindī to Theodore, Frederick II's Astrologer, Concerning Gog and Magog, the Enclosed Nations, and the Scourge of the Mongols', *Viator*, 15 (1984), 151–67

*7. 'Arabic, Greek and Latin Works on Astrological Magic attributed to Aristotle', in *Pseudo-Aristotle in the Middle Ages: The Theology and Other Texts*, ed. by Jill Kraye, William Francis Ryan, and Charles B. Schmitt, Warburg Institute Surveys and Texts, 11 (London: Warburg Institute, 1986), pp. 84–96

*8. 'Adelard, Ergaphalau and the Science of the Stars', in *Adelard of Bath: An English Scientist and Arabist of the Early Twelfth Century*, ed. by Charles Burnett, Warburg Institute Surveys and Texts, 14 (London: Warburg Institute, 1987), pp. 133–45

*9. 'El kitab al-Istamatis i un manuscrit Barceloní d'obres astrològiques i astronòmiques', *Llengua i literatura*, 2 (1987), 431–51; revised and translated as 'The *kitāb al-Istamāṭīs* and a Manuscript of Astrological and Astronomical Works from Barcelona (Biblioteca de Catalunya, 634)', in *Magic and Divination in the Middle Ages: Texts and Techniques in the Islamic and Christian Worlds*, Variorum Collected Studies Series, C557 (Aldershot: Variorum, 1996), Article VII

*10. 'The Earliest Chiromancy in the West', *Journal of the Warburg and Courtauld Institutes*, 50 (1987), 189–95

*11. 'The Eadwine Psalter and the Western Tradition of the Onomancy in Pseudo-Aristotle's *Secret of Secrets*', *Archives d'histoire doctrinale et littéraire du moyen âge*, 55 (1988), 143–67

*12. 'A Note on Two Astrological Fortune-Telling Tables', *Revue d'histoire des textes*, 18 (1988), 257–62

*13. 'Divination from Sheep's Shoulder Blades: A Reflection on Andalusian Society', *Cultures in Contact in Medieval Spain: Historical and Literary Essays Presented to L. P. Harvey*, ed. by David Hook and Barry Taylor (London: King's College, 1990), pp. 29–45

*14. 'The Astrologer's Assay of the Alchemist: Early References to Alchemy in Arabic and Latin Texts', *Ambix*, 39 (1992), 103–09

15. 'The Prognostications of the Eadwine Psalter', in *The Eadwine Psalter: Text, image, and monastic culture in twelfth-century Canterbury*, ed. by Margaret T. Gibson, T. A. Heslop, and Richard William Pfaff (London – University Park, PA: Modern Humanities Research Association – Pennsylvania State University Press, 1992), pp. 165–67

16. 'An Unknown Latin Version of an Ancient *Parapegma*: The Weather-Forecasting Stars in the *Iudicia* of Pseudo-Ptolemy', in *Making Instruments Count: Essays on Historical Scientific Instruments presented to Gerard L'Estrange Turner*, ed. by Robert Geoffrey William Anderson, James A. Bennett, and William Francis Ryan (Aldershot: Variorum, 1993), pp. 27–41

*17. 'An Islamic Divinatory Technique in Medieval Spain', *The Arab Influence in Medieval Europe: Folia scholastica mediterranea*, ed. by Dionisius A. Agius and Richard Hitchcock, Middle East Cultures Series, 18 (Reading: Ithaca, 1994), pp. 100–35

*18. 'The Scapulimancy of Giorgio Anselmi's *Divinum opus de magia disciplina*', *Euphrosyne* 23 (1995), 63–81

*19. 'Talismans: Magic as Science? Necromancy among the Seven Liberal Arts', in *Magic and Divination in the Middle Ages: Texts and Techniques in the Islamic and Christian Worlds*, Variorum Collected Studies Series, C557 (Aldershot: Variorum, 1996), Article I

*20. 'The Conte de Sarzana Magical Manuscript', in *Magic and Divination in the Middle Ages: Texts and Techniques in the Islamic and Christian Worlds*, Variorum Collected Studies Series, C557 (Aldershot: Variorum, 1996), Article IX

21. 'Al-Kindī on Finding Buried Treasure', with Keiji Yamamoto and Michio Yano, *Arabic Sciences and Philosophy*, 7 (1997), 57–90

22. 'The Establishment of Medieval Hermeticism', in *The Medieval World*, ed. by Peter Linehan and Janet L. Nelson (London: Routledge, 2001), pp. 111–30

23. 'Remarques paléographiques et philologiques sur les noms d'anges et d'esprits dans les traités de magie traduits de l'arabe en latin', in *Les anges et la magie au moyen âge*, ed. by Jean-Patrice Boudet, Henri Bresc, and Benoît Grévin, *Mélanges de l'École française de Rome*, 114/2 (2002), pp. 657–68

24. 'The Arabic Hermes in the Works of Adelard of Bath', in *Hermetism from Late Antiquity to Humanism*, ed. by Paolo Lucentini, Ilaria Parri, and Vittoria Perrone Compagni, Instrumenta Patristica et Mediaevalia, 40 (Turnhout: Brepols, 2003), pp. 369–84

25. 'Images of Ancient Egypt in the Latin Middle Ages', in *The Wisdom of Egypt: Changing Visions through the Ages*, ed. by Peter J. Ucko and Timothy C. Champion (London: Routledge, 2003), pp. 65–99

26. 'Medieval Latin Translations of Greek Texts on Astrology and Magic', in *The Occult Sciences in Byzantium*, ed. by Maria Mavroudi and Paul Magdalino (Geneva: La Pomme d'or, 2006), pp. 325–59

27. 'Ṯābit ibn Qurra the Ḥarrānian on Talismans and the Spirits of the Planets', *La Corónica*, 36 (2007), 13–40

28. '*Nīranj*: A Category of Magic (Almost) Forgotten in the Latin West', in *Natura, scienze e società medievali. Studi in onore di Agostino Paravicini Bagliani*, ed. by Claudio Leonardi and Francesco Santi, Micrologus Library, 28 (Florence: SISMEL–Edizioni del Galluzzo, 2008), pp. 37–66

29. 'The Theory and Practice of Powerful Words in Medieval Magical Texts', in *The Word in Medieval Logic, Theology and Psychology*, ed. by Tetsuro Shimizu and Charles Burnett (Turnhout: Brepols, 2009), pp. 215–31

30. 'Introduction' to David Pingree, 'Between the *Ghāya* and the *Picatrix* II: The *Flos Naturarum* ascribed to *Jabir*', *Journal of the Warburg and Courtauld Institutes*, 72 (2009), 41–44

31. 'Divination', with Guido Giglioni, in *The Classical Tradition*, ed. by Anthony Grafton, Glen W. Most, and Salvatore Settis (Cambridge, MA: Belknap, 2010), pp. 275–79

32. 'Geomancy in the Islamic World and Western Europe', *Zhouyi Studies (English Version)*, 7 (2011), 176–80

33. 'A Judaeo-Arabic Version of Ṯābit ibn Qurra's *De imaginibus* and Pseudo-Ptolemy's *Opus imaginum*', with Gideon Bohak, in *Islamic Philosophy, Science, Culture, and Religion: Studies in Honor of Dimitri Gutas*, ed. by Felicitas Opwis and David Reisman (Leiden: Brill, 2012), pp. 179–200

34. 'East (and South) Asian Traditions in Astrology and Divination as Viewed from the West', in *Les Astres et le destin: astrologie et divination en Asie Orientale*, ed. by Jean-Noël Robert and Pierre Marsone, Extrême orient-Extrême occident, 35 (Saint-Denis: Presses Universitaires de Vincennes, 2013), pp. 285–93

35. 'The Latin Versions of Pseudo-Aristotle's *De signis*', in *Translating at the Court: Bartholomew of Messina and Cultural Life at the Court of Manfred, King of Sicily*, ed. by Pieter De Leemans, Mediaevalia Lovaniensia, Series 1, Studia 45 (Leuven: Leuven University Press, 2014), pp. 285–301

36. 'Magic in the Court of Alfonso el Sabio: The Latin Translation of the *Ghāyat al-Ḥakīm*', in *De Frédéric II à Rodolphe II: Astrologie, divination et magie dans les cours (XIIIe-XVIIe siècle)*, ed. by Jean-Patrice Boudet, Martine Ostorero, and Agostino Paravicini Bagliani, Micrologus Library, 85 (Florence: SISMEL–Edizioni del Galluzzo, 2017), pp. 37–52

37. 'Hermetic Geomancy, "Ratione certi experimenti usitata"', in *Geomancy and Other Forms of Divination*, ed. by Alessandro Palazzo and Irene Zavattero, Micrologus Library, 87 (Florence: SISMEL–Edizioni del Galluzzo, 2017), pp. 135–41

38. 'Arabic Magic: The Impetus for Translating Texts and their Reception', in *The Routledge History of Medieval Magic*, ed. by Sophie Page and Catherine Rider (London – New York: Routledge, 2019), pp. 71–84

39. 'The Three Divisions of Arabic Magic', in *Islamicate Occult Sciences in Theory and Practice*, ed. by Liana Saif, Francesca Leoni, Matthew Melvin-Koushki, and Farouk Yahya (Leiden: Brill, 2020), pp. 43–56

40. Pseudo-Ptolemy, *Liber figure*, transcribed by Charles Burnett (update 25 May 2021), <https://ptolemaeus.badw.de/ms/239/343/transcription/1> [accessed 19 April 2023]

41. '*Alia littera*: Editorial Strategies in Copies of a Medieval Latin Text on Talismans by Thābit ibn Qurra', in *Le Moyen Âge et les sciences*, ed. by Danielle Jacquart and Agostino Paravicini Bagliani, Micrologus Library, 100 (Florence: SISMEL–Edizioni del Galluzzo, 2021), pp. 595–616

42. 'Thābit ibn Qurra's *On Talismans* between the Middle Ages and the Renaissance, and between the Science of the Stars and Magic', *Bruniana & Campanelliana*, 27 (2021), 23–50

43. '*Inscriptio characterum*: Solomonic Magic and Paleography', in *Unveiling the Hidden – Anticipating the Future: Divinatory Practices Among Jews Between the Qumran and the Modern Period*, ed. by Josefina Rodríguez-Arribas and Dorian Gieseler Greenbaum, Prognostication in History, 5 (Leiden: Brill, 2021), pp. 311–32

44. 'Society and the Environment: The Social Position of the Alchemist and Alchemy in the Court, in the Church, and in Society', with Antoine Calvet and Justine Bayley, in *A Cultural History of Chemistry in the Middle Ages*, ed. by Charles Burnett and Sébastien Moureau, A Cultural History of Chemistry, 6 vols (London: Bloomsbury Academic, 2022), II: 93–106

G. Anglo-Norman Science and Learning in the Twelfth Century

‡1. 'Adelard of Bath and the Arabs', *Rencontres de cultures dans la philosophie médiévale: Traductions et traducteurs de l'antiquité tardive au XIVe siècle*, ed. by Jacqueline Hamesse and Marta Fattori, Rencontres de philosophie médiévale, 1 (Louvain-la-Neuve: L'Institut d'Études Médiévales, 1990), pp. 89–107

2. '*Omnibus convenit Platonicis*: An Appendix to Adelard of Bath's *Quaestiones naturales?*', in *From Athens to Chartres Studies in Honour of Edouard Jeauneau*, ed. by Haijo Jan Westra, Studien und Texte zur Geistesgeschichte des Mittelalters, 35 (Leiden: Brill, 1992), pp. 259–81

3. 'Adelard of Bath's Use of Arabic Astrology and Astronomy', *Kokusai gengogaku kenkyusho shohou* (Journal of the International Institute of Linguistic Science, Kyoto Sangyo University), 14 (1993) (in Japanese)

4. 'Ocreatus', in *Vestigia mathematica: Studies in Medieval and Early Modern Mathematics in Honour of H. L. L. Busard*, ed. by Menso Folkerts and Jan P. Hogendijk (Amsterdam: Rodopi, 1993), pp. 69–77

5. 'The Introduction of Arabic Learning into British Schools', *The Introduction of Arabic Philosophy into Europe*, ed. by Charles E. Butterworth and Blake Andrée Kessel, Studien und Texte zur Geistesgeschichte des Mittelalters, 39 (Leiden: Brill, 1994), pp. 40–57

6. 'Mathematics and Astronomy in Hereford and its Region in the Twelfth Century', in *Medieval Art, Architecture and Archaeology at Hereford*, ed. by David Whitehead, British Archaeological Association Conference Transactions, 15 (Leeds: British Archaeological Association, 1995), pp. 50–59

7. 'Adelard of Bath's Doctrine on Universals and the *Consolatio Philosophiae* of Boethius', *Didascalia*, 1 (Sendai) (1995), 1–13

8. 'Give him the White Cow: Notes and Note-taking in the Universities in the Twelfth and Thirteenth Centuries', *History of Universities*, 14 (1995–1996), 1–30

9. 'The Introduction of Aristotle's Natural Philosophy into Great Britain: A Preliminary Survey of the Manuscript Evidence', in *Aristotle in Britain during the Middle Ages*, ed. by John Marenbon, Rencontres de philosophie médiévale, 5 (Turnhout: Brepols, 1996), pp. 21–50

10. 'The Instruments which are the Proper Delights of the *Quadrivium*: Rhythmomachy and Chess in the Teaching of Arithmetic in Twelfth-Century England', *Viator*, 28 (1997), 175–201

11. 'Hildegard in England: A Note on Hildegard's Texts in the Library of the Austin Friars in York', in *Hildegard of Bingen: The Context of Her Thought and Art*, ed. by Charles Burnett and Peter Dronke, Warburg Institute Colloquia, 4 (London: Warburg Institute, 1998), pp. 63–64

12. 'The Introduction of Scientific Texts into Britain, c. 1100–1250', in *The Cambridge History of the Book in Britain*, 7 vols (Cambridge: Cambridge University Press, 1998–2019), II (2008): *1100–1400*, ed. by Nigel J. Morgan and Rodney M. Thomson, 446–53

13. '‘Abd al-Masīḥ of Winchester', in *Between Demonstration and Imagination: Essays on the History of Science and Philosophy Presented to John D. North*, ed. by Lodi Nauta and Arie Johan Vanderjagt (Leiden: Brill, 1999), pp. 159–69

14. 'Avranches, B. M. 235 et Oxford, Corpus Christi Collège, 283', in *Science antique, science médiévale (autour d'Avranches 235)*, ed. by Louis Callebat and O. Desbordes (Hildesheim: Olms-Weidmann, 2000), pp. 63–70

15. 'Les langues d'Angleterre dans les ouvrages d'Adélard de Bath', in *Il latino e l'inglese: Una storia di lunga durata. Atti del convegno di Treviso, 25 novembre 2005* (Paris: Unione Latina, 2006), pp. 47–52

16. 'William of Conches and Adelard of Bath', in *Guillaume de Conches: Philosophie et science au XIIe siècle*, ed. by Barbara Obrist and Irene Caiazzo, Micrologus' Library, 42 (Florence: SISMEL–Edizioni del Galluzzo, 2011), pp. 67–77

17. 'The Arrival of the Pagan Philosophers in the North: A Twelfth-Century *Florilegium* in Edinburgh University Library', in *Knowledge, Discipline and Power in the Middle Ages, Essays in Honour of David Luscombe*, ed. by Joseph Canning, Edmund King, and Martial Staub (Leiden: Brill, 2011), pp. 79–93

18. Prologue and Epilogue to a reprint of *Adelard of Bath: The First English Scientist*, by Louise Cochrane (Bath: Bath Royal Literary and Scientific Institution, 2013), pp. 11–13 and 159–66

19. 'The Introduction of Arabic Words in Medieval British Latin Scientific Writings', in *Latin in Medieval Britain*, ed. by Richard Ashdowne and Carolinne White, Proceedings of the British Academy, 206 (Oxford: Oxford University Press, 2017), pp. 198–210

H. Peter Abelard and the French Schools

1. 'The Contents and Affiliation of the Scientific Manuscripts Written at, or Brought to, Chartres in the Time of John of Salisbury', in *The World of John of Salisbury*, ed. by Michael Wilks, Studies in Church History–Subsidia, 3 (Oxford: Basil Blackwell, 1984), pp. 127–60

2. 'Peter Abelard, *Soliloquium*: A Critical Edition', *Studi Medievali*, Series 3, 25 (1984), 857–94

3. 'Notes on the Tradition of the Text of the *Hymnarius Paraclitensis* of Peter Abelard', *Scriptorium*, 38 (1984), 295–302

4. 'The *Expositio Orationis Dominicae "Multorum legimus orations"*: Abelard's Exposition of the Lord's Prayer?', *Revue bénédictine*, 95 (1985), 60–72

5. 'Les Épitaphes d'Abélard et d'Héloïse au Paraclet et au Prieuré de Saint-Michel, à Chalon-sur-Saône', with Constant Mews, *Studia Monastica*, 27 (1985), 61–67

6. 'Peter Abelard, *Confessio fidei "Universis"*: Abelard's Reply to Criticisms of Heresy', *Mediaeval Studies*, 48 (1986), 111–38

7. '*Confessio fidei ad Heloisam*: Abelard's Last Letter to Heloise? A Discussion and Critical Edition of the Latin and Medieval French Versions', *Mittellateinisches Jahrbuch*, 21 (1986), 147–55

8. 'A New Text for the "School of Peter Abelard" Dossier?', *Archives d'histoire doctrinale et littéraire du moyen âge*, 55 (1988), 7–21 (on Albertus Vangaditiensis, *Summa sententiarum*)

9. 'L'Astronomie à Chartres au temps de l'évêque Fulbert', in *Le Temps de Fulbert. Actes de l'Université d'été du 8 au 10 juillet 1996* (Chartres: Société archéologique d'Eure-et-Loir, 1996), pp. 91–103

10. 'John of Salisbury and Aristotle', *Didascalia* (Sendai), 2 (1996), 19–32

11. 'Vincent of Beauvais, Michael Scot and the "New Aristotle"', in *Lector et compilator: Vincent de Beauvais, frère prêcheur. Un intellectuel et son milieu au XIIIe siècle*, ed. by Serge Lusignan, Monique Paulmier-Foucart, and Marie-Christine Duchenne (Grâne: Editions Créaphis, 1997), pp. 189–213

12. 'La réception des mathématiques, de l'astronomie et de l'astrologie arabes à Chartres', in *Aristote, L'école de Chartres et la cathédrale. Actes du colloque européen des 5 et 6 juillet 1997* (Chartres: Association des Amis du Centre Médiéval européen de Chartres, 1997), pp. 101–07

13. 'A New Student for Peter Abelard: The Marginalia in British Library MS Cotton Faustina A. X', with David Luscombe, in *Itinéraires de la raison. Études de philosophie médiévale offertes à Maria Cândida Pacheco*, ed. by José Meirinhos, Textes et études du Moyen Âge, 32 (Louvain-la-Neuve: Fédération internationale des instituts d'études médiévales, 2005), pp. 163–86

I. Music

1. 'The Use of Geometrical Terms in Medieval Music: *el muahim* and *elmuarifa* and the Anonymus IV', *Sudhoffs Archiv*, 70 (1986), 198–205

2. 'Adelard, Music and the Quadrivium', in *Adelard of Bath: An English Scientist and Arabist of the Early Twelfth Century*, ed. by Charles Burnett, Warburg Institute Surveys and Texts, 14 (London: Warburg Institute, 1987), pp. 69–86

3. 'Sound and its Perception in the Middle Ages', in *The Second Sense: Studies in Hearing and Musical Judgement from Antiquity to the Seventeenth Century*, ed. by Charles Burnett, Michael Fend, and Penelope Gouk, Warburg Institute Surveys and Texts, 22 (London: Warburg Institute, 1991), pp. 43–69

4. 'European Knowledge of Arabic Texts Referring to Music: Some New Material', *Early Music History*, 12 (1993), 1–17; Italian version: 'Teoria e pratica musicali arabe in Sicilia e nell'Italia meridionale in età normanna e sveva', *Nuove effemeridi*, 11 (1990), 79–89

5. 'Boethius on Vibrational Frequency and Pitch: Correspondence between Dr Charles Burnett … and Professor H. Floris Cohen, University of Twente, Enschede, occasioned by a statement in Cohen's review of *The Second Sense: Studies in Hearing and Musical Judgement from Antiquity to the Seventeenth Century*, edited by C. Burnett, M. Fend and P. Gouk (London: Warburg Institute, 1991)', *Annals of Science*, 52 (1995), 303–05

6. 'Hearing and Music in Book XI of Pietro d'Abano's *Expositio Problematum Aristotelis*', in *Tradition and Ecstasy: The Agony of the Fourteenth Century*, ed. by Nancy van Deusen, Musicological studies, 62/3 (Ottawa: The Institute of Mediaeval Music, 1997), pp. 153–90

7. *The Cashel Music Treatise*, ed. with Michael Lundell for database *Thesaurus Musicarum Latinarum: Canon of Data Files*, ed. by Thomas J. Mathiesen (Lincoln: University of Nebraska Press, 1999), catalogue entry p. 21, edition at <chmtl.indiana.edu/tml/13th/ANOTRAC_MCBOLT1> [accessed 19 April 2023]

8. 'Music and Healing in Islam', *Avidi Lumi: Quadrimestrale di culture musicali de Teatro Massimo di Palermo*, 5 (February 1999), 36–40, 83–84 and 108–09 (Italian, English and Arabic versions)

9. '"Spiritual Medicine": Music and Healing in Islam and its Influence in Western Medicine', in *Musical Healing in Cultural Contexts*, ed. by Penelope Gouk (Aldershot: Ashgate, 2000), pp. 85–91

10. 'Music and Magnetism, from Abū Maʿshar to Kircher', in *Music and Esotericism*, ed. by Laurence Wuidar, Aries Book Series, 9 (Leiden: Brill, 2010), pp. 13–22

11. 'Musical Instruments as Conveyors of Meaning from One Culture to Another: The Example of the Lute', in *The Power of Things and the Flow of Cultural Transformations: Art and Culture between Europe and Asia*, ed. by Lieselotte E. Saurma-Jeltsch and Anja Eisenbeiss (Berlin: Deutscher Kunstverlag, 2010), pp. 156–69

12. 'Music and the Stars in Cashel, Bolton Library, MS 1', in *Music and the Stars: Mathematics in Medieval Ireland*, ed. by Mary Kelly and Charles Doherty (Dublin: Four Courts Press, 2013), pp. 142–58, and Plates 6 and 7

J. Contacts between the West and the Far East

1. 'Attitudes towards the Mongols in Medieval Literature: the XXII Kings of Gog and Magog from the Court of Frederick II to Jean de Mandeville', with Patrick Gautier Dalché, *Viator*, 22 (1991), 153–67
2. Summary of lecture 'Antipelargesis: A Jesuit Latin Play in a Japanese Setting', *Chuo University Bulletin*, Tokyo (1992) (in Japanese)
3. 'Common Sources of Astrology and Astronomy in West and East', in *The Mutual Encounter of East and West, 1492–1992*, ed. by Peter Milward (Tokyo: Renaissance Institute, 1993), pp. 81–87
4. 'Humanism and the Jesuit Mission to China: The Case of Duarte de Sande (1547–1599)', *Euphrosyne*, n.s., 24 (1996), 425–71
5. 'The Navigational Instruments in Duarte de Sande's *Dialogus de missione legatorum Iaponensium ad Romanam Curiam (1590)*', in *História das ciências matemáticas, Portugal e o Oriente / History of Mathematical Sciences, Portugal and East Asia*, ed. by Luis Saraiva (Camarate: Fundação Oriente, 2000), pp. 263–74
6. 'The Freising Titus Play', in *Mission und Theater: Japan und China auf den Bühnen der Gesellschaft Jesu*, ed. by Adrian Hsia and Ruprecht Wimmer, Jesuitica, 7 (Regensburg: Schnell & Steiner, 2005), pp. 413–97

K. Miscellaneous

1. 'Marie-Thérèse d'Alverny et l'islamologie', *Cahiers de civilisation médiévale*, 35 (1992), 290–91
2. 'Wobagu Kenkyujo ni okeru Arabia kagaku no kenkyu' (in Japanese: 'Arabic studies at the Warburg'), in *Tozai Namboku*, Machida, Tokyo (2001), 108–15
3. 'Marie-Thérèse d'Alverny (1903–1991): The History of Ideas in the Middle Ages in the Mediterranean Basin', in *Women Medievalists and the Academy*, ed. by Jane Chase (Madison: University of Wisconsin Press, 2005), pp. 585–97
4. 'John North', in *Biographical Memoirs of Fellows of the British Academy*, XI (Oxford: Oxford University Press, 2013), pp. 493–99
5. In memoriam Peter Dronke, *Mediterranea*, 6 (2021), 157–61
6. In memoriam Helena Avelar de Carvalho, *Mediterranea*, 6 (2021), 189–90

IV. Editions of Arabic and Latin (and some Greek and Hebrew) texts in the books and articles catalogued

The language of the editions is indicated as follows:
H = Hebrew, G = Greek, A = Arabic, L = Latin

Abdalaben Zeleman, *De spatula* 24 L
Abraham Ibn Ezra (Avenezra), *De revolutionibus annorum mundi* 27 L
Abraham Ibn Ezra, *Sefer ha-Midot* 19 HL
Abraham Iudaeus, *Horoscope* D38 L
Abū Maʿšar, *Abbreviation of the Introduction* (*Ysagoga minor*) 6, 12 AL
Abū Maʿšar, *On Historical Astrology* 15 AL
Abū Maʿšar, *The Great Introduction to Astrology* 26 GA
Adelard of Bath, *De avibus tractatus* 13 L
Adelard of Bath, *De eodem et diverso* 13 L
Adelard of Bath, *Questiones naturales* 13 L
Alcabitius, *De coniunctionibus planetarum* 17 L
Alcabitius, *Liber introductorius* 17, D29 (preface) GAL
Alkindi, *De cometis* (fragment) B14 German
Alkindi, *De mutatione temporum* 14 HA
Alkindi, *De thesauris* and *De thesauri reperto* F21 AL
Alkindi, *Forty Chapters (Iudicia)* 5, D2 L
Alkindi, *Forty Chapters (Iudicia)* (excerpts) 14, 20 AL
Alkindi, *Kitāb fī ʿilm al-katif* 16 A
Alkindi, *Liber de iudiciis revolutionum mundi* 27 L
Alkindi, *On the Rule of the Arabs and Its Duration* 15 A
Alkindi, *Risāla fī aḥdāth al-jaww* 14 A
Alkindi, *Tehran Chapters (on Weather)* 14 A
Pseudo-Alkindi, *Epistola prudenti viro* 9, F6 L
Amitegni D40 L
Antimaquis 16 L
Antipelargesis (Jesuit play) 7 L
Anxiomata artis aritmetice 8 L
Apertio portarum 14 L
Pseudo-Aristotle, *De Luna* 9, F7, F26, F28 L
Pseudo-Aristotle, *Secreta secretorum* F11 L
Ascelin of Augsburg, *Compositio astrolabii* 21, D8 L
Averroes Junior, *De intellectu* B12, B15 AHL
Axiomata artis aritmetice 8 L

Bartholomew of Parma, *Tractatus spere, pars tercia* D10 L
Pseudo-Bede, *De mundi celestis terrestrisque constitutione* 3 L
Bernardus Silvestris, *Experimentarius* F2 L
Borotin, Introduction to Commentary on Alcabitius's *Introduction to Astrology* D35 L

The Cashel Arithmetical Treatises C5 L
The Cashel Music Treatise I7 L
Chiromantia parva ('Adelard Chiromancy') 9 L
Constantine the African, *De oblivione* E6 L
Constantine the African, *Pantegni*, chapter on the elements E8, chapter on the spirits E5 L

David Ben Yom Tov, *Medical Astrology* 42 HL
De amicitia vel inimicitia planetarum 16, F9 L
De cibis 21, B17 L
De elementis 21, B17 L
De metallis 21, B17 L
De saporibus B24 L
De spatula 16 L
De spatula, liber alius 16, 24 L
Divisio scientiarum (Chartrian) B5 L
Duarte de Sande, trans., *De missione legatorum Iaponensium* (chapters), J4, J5 L

Eadwine Chiromancy 9, F11 L

al-Fārābī, *Commentary on Euclid's Elements*, Book 6, C7 L
al-Farghānī, *On the Science of the Stars*, preface to Jacob Golius's edition and translation D49 L
Freising Titus Play J6 L

Gerard of Cremona, Preface to *Almagest* D27 L
Gerard of Cremona, *Vita, Commemoratio librorum* and *Eulogium* 21, A23 L
Giorgio Anselmi, *Notitia spatulæ* 9, F18 L

Haliabas, *Regalis dispositio* (chapter on travel) A38 L; see also Stephen of Antioch
Hermann of Carinthia, *De essentiis* 1 L
Hermann of Carinthia, Preface to Ptolemy, *Planisphere* 1, A2 L
Hugo Sanctelliensis (Hugo of Santalla), *Liber Aristotilis* 10 L
Hugo Sanctelliensis, Preface to *Liber novem iudicum* D16 L
Hugo Sanctelliensis/Hermann of Carinthia, Preface to *Liber trium iudicum* A1 L

Ibn Abī al-Rijāl, *Kitāb al-Bāriʿ* (chapters) 14, 15 A(L)
Iohannes Hispalensis, *Tractatus pluviarum et aeris mutationis* D13, D22 L

Jabir, *Flos naturarum* F30 L
Jafar Indus, *De pluviis* D13 L
John of Seville and Limia, Preface to Pseudo-Aristotle, *Secreta secretorum* 21, A15 L
John of Seville and Limia, Preface to Thābit, *De imaginibus* 21, A15 L

Kankaf Indus, Dixit Kankaf Indus 15 L
Kitāb al-Bulhān (lunar mansions) F7 A

The Leicester Iudicia D47 L
Liber Alchorismi, Part II, 23, C12 L
Liber novem iudicum (chapters on war) D31 L
Liber runarum 9, F4 L
Liber ysagogarum Alchorismi, Book 4, C15 L
'Lineae naturales ...' (Eadwine Chiromancy) 9, F11 L

Maimonides, *De coitu* 25 AHLSlavonic
Māshāʾallāh, *De revolutione annorum mundi* (partial) 15 L
Māshāʾallāh, *On the Accession of the Caliphs* 15 A
Michael Scot, *Divisio philosophiae* H11 L

Nemesius, *On the Nature of Man* (chapter on the elements) B17 L

Ocreatus, *Helcep Sarracenicum* 8 L
'Omnibus convenit Platonicis ...' G2 L
On the Victorious and the Vanquished 9, C3, F2, F11 L
'Oxford Gloss' to the first chapter of Aristotle's *Physics* B7 L

Peter Abelard, *Abbreviatio Expositionis Hexameron* 18 L
Peter Abelard, *Confessio fidei ad Heloisam* H7 L
Peter Abelard, *Confessio fidei 'Universis'* H6 L
Peter Abelard, Epitaphs H5 L
Peter Abelard, *Expositio Orationis Dominicae 'Multorum legimus orationes'* H4 L
Peter Abelard, *Soliloquium* H2 L
Petrus Alfonsi, Prologue to *Zīj* of al-Khwārizmī A16 L
Pietro d'Abano, *Expositio Problematum* (chapters on hearing) I6 L
Pietro d'Afeltro, *Expositio Proemii Averroys* (preface and section on method) B7 L
Prologue to Boethius, *De arithmetica* in Cambridge, Trinity R.15.16 B5 L
Proverbia Senece et versus Ebrardi super eadem B11 L
Ptolemy, *Almagest*, first chapter D27 L
Pseudo-Ptolemy, *Liber figure* F40 L
Pseudo-Ptolemy, *On Images* (*De ymaginibus super facies signorum*) (partial) 28, F33 AL
Pseudo-Ptolemy, *Parapegma* F16 L
Pseudo-Ptolemy, *Proverbia* B11 L

Raymond of Marseilles, *Liber cursuum planetarum* 22 L
Raymond of Marseilles, *Tractatus astrolabii* 22, D32 L
Razes (Ibn Zakarīyāʾ al-Rāzī), *Dubitationes contra Galenum* (fragment) B10 L
Rememoratio spatule et expositio eius 9, F17 L
Robert of Ketton, Prologue to Alkindi, *Iudicia* 5, D2 L
Rhythmomachia G10 L

'Sanctus Francis Xaverius' (Jesuit play) 7 L
'Sapientes Indiae ...' D13 L
'Saturnus in Ariete ...' 14 L
'Septem sunt genera metalli ...': see *De metallis*
Sloane Chiromancy 9 L
Sortes regis Amalrici F2, D7 L
Stelle fixe aerem turbantes in singulis mensibus F26 L
Stephen of Antioch (Pisa), Prefaces to *Regalis dispositio* and *Liber Mamonis* 21, A22 L
Summa de saporibus et odoribus B23 L

Tables of Lucca C3 L
Thābit ibn Qurra (Thebit Ben Corah), *On Talismans (De imaginibus)* 28, F42 AL
Thābit ibn Qurra, *Tashīl al-Majisṭī* ('making the Almagest easy', excerpt) F27 L
Theodore of Antioch, *Epistola ad imperatorem Fridericum de conservanda sanitate* 21, A14 L
Theodore of Antioch, Prefaces to *Moamin* and *Gatriph* 21, A14 L
Toledan *Regule* 23, C12 L

'Umar ibn al-Farrukhān, *Kitāb mukhtaṣar al-masā'il (Liber Aomaris)* (chapters on weather forecasting) 14 AL
'Ut testatur Ergaphalau ...' F8 L

Victorious and Vanquished: see *On the Victorious and the Vanquished*

William Scot, *De coniunctionibus planetarum transitis et futuris* D5 L

Zadanfarrūkh al-Andarzaghar on Anniversary Horoscopes 4 AL

Index of Names

Abraham Ibn Ezra: 46, 49
Abū Maʿshar (Albumasar): 13, 19–52, 63, 74 n. 43, 112 n. 23
Adelard of Bath: 14, 32, 83–85, 88, 92–93, 95–103
Adrian Zeelst: 17, 231, 243, 245–50
Afrīdūn the Nabatean: 61
Aggere, Petrus Ab: 239, 241
Aḥmad ibn Yūsuf ibn al-Dāya: 11 n. 1
Albertus Magnus: 107, 108 n. 1, 116, 118, 152, 153 n. 27, 154, 164 n. 68, 166, 169
Albumasar, *see* Abū Maʿshar
Alexander of Hales: 170 n. 92
Alexander the Great, King: 60, 184, 193
Alfonso X, King: 186, 238, 239 n. 13
Alfred of Shareshill: 109
Alpetragius: 110
Amenus: 185
Ananias of Širak: 89, 92
Andarzaghar: 31, 42 n. 85, 43–46, 48 n. 110, 51
Antiochus of Athens: 22 n. 8, 24, 26–30, 33, 35 n. 60, 43
Apuleius: 225
Ardāshir b. Bābakān: 61
Aristotle: 15, 70, 100 n. 40, 108, 109 n. 9-10, 110, 112, 143–151, 152 n. 21, 153–158, 160–61, 164–72, 182, 184–85, 193, 201
Arsenius, Gualterus: 239, 245, 246
Arzachel, *see* al-Zarqālī
Augustine: 15, 101, 143–44, 150–52, 153 n. 22, 209, 221, 225
Averroes (Ibn Rushd): 15, 107, 108, 109 n. 13, 110, 114, 115 n. 31, 118, 144, 153 n. 22, 166, 171, 192
Avicenna (Ibn Sīnā): 14–15, 70, 107–17, 125, 127 n. 2, 143–44, 149 n. 9, 150, 152, 153 n. 22, 158, 160–61, 165, 171
Azarquiel, *see* al-Zarqālī

Bacon, Francis: 103
Behentater: 186
Benoît de Sainte-Maure: 219
al-Bīrūnī: 24 n. 17, 48, 61, 240
Blagrave, John: 240, 241
Boethius: 225 n. 75
Boethius of Dacia: 146, 148 n. 3, 156, 167–69, 170 n. 91
Bonaventure: 145, 149, 155, 160–61, 165, 171
Brahmagupta: 68
Brethren of Purity, *see* Ikhwān al-Ṣafāʾ
Buzurjmihr: 46

Calcidius: 99 n. 32, 100 n. 37, 116
Caraphzebiz: 185
Cicero: 97
Circe: 218, 225
Claude Tholosan: 225
Clement Canterbury: 210 n. 18
Cleopatra: 193
Constantine VII Porphyrogenitus: 29

Dee, John: 209
Dido: 218, 219
Dorotheus of Sidon: 24–25, 29–30, 43–45, 51, 184

Empedocles: 184
Enoch: 184
Erichtho: 221
Ernst of Bavaria: 17, 245–46, 249

al-Fārābī: 12 n. 2, 108–09
Firmicus Maternus: 23 n. 12, 24, 27, 29, 30, 34
Fonseca, Pedro: 108
Frederick II Hohenstaufen: 15, 109
Frisius, Gemma: 238–39

Galen: 182, 184
Geber, *see* Jābir ibn Ḥayyān
Geoffrey of Aspall: 164–65
Geoffrey of Monmouth: 219
Gerard of Cremona: 108–09
al-Ghazālī: 152, 153 n. 22, 162
Giles of Rome: 146, 153, 155, 156 n. 37, 160, 163, 167, 168
Godfrey of Fontaines: 158, 159 n. 45, 160
Guido Bonatti: 46, 50
Guido Terreni: 153 n. 26, 154 n. 27, 166, 169, 172

Habermel, Erasmus: 245
Hans Dorn: 239
Hans Fründ: 224
Hartlieb, Johannes: 224
al-Ḥasan al-Marrākushī: 240
Heliodorus: 27 n. 32
Hemmerlin, Felix: 224
Henricus Aristippus: 108–09
Henry of Ghent: 158, 160–61
Henry of Harclay: 159, 163
Hephaestio of Thebes: 23 n. 9, 24–25, 30, 44 n. 92
Hermann of Carinthia: 32
Hermes Trismegistus: 25, 28, 30–31, 34, 43, 51, 184, 186
Hipparchus: 66, 112
Hippocrates: 184
Ḥunayn ibn Isḥāq (Johannitius): 184, 201

Ibn al-Jazzār: 211, 240
Ibn al-Sarrāj :241
Ibn Khalaf: 238–39
Ibn Rushd, *see* Averroes
Ibn Sīnā, *see* Avicenna
Ibn Ṭufayl: 93, 95, 102
Ibn Waḥshīyya: 200–01
Ibn Zakarīyā᾽ al-Rāzī, *see* al-Rāzī
Ikhwān al-Ṣafā᾽ ('Brethren of Purity'): 13–14, 58–66, 67 n. 26, 68–77, 112
Isidore of Seville: 218

Jābir ibn Ḥayyān (Geber): 182, 184
Jacob ben Makir ibn Tibbon (Profacius): 238
Jean de Lignères: 243
Jean Vincent: 223
Johannitius, *see* Ḥunayn ibn Isḥāq
John of Brexia: 238
John of Jandun: 166, 170 n. 91, 171 n. 93
John of Salisbury: 219 n. 50
John of Seville: 32 n. 46, 33, 46, 50, 57 n. *, 83 n. *
John Pecham: 160, 161 n. 59, 162
John Quidort: 158, 159 n. 45, 160–61
Jubert of Bavaria: 225

Keyser, Conrad: 215
al-Khwārizmī: 183

Leopold of Austria: 50
Lucan: 221

Maelcote, Odo van: 241
Maimonides (Moses ben Maimon): 149, 152, 154, 162
al-Maʾmūn: 17, 238
Marcus Graecus: 215
Marinus of Tyre: 89
Māshāʾallāh: 45, 46, 72 n. 37
Maslama ibn Qāsim al-Majrīṭī al-Qurṭubī: 16, 178, 189, 203 n. 29
al-Masʿūdī: 63 n. 18, 68–69, 93 n. 21, 95 n. 22
Matthew of Aquasparta: 152 n. 18, 153, 160–62
Mercator, Gerard: 239 n. 13, 245–46
Mercurius: 184
Michael Scot: 15, 109–10, 114–17
Molitoris, Ulrich: 225
More, Thomas: 103
Moses ben Maimon, *see* Maimonides

Nechepso: 20, 21 n. 7, 24
Nemesius: 100 n. 40
Nicole Oresme: 212, 220–22, 225

Olympiodorus: 22 n. 8, 24, 27–30, 34, 35 n. 60, 38, 40, 41 n. 81, 42, 46, 112
Ovid: 221

Pappus of Alexandria: 89
Paulus of Alexandria: 22–24, 27–30, 34–38, 40–41, 42 n. 89, 43, 46, 51
Peter Lombard: 145, 147 n. 2
Peter of Auvergne: 153 n. 26, 163
Peter of Tarentaise: 165, 168
Petosiris: 20, 21 n. 7, 24
Petrus Alfonsi: 14, 83–85, 86 n. 6, 87, 88 n. 8-9, 92–93, 98, 100–03
Philip the Chancellor: 170 n. 92
Philoponus: 150 n. 14, 161 n. 58, 162 n. 62
Pietro d'Abano: 108
Plato: 42 n. 88, 89 n. 13, 99, 100 n. 37, 107, 111–12, 116–17, 124, 182, 184, 192, 200, 207 n. 1
Plato of Tivoli: 11 n. 1
Pliny: 184
Pomponazzi, Pietro: 108
Porphyry: 150
Proclus: 150 n. 14, 153 n. 22
Profacius, *see* Jacob ben Makir ibn Tibbon
Ps.-Aristotle: 164
Ps.-Ptolemy: 11 n.1
Ptolemy: 24, 27, 31, 38 n. 68, 43, 48, 60–61, 66, 88, 89, 184, 192 n. 25
Pythagoras: 182, 184

al-Qabīṣī: 45 n. 100, 47, 48
Qummī: 63

al-Rāzī (Ibn Zakarīyāʾ al-Rāzī): 182, 184
Rhetorius: 22 n. 8, 24, 26–28, 30, 33, 35 n. 60, 43–46, 51
Richard of Middleton: 158, 162
Rojas, Juan de: 238
Roland of Cremona: 218, 225

Sacrobosco: 116
Sahl ibn Bishr: 24 n. 17, 42 n. 85, 44–45, 46 n. 102, 51
Samuel ibn Tibbon: 110
Schöner, Johannes: 238
Socrates: 184
Solomon: 215
Stempel, Gerard: 245, 247
Stöberl, Andreas: 238
Sulayman b. Dāwūd: 61

al-Ṭabarī: 184, 199
Thābit ibn Qurra (Thebit ben Corat): 185
Theophilus of Edessa: 31, 43, 45–46, 74
Thomas Aquinas: 116–17, 146, 147 n. 2, 148 n. 5, 149–55, 158–61, 163, 164 n. 68, 166–67, 169, 170 n. 92, 171 n. 93, 172
Thomas of Willesborough: 209
Thoos: 185
Tiberio Russiliano: 108
Tintinz the Greek: 185
Trombetta, Antonio: 108
Tubbaʿ al-Ḥimyarī: 61
Tymtym: 185

Vettius Valens: 22–25, 28–31, 34–35, 39, 40 n. 71, 41, 43–45, 51

Whitwell, Charles: 240
William de la Mare: 159
William of Auvergne: 218
William of Baglione: 164, 168, 169 n. 86
William of Conches: 116
William the Englishman: 238
Winstanley, Gerrard: 103
Witch of Endor: 218

Yehuda ben Mose: 238

Zadealis: 185
al-Zarqālī (Arzachel, Azarquiel): 17, 238, 239, 243
Zeherith: 183
Ziegler, Jacob: 238

Contact and Transmission

Intercultural Encounters from Late Antiquity to the Early Modern Period

All volumes in this series are evaluated by an Editorial Board, strictly on academic grounds, based on reports prepared by referees who have been commissioned by virtue of their specialism in the appropriate field. The Board ensures that the screening is done independently and without conflicts of interest. The definitive texts supplied by authors are also subject to review by the Board before being approved for publication. Further, the volumes are copyedited to conform to the publisher's stylebook and to the best international academic standards in the field.

Titles in Series

Isaac Lampurlanés Farré, *Excerptum de Talmud: Study and Edition of a Thirteenth-Century Latin Translation* (2020)

Premodern Translation: Comparative Approaches to Cross-Cultural Transformations, ed. by Sonja Brentjes and Alexander Fidora (2021)

Narratives on Translation across Eurasia and Africa: From Babylonia to Colonial India, ed. by Sonja Brentjes in cooperation with Jens Høyrup and Bruce O'Brien (2022)